I0823305

# Digital Payments with SAP®

Saravana Kumar Kuppusamy

# Digital Payments with SAP®

## Credit Card Processing, PSP Integration, and PCI Compliance

**Editor** Meagan White
**Acquisitions Editor** Emily Nicholls
**Copyeditor** Melinda Rankin
**Cover Design** Abigail Damke
**Photo Credit** iStock: 2106574580/© Elena Uve
**Layout Design** Vera Brauner
**Production** Hannah Lane
**Typesetting** III-satz, Germany
**Printed and bound in** Canada, on paper from sustainable sources

**ISBN 978-1-4932-2836-2**
1st edition 2026


Rheinwerk Publishing, Inc.
2 Heritage Drive, Suite 305
Quincy, MA 02171
USA
info@rheinwerk-publishing.com
+1.781.228.5070

**Represented in the E.U. by:**
Rheinwerk Verlag GmbH
Rheinwerkallee 4
53227 Bonn
Germany
service@rheinwerk-verlag.de
+49 (0) 228 42150-0

**Library of Congress Cataloging-in-Publication Control Number:** 2026018064

# Contents at a Glance

# Contents

# Preface

Enterprise payment processing presents a peculiar paradox: It's invisible when it works, yet catastrophic when it fails. Nobody celebrates when enterprise payments process smoothly; your customers and your customer service and finance teams expect it to work flawlessly, and executives notice only when it fails and revenue stops flowing. This book documents my journey solving that invisible-until-broken challenge, in which discontinued legacy payment systems and rushed deployments created crises that I learned to prevent. When I began implementing the SAP digital payments add-on, I found powerful SAP technology, but guidance was scarce and scattered. I pieced together disparate information sources to create a comprehensive implementation guide that bridges SAP's technical documentation and real-world implementation experience—knowledge I now share with you.

Every enterprise-grade implementation of a production-ready system carries implementation scars—lessons learned through mistakes during high-pressure situations. In this book, I share those scars: my trial and error experience of what worked, what failed, and, most importantly, why, so that you can be successful in your digital payments add-on implementation. Many organizations struggle with legacy payment systems that vendors have discontinued or replaced with cloud-based solutions that are incompatible with their existing infrastructure or roadmap. Implementation teams face pressure to deploy quickly while avoiding the consequences of payment-processing failures: unauthorized charges, delayed settlements, or compliance violations that erode customer trust and expose organizations to regulatory penalties.

Modern payment complexity demands sophisticated handling of credit card processing, payment service provider (PSP) integration, and regulatory compliance. The digital payments add-on addresses these requirements by orchestrating payment operations across channels, banks, and geographies while reducing Payment Card Industry Data Security Standard (PCI DSS) compliance scope through tokenization. However, the add-on's power comes with implementation complexity. Configuration options are vast, integration points numerous, and consequences of misconfiguration severe. This book provides practical guidance for implementing, configuring, and operating the digital payments add-on in enterprise environments, helping you avoid common pitfalls along the way and achieve a successful production deployment.

## Target Audience

This book serves diverse professionals involved in enterprise payment processing within SAP environments. Technical consultants, functional leads, project managers, and business stakeholders will find detailed guidance applicable to their specific roles and responsibilities.

SAP consultants specializing in finance and sales and distribution will deepen their expertise in payment processing through practical configuration guidance that complements formal training. The book examines not only configuration mechanics but also design decisions and their business implications.

Solution architects and technology leads designing payment solutions spanning multiple systems will understand how the digital payments add-on fits within broader technical landscapes. The architectural perspective supports decisions about centralization versus distribution, real-time versus batch processing, and build versus buy trade-offs for payment infrastructure.

SAP developers and integration specialists building custom enhancements or external system interfaces will gain detailed knowledge of data structures, APIs, and extension points. Chapter 5, which focuses extensively on SAP ERP, provides sample code, debugging techniques, and integration patterns for developers implementing payment processing in SAP ERP environments.

Project managers and business analysts translating business requirements into system specifications will understand effort drivers, realistic timelines, and common implementation challenges. The book provides project planning guidance that bridges the gaps between business stakeholders and technical teams.

Finance and treasury leaders making strategic decisions about payment infrastructure investments will find business context and compliance implications relevant to leadership decisions. Understanding PCI DSS scope reduction and security architecture informs investment justification and risk assessment.

IT operations and support teams responsible for production payment processing will learn about operational monitoring, troubleshooting procedures, and maintenance tasks that ensure long-term system health. The operational chapters provide runbooks and support procedures for day-to-day payment operations.

The book assumes working knowledge of SAP, including its navigation, organizational structure, and transaction processing, along with basic accounting concepts and financial process understanding. No prior payment processing or digital payments add-on experience is required.

## How to Read This Book

The chapters are structured to build knowledge progressively, with concepts introduced in early chapters applied in subsequent implementation sections. Reading sequentially provides optimal learning progression, though the book also serves as a reference for specific topics, as follows:

- **For first-time implementations**
  Begin with Chapter 1 and Chapter 2, which establish the business context and technical architecture. Proceed through Chapter 3 and Chapter 4, covering SAP Business Technol-

ogy Platform (SAP BTP) setup and PSP integration. Then focus on either Chapter 5 (SAP ERP) or Chapter 6 (SAP S/4HANA), depending on your system landscape. Complete your reading with Chapter 7 on security and compliance before deployment, using Chapter 8 through Chapter 10 for testing, go-live, and operational procedures.

- **For SAP ERP environments**
  After the foundation chapters, concentrate on Chapter 5, which provides extensive development guidance, SAP Note implementation procedures, and function modules required for SAP ERP integration. This chapter is particularly relevant for organizations that require custom payment processing logic that is unavailable in standard SAP ERP functionality.
- **For SAP S/4HANA environments**
  After the foundation chapters, focus on Chapter 6. SAP S/4HANA implementation leverages standard configuration and delivered SAP Fiori applications, requiring significantly less custom development than SAP ERP. Chapter 6 explains activation procedures and out-of-the-box integration approaches.
- **Hands-on practice**
  Replicate the configuration examples in sandbox or training systems as you read. Practical experience with transaction codes, menu navigation, and configuration settings reinforces conceptual understanding more effectively than passive reading. Experiment with variations to understand dependencies and the impacts of configuration choices.
- **Reference usage**
  Although it is designed for sequential reading, the book's chapter organization supports using it as an implementation reference. Cross-references guide you to related information, and the table of contents and index enable quick location of specific topics during active implementation projects.

## How This Book Is Organized

This book progresses from the business context and architecture through detailed implementation guidance to operational procedures and business user training. Each chapter builds on the knowledge from previous sections while remaining accessible to readers focusing on specific topics:

- **Chapter 1: Introduction to Payment Modernization with SAP**
  This chapter establishes why organizations need payment infrastructure modernization and introduces the digital payments add-on as the solution. It examines legacy payment system limitations, including manual processes, the PCI compliance burden, and an inability to support modern payment methods like digital wallets and buy now, pay later options. You will explore the business drivers for transformation and how the add-on addresses these challenges through standardization, security, and flexibility.

- **Chapter 2: SAP Digital Payments Add-on Architecture**
  This chapter describes the technical foundation of the digital payments add-on within the SAP ecosystem. It explains how the add-on functions as secure middleware between SAP applications and payment service providers, standardizing payment operations while reducing PCI compliance scope. You will learn about component architecture, communication flows between SAP and PSPs, security mechanisms that protect payment data, and high-availability design. The chapter establishes the technical knowledge required for subsequent implementation chapters.
- **Chapter 3: Setup in SAP BTP**
  This chapter provides step-by-step instructions for configuring the digital payments add-on in SAP BTP. It covers the complete setup from initial subscription through security configuration to validation testing. You will learn how to configure communication between the add-on and SAP systems, establish proper security settings, manage user roles and authorizations, and validate installation completeness. The chapter includes troubleshooting guidance for common setup issues encountered during SAP BTP configuration.
- **Chapter 4: Integration with Payment Service Providers**
  This chapter provides comprehensive guidance for integrating PSPs with the digital payments add-on, using Stripe as the primary implementation example throughout. It covers PSP integration architecture, detailed configuration steps using Stripe's certified adapter, and setup of payment methods.
- **Chapter 5: Implementation in SAP ERP**
  This chapter addresses unique requirements for implementing the digital payments add-on in SAP ERP environments that require custom development. It covers SAP Notes that provide payment-processing functionality, ABAP classes and function modules for authorization and settlement, payment card configuration in cross-application components, SAP ERP Sales and Distribution integration for order processing, batch job configuration for automated settlement, and custom report development for reconciliation.
- **Chapter 6: Implementation in SAP S/4HANA**
  This chapter focuses on implementing the digital payments add-on in SAP S/4HANA, where the integration is streamlined and largely delivered out of the box. It covers simplified configuration activation procedures and standard SAP Fiori applications for payment monitoring. You will learn the differences between the cloud and on-premise deployment options and explore points to consider when migrating from SAP ERP to SAP S/4HANA.
- **Chapter 7: Security and Compliance**
  This chapter addresses critical security requirements and compliance mandates for payment processing. It explains how the digital payments add-on reduces the PCI DSS compliance scope through tokenization, removing sensitive cardholder data from SAP systems entirely. You will learn about tokenization implementation and token lifecycle

management, encryption architecture and certificate handling, audit logging requirements and monitoring procedures, General Data Protection Regulation compliance for data privacy, and security testing validation approaches. The chapter includes practical guidance for compliance documentation and audit preparation.

- **Chapter 8: Testing and Troubleshooting**
  This chapter provides comprehensive testing approaches and troubleshooting techniques for digital payments add-on implementations. It covers a test environment setup using Stripe's test mode, test scenario development spanning authorization through settlement and refunds, performance testing to validate the transaction volume capacity, systematic troubleshooting methodologies, and application log analysis procedures.
- **Chapter 9: Go-Live and Administration**
  This chapter ensures successful production deployment and sustainable payment operations. It provides detailed cutover planning procedures, migration strategies from legacy payment systems with parallel operations approaches, production deployment execution with clear rollback criteria, post-go-live support, and operational monitoring and maintenance procedures, including batch job monitoring and reconciliation.
- **Chapter 10: Processing Digital Payments**
  This chapter addresses how business users interact with the digital payments add-on for daily payment operations. It covers customer service procedures for processing sales orders with payment cards and handling authorization failures, finance operations including refund processing and monthly reconciliation, dispute and chargeback management procedures, and training approaches for role-based user enablement.

## Conclusion

Whether you are implementing the digital payments add-on for the first time, migrating from a legacy payment system, or optimizing an existing deployment, this book provides practical guidance grounded in production experience. The combination of technical depth, business context, and operational procedures supports successful outcomes across diverse organizational scenarios and SAP system landscapes.

Successful implementation requires understanding not only configuration mechanics but also business processes, integration patterns, security requirements, and operational procedures. This book provides that complete perspective, serving as comprehensive guide for implementation teams and operational reference for production support.

Organizations successfully implementing the digital payments add-on report payment processing reliability, reduced compliance audit effort, improved cash flow visibility, and the business confidence to expand digital commerce capabilities. The journey from payment processing confusion to operational clarity is challenging but achievable. This book provides the roadmap.

## Acknowledgements

I am truly grateful to my family, especially to my wife Vidhya and my son Arnav, for their continuous encouragement, love, and support throughout this journey. Without my wife's patience, understanding, and belief in the value of this work, I could not have dedicated weekends and late nights to developing this book. Special thanks go to my editor Meagan White for her tremendous support, guidance, and patience throughout the writing process. Emily Nicholls from Rheinwerk Publishing was instrumental in the early development of this book, helping shape its vision and structure. I thank them both, along with the entire Rheinwerk Publishing team, for providing me with this opportunity that I will cherish.

Many of the insights shared in this book are based on my hands-on experience, complemented by the perspectives gained through collaborative discussions with the SAP digital payments add-on consulting team. Their willingness to share lessons learned from other implementations and their collaborative approach to solving challenges enriched the deployment. I thank my team members for their contributions to the development, testing, and deployment of the digital payments add-on. Their dedication during challenging implementation phases and their collaborative problem-solving transformed complex requirements into operational reality. I appreciate my colleagues' support during the implementation, from the initial evaluation, architecture review, and requirements gathering to the SAP BTP environment setup and PSP integration configuration. Their constructive feedback and readiness to question assumptions contributed to enhancing the implementation.

## Chapter 1
# Introduction to Payment Modernization with SAP

*Before we dive into the technical architecture, configuration, and implementation details covered in this book, it is essential to understand why payment modernization matters and what business problems it solves. This chapter establishes that foundation by tracing the evolution of enterprise payment processing; examining the compliance, security, and integration challenges organizations face today; and introducing the SAP digital payments add-on as the platform that the rest of this book will guide you to implement.*

Enterprise payment processing is undergoing a fundamental transformation driven by cloud technology, real-time settlement demands, new payment methods, and increasingly complex compliance requirements. For organizations running SAP, this transformation leverages the SAP digital payments add-on, a cloud-based middleware solution hosted on SAP Business Technology Platform (SAP BTP). This chapter establishes the foundational context you need before exploring the following chapters.

This chapter is organized into four main sections that build upon each other. We begin with the evolution of enterprise payment processing, tracing the journey from mainframe-era batch systems to today's cloud-based and real-time landscape. We then examine current challenges in payment processing, including compliance, security threats, integration complexity, and manual processes. From there, we introduce the SAP digital payments add-on and its core capabilities, showing how the solution's architecture directly addresses these challenges. Finally, we build a business case for payment modernization, connecting technical capabilities back to measurable cost savings, strategic benefits, and risk reduction.

## 1.1 Evolution of Enterprise Payment Processing

Grasping the evolution of enterprise payment processing gives valuable insight into today's payment setups in SAP. By understanding this history, you can see why legacy systems work the way they do, learn about the architectural choices behind current solutions, and map out steps for modernizing payments in the future. For SAP professionals, this context is crucial for making informed design, implementation, and strategic decisions. When you must justify the significant investments required for payment modernization projects, this background supports smarter, more effective choices.

The evolution from paper-based manual processing in the past to today's instant mobile payments (see Figure 1.1) represents a continuing transformation. What makes this particularly relevant for SAP implementations is that on any given day, enterprise systems might be processing modern cloud-based payment services as well as running batch payment programs that were configured in the 1990s. This isn't a limitation; it's the reality of production systems that must continue operating while they evolve.

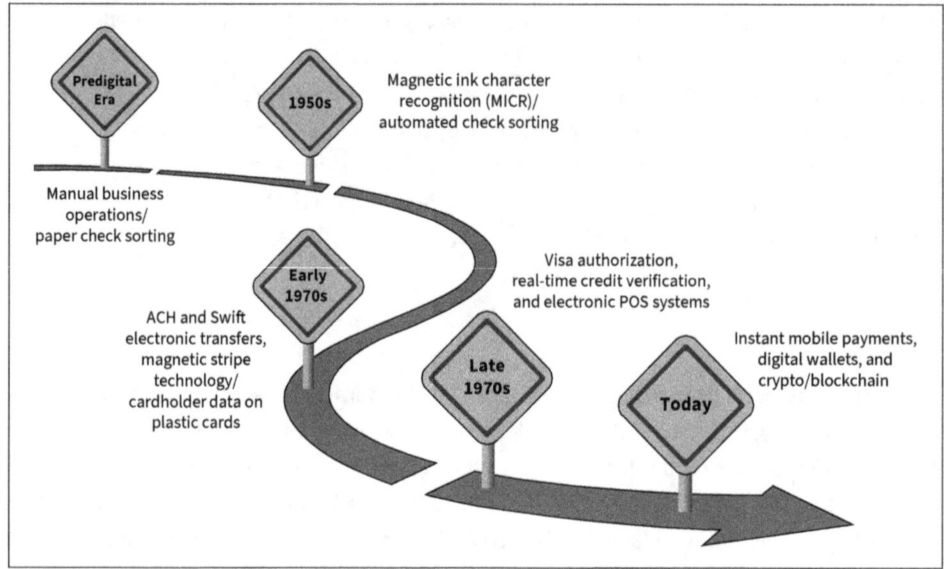


*Figure 1.1: Enterprise Payment Processing Evolution*

### 1.1.1 Background

Before digital payment systems, businesses relied on manual operations in which employees sorted paper checks by hand. The introduction of magnetic ink character recognition (MICR) in the 1950s made automated check sorting possible. The ACH Network and the Swift standards introduced in the early 1970s offered an electronic alternative to paper checks. In the late 1970s, the Visa authorization system provided a method for real-time credit verification. IBM introduced magnetic stripe technology in 1970 through partnerships with American Airlines and American Express. This technology embedded cardholder data directly on plastic cards. Soon after, electronic data-capturing point-of-sale (POS) systems emerged, eliminating paper sales drafts and carbon copy imprinters. These systems laid the technical foundation for the card-based economy.

IBM mainframes became the backbone of enterprise payment processing. Interestingly, most credit card transactions worldwide still run on mainframes; COBOL handles most ATM transactions, and core banking systems at major financial institutions continue to operate on mainframe architecture. For SAP practitioners, understanding mainframe-era concepts, batch-processing orientation, transaction-logging rigor, and data integrity controls remains relevant because these patterns influenced subsequent system designs.

The 1990s saw significant changes as businesses shifted from mainframe-centric computing to distributed client-server systems. This change allowed for tighter integration between payment processing and other business functions, moving payments from isolated back-office operations to essential parts of enterprise resource planning (ERP) systems.

Table 1.1 highlights the design choices made in payment processing over the years.

| Mainframe Era (1970s–1980s) | Client/Server Era (1990s–2000s) | Cloud and Real-Time Era (2010s–Current) |
|---|---|---|
| ■ Centralized processing (dumb terminals)<br>■ Batch-oriented<br>■ Took days to settle<br>■ Limited scalability | ■ Distributed architecture<br>■ Real-time capable<br>■ ERP integrated<br>■ Horizontal scaling | ■ Microservices<br>■ API orchestration<br>■ Takes seconds to settle; tokenization enabled<br>■ Elastic scalability<br>■ Multichannel and PCI compliant |

*Table 1.1: Enterprise Payment Processing Eras*

Figure 1.2 illustrates the fundamental shift in how payment systems were designed. It illustrates the architectural shift from left to right:

- **Mainframe era**
  Centralized processing with batch jobs and dumb terminals
- **Client-server era**
  Three-tier architecture with the presentation, application, and database layers separated
- **Cloud era**
  Distributed services with API integration, microservices, and real-time processing

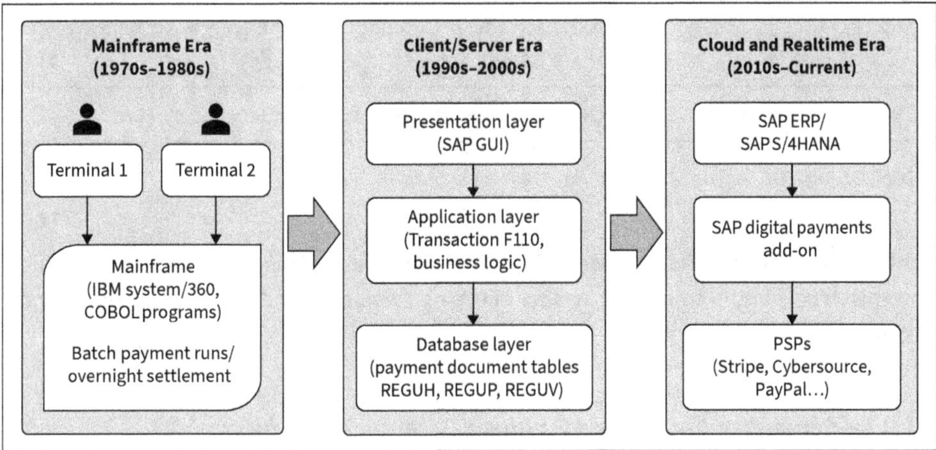


*Figure 1.2: Architectural Shift Through Various Eras*

When SAP launched SAP R/2 in 1979, it operated as a two-tier mainframe system in which all processing occurred on central hardware. SAP R/3, launched in 1992, pioneered a revolutionary three-tier, client-server architecture that separated the presentation, application logic, and database layers. SAP introduced the Transaction F110 automatic payment program in SAP R/3, which continues to be the foundation of SAP payment processing today, more than three decades later. The Transaction F110 architecture established key data structures that SAP practitioners continue working with today, including in SAP S/4HANA. Table 1.2 lists these core database tables.

| Table Name | Description |
|---|---|
| REGUH | Payment run header data (company code, run date, etc.) |
| REGUP | Payment line items (invoice document numbers, amounts, cash discounts, due dates, etc.) |
| REGUV | Administration data (processing status, error messages, payment medium creation status, etc.) |

*Table 1.2: SAP Data Model: Core Payment Processing Tables*

As electronic data interchange (EDI) became the standard protocol for business-to-business payment communications, SAP R/3 incorporated IDoc-based payment integration in order to support electronic bank communication. Table 1.3 highlights key IDocs used in payment integration.

| Message Type | IDoc Type | Purpose/Context/Trigger |
|---|---|---|
| PAYEXT | PEXR2002 | Payment instructions sent from your SAP system to the bank<br>Generated during the payment run via Transaction F110 |
| REMADV | PEXR2002 | Remittance advice to vendors providing details on the invoices that are being paid |
| FIDCCP | FIDCCP02 | Exchange credit card payment data between systems |

*Table 1.3: Key EDI Messages Used in Payment Integration*

Configuration through Transaction WE20 partner profiles enabled electronic communication with banks and trading partners, eliminating manual file transfers and reducing processing errors.

### 1.1.2 E-Commerce Growth and Payment Gateway Innovation

The rapid growth of e-commerce in the late 1990s and the subsequent shift to mobile and social e-commerce significantly reshaped enterprise payment requirements as the growth in e-commerce required a new payment infrastructure, one capable of processing card-not-present transactions securely and efficiently.

Payment gateways helped to bridge the gap between e-commerce platforms and traditional card-processing networks. Cybersource was one of the first e-commerce payment-management companies that offered merchants a way to accept credit cards online without building their own processing infrastructure. Authorize.net soon thereafter introduced its gateway that did not require preregistration before processing test transactions. Payment gateways soon became critical e-commerce infrastructure rather than optional add-ons. SAP also enhanced its payment capabilities during this period. SAP ERP, released in 2004, brought revised technical architecture and expanded functionality.

Table 1.4 compares the payment solutions SAP introduced in SAP ERP.

| Solution | Key Capabilities | Primary Use Cases |
|---|---|---|
| Payment Medium Workbench | Configuration-based formats; enhanced performance for 50,000+ payments | Mass payment processing; format customization |
| SAP Bank Communication Management | Payment batching; multilevel approval; status monitoring (Transaction BNK_MONI) | Treasury operations; payment control |
| SAP In-House Cash | POBO (payment on behalf of)/ROBO (receipt on behalf of); virtual internal bank | Shared service centers; corporate treasury |
| Cross-Application Payment Card Interface (CA-PCI) | Credit card authorization and capture | E-commerce; call center sales |

*Table 1.4: SAP ERP Payment Capabilities*

SAP ERP credit card processing used several key database tables and transaction codes (see Table 1.5).

| Database Tables | Transaction Codes |
|---|---|
| ▪ `FPLTC`: Payment card transaction data in Sales and Distribution<br>▪ `BSEGC`: Payment card transaction data in Financial Accounting<br>▪ `BAPIACCARD`: BAPI structure for card authorization<br>▪ `BAPIACCARDC`: BAPI structure for card capture | ▪ Transaction VA01: Sales order entry with payment card capture<br>▪ Transaction VCC1: Manage credit card orders on hold<br>▪ Transaction VFX3: Manage blocked invoices awaiting payment authorization |

*Table 1.5: SAP ERP: Key Payment Tables and Transaction Codes*

Now, there is a fundamental architectural shift toward cloud-based services, mobile-enabled payment methods, and real-time settlement capabilities. SAP's modern payment evolution has centered on SAP S/4HANA and SAP BTP to align with technological advances and evolving consumer expectations shaped by smartphones and on-demand service models. Table 1.6 describes the modern payment solutions offered by SAP.

| Solution | Capabilities |
|---|---|
| SAP S/4HANA | ■ Manage automatic payments; schedule payment proposals |
| SAP Multi-Bank Connectivity | ■ Direct Swift connectivity; native ISO 20022 format support<br>■ Automatic bank statement imports; intelligent message routing<br>■ Simplified bank communication compared to point-to-point approach |
| SAP digital payments add-on | ■ Replaces CA-PCI as a software-as-a-service cloud product on SAP BTP<br>■ Payment Card Industry Data Security Standard (PCI DSS)–compliant tokenization; unified payment processing across multiple sales channels; multi-PSP support<br>■ Native integration with SAP S/4HANA, SAP Commerce Cloud, SAP Subscription Billing, and so on |
| SAP S/4HANA Cloud for advanced payment management | ■ Centralized payment processing across corporate groups<br>■ Support for internal payments between group entities, payments made in the name of a subsidiary with bank forwarding, and routing optimization<br>■ Supports POBO scenarios in which the treasury executes payments for multiple entities, centralized incoming payments through ROBO, and in-house banking for SAP S/4HANA Cloud for advanced payment management<br>■ Real-time balance updates; configuration-based payment routing |

*Table 1.6: SAP's Modern Payment Solution Offerings*

SAP S/4HANA offers several key architectural improvements for payment processing:

- In-memory database processing via SAP HANA enables real-time analytics on payment data.
- Universal Journal table `ACDOCA` replaces cluster tables for a simplified data model.

- SAP Fiori–based user interfaces improve the user experience without changing the underlying logic.
- Native integration with SAP BTP services enhances cloud capabilities.
- Support for ISO 20022 messaging standards allows for international payment standardization.

### 1.1.3 Global Payment Schemes and Regional Innovation

As enterprises expanded across borders through organic or inorganic growth, they needed to support diverse regional payment schemes alongside international standards. Rather than converging toward a single global standard, payment processing has become increasingly fragmented, with strong regional schemes emerging to serve specific markets. Table 1.7 summarizes the major regional payment initiatives.

| Region | Key Features |
|---|---|
| Europe | ■ Single Euro Payments Area (SEPA) Credit Transfer: Support for electronic bank transfer for euro payments within the SEPA zone<br>■ SEPA Direct Debit: Automated pull payment method for recurring payments with support for customer mandate<br>■ SEPA Instant Credit Transfer: Immediate electronic payments with instant settlement and fund availability within 10 seconds |
| China | ■ Alipay: Dominant payment platform in China that allows users to link credit or debit cards and facilitate transactions using QR codes<br>■ WeChat Pay: Another dominant digital payment platform that offers payment features integrated directly into WeChat, China's most popular instant messaging and social media app, allowing users to send money in chats without having to leave the chat app |
| India | ■ UPI: Real-time mobile payment system that supports instant, real-time fund transfers, peer-to-peer (P2P) transfers, merchant payments via QR codes, and automatic bill payments |
| Brazil | ■ PIX: Real-time payment platform that supports P2P, business-to-business (B2B), government, and in-store payments (QR codes, PIX key support) |
| United States | ■ Zelle: Instant P2P payment service that allows users to send and receive money directly between eligible US bank accounts, integrated into mobile banking apps |

*Table 1.7: Regional Payment Schemes Overview*

Since the early 2000s, buy now, pay later (BNPL) has emerged as a new payment category. It follows the model of splitting purchases into installment payments without traditional credit card interest charges. Klarna, Affirm, Afterpay, PayPal, and Zip are some of the well-known names in BNPL, and customer adoption is steadily growing. BNPL integration requirements for SAP customers include real-time credit-decisioning API integration, split payment handling across multiple settlement dates, merchant discount fee calculation and accounting, reconciliation of partial payments with order fulfillment, and integration with the digital payments add-on or custom gateway connections. The digital payments add-on represents the latest architectural response from SAP to manage the evolution and complexity of enterprise payment methods.

## 1.2 Current Challenges in Payment Management

Organizations managing payment operations face issues like compliance complexity, operational inefficiency, security vulnerabilities, and integration difficulties. PCI DSS compliance demands extensive security controls and ongoing validation, with the scope expanding as organizations process more cardholder data. Payment method diversification requires supporting credit cards, digital wallets, and regional payment schemes, each with distinct integration needs. Manual payment processes that worked at lower volumes create bottlenecks as transaction counts grow, while legacy system integrations multiply the maintenance effort across multiple payment gateways and challenges. Each challenge comes with costs, including financial penalties, operational expenses, customer dissatisfaction, and competitive disadvantage.

### 1.2.1 PCI DSS Compliance

PCI DSS compliance is one of the most significant challenges for enterprises processing credit card payments. PCI DSS version 4.0 has over 50 new requirements compared to past versions, which substantially increases the compliance complexity. Table 1.8 outlines the key challenges of PCI DSS 4.0 implementation.

| Area | PCI DSS 4.0 Requirement | Impact |
|---|---|---|
| Scope determination | Identify all systems, processes, people, and technologies touching cardholder data | Requires comprehensive data flow mapping across complex enterprise architectures |
| Multifactor authentication (MFA) | MFA required for all access to cardholder data environment | Must deploy MFA infrastructure across all access points |
| Continuous monitoring | Shift from periodic assessments to ongoing threat detection | Requires advanced security monitoring tools and dedicated personnel |

*Table 1.8: PCI DSS 4.0 Challenges*

| Area | PCI DSS 4.0 Requirement | Impact |
|---|---|---|
| Customized controls | Flexibility to implement alternative controls that meet security objectives | Demands higher security expertise to design and justify alternatives |
| Enhanced encryption | Stronger encryption protocols for data transmission and storage | May require infrastructure upgrades and cryptographic key management |
| Service provider standards | Increased accountability and more frequent audits | Heightened documentation and validation requirements |
| Risk analysis | Ongoing risk analyses to identify and mitigate threats | Must establish formal risk-management processes |

*Table 1.8: PCI DSS 4.0 Challenges (Cont.)*

Scope reduction through architectural changes is the most effective approach to managing the PCI compliance burden. Tokenization, point-to-point encryption, and cloud-based payment processing that keeps sensitive data entirely outside enterprise systems eliminate most compliance obligations, transferring responsibility to specialized payment service providers (PSPs) that are already maintaining compliance at scale.

### 1.2.2 Payment Method Expansion

Consumer payment preferences have diversified a lot, creating integration and support challenges for enterprises. Organizations once needed to support credit cards, debit cards, and perhaps ACH transfers. Today's payment processing demands accommodation of digital wallets, BNPL options, real-time payments, account-to-account transfers, and region-specific schemes—each with distinct technical requirements, settlement timing, and failure scenarios, as follows:

- Customers may abandon transactions when their preferred payment method is unavailable. This creates a direct revenue impact as organizations limiting payment options lose sales to competitors offering broader choices.
- BNPL presents additional integration complexity. Providers like Klarna, Affirm, and Afterpay each maintain proprietary APIs with different integration requirements. BNPL transactions require handling split payments across multiple settlement dates, calculating merchant discount fees correctly, reconciling partial payments with order fulfillment, and managing credit-decisioning workflows that occur during checkout. For SAP customers, this means custom development or middleware must connect Transaction F110 payment processing with BNPL provider APIs.
- Digital wallets introduce token-management complexity. When customers pay via Apple Pay or Google Pay, merchants receive network tokens rather than actual card numbers.

These tokens require different validation, have distinct lifecycle management needs, and involve additional parties in the payment chain. Organizations must integrate with Visa Token Service, Mastercard Digital Enablement Service, and other token service providers while maintaining backward compatibility with traditional card processing.

### 1.2.3 Manual Processes

Despite automation progress, many enterprises have manual payment processes that increase operational costs and introduce errors. These manual workflows exist due to custom requirements, legacy system limitations, or deferred modernization investments. The accumulated inefficiency creates a financial impact that often exceeds the cost of modernization. Let's look at the negative impacts of manual processes:

- Manual data entry introduces increased error rates. In accounts payable operations, these errors show up as duplicate payments, incorrect payment amounts, payments to wrong vendors, and missed early payment discounts. Processing delays are another hidden cost. Manual payment approval workflows often require moving documents between departments via email or physical routing. This creates delays in payment cycles that negatively impact supplier relationships and miss early payment discount opportunities.
- Cash flow visibility suffers with manual processes. Finance teams lack real-time insight into payment status when tracking occurs via spreadsheets and email threads. This makes cash forecasting difficult, complicates working capital management, and creates surprises when large batches of payments clear simultaneously. Organizations with strong cash flow forecasting typically hold less idle cash, reducing opportunity costs while maintaining adequate liquidity.
- Security and fraud risks increase with manual handling. Paper checks can be lost, stolen, or altered. Payment information written on physical documents or stored in spreadsheets lacks encryption and access controls. Manual processes do not include automated validation checks that detect suspicious patterns like payment amount anomalies, duplicate vendor records, or unusual payment timing. Without these controls, fraudulent invoices and unauthorized payments may escape detection until identified during audits.
- Manual processes that work well at lower volumes break down when transaction counts increase. Organizations experiencing rapid growth often find accounts payable departments overwhelmed, creating backlogs that damage vendor relationships and lead to missed payment deadlines.

### 1.2.4 Integration Complexity

Enterprise payment operations do not happen in isolation. Payment processing must integrate with ERP systems, customer relationship management platforms, e-commerce applications, order management systems, subscription billing platforms, and treasury management systems. Each integration point introduces complexity, creates potential failure points, and demands ongoing maintenance as systems evolve, as follows:

- Each gateway maintains proprietary APIs with different authentication mechanisms, data formats, error-handling approaches, and timeout behaviors. Organizations supporting multiple gateways need separate integration code for each, multiplying development and maintenance effort.
- E-commerce platforms need payment integration that differs from that of call center applications or customer service systems using custom UIs. Each channel may integrate with different payment gateways based on deployment timing, regional requirements, or business unit preferences. Maintaining consistent payment-processing logic, reconciliation procedures, and exception handling across these varied integrations requires significant technical effort.
- Data synchronization introduces additional complexity. Payment transactions generate data that must flow to multiple systems. Successful payment authorizations trigger order processing, inventory allocation, and fulfillment workflows. Captured payments update accounts receivable, trigger revenue recognition, and affect cash forecasting. Refunds require coordination between payment processing, order management, inventory, and accounting. Manual data entry to synchronize these systems introduces delays and errors, while automated integration requires careful design to handle edge cases and maintain consistency.

Legacy SAP implementations process payments in batch cycles: daily Transaction F110 runs, nightly bank statement imports, and weekly reconciliation jobs. Modern payment methods providing instant settlement and immediate confirmation don't align with these batch windows. Bridging real-time payment events to batch-oriented financial systems requires middleware that queues transactions, manages timing differences, and handles scenarios in which the real-time payment succeeds but the batch processing encounters errors.

Regulatory requirements vary by jurisdiction, complicating global implementations. Strong customer authentication mandates in Europe under PSD2, 3D Secure requirements, data localization laws, and reporting obligations differ across regions. Payment integration must accommodate these variations while maintaining consistent user experiences and operational processes. Organizations expanding internationally may find that their payment integration designed for one market doesn't satisfy requirements in others, forcing costly rework.

### 1.2.5 Security and Fraud

Payment processing is prone to fraud and cyberattacks. Card-not-present fraud increases substantially with e-commerce. Customers using stolen card numbers for online purchases face lower detection risk than those committing in-person fraud. EMV chip adoption reduces card fraud at physical points of sale, but online channels lack chip technology's protections. Organizations accepting online payments experience higher fraud rates than those processing primarily in-person transactions, driving the need for sophisticated fraud detection and prevention tools. Beyond basic fraud prevention, organizations must address key security and operational risks that impact payment-processing integrity and financial stability—for example:

- Account takeover attacks target customer accounts with stored payment methods. Once authenticated to customer accounts, attackers make fraudulent purchases using stored payment cards or change payment settings to redirect funds. Organizations must implement strong authentication, unusual activity detection, and transaction verification to mitigate these risks.
- Business email compromise attacks target payment operations directly. Attackers impersonate executives or vendors through compromised email accounts, requesting urgent payments to fraudulent accounts. Finance teams receiving authentic-appearing emails from executive accounts or known vendors may process payments without sufficient verification, resulting in funds being transferred to attacker-controlled accounts.
- Malware specifically targeting payment systems poses a major threat. These technical attacks require robust security controls at payment processing endpoints—controls that may be missing or outdated in legacy implementations.
- Organizations must implement segregation of duties, access logging, and monitoring of unusual employee activity within payment systems. In addition, organizations must assess third-party security postures, monitor for breaches affecting partners, and architect payment flows to minimize data sharing with external parties.
- Currency conversion and foreign exchange (FX) expose organizations to additional complexity and risk. Payments made in foreign currencies create FX exposure between the transaction and settlement dates. Organizations must decide whether to accept this exposure, hedge through derivative instruments, or use payment processors offering dynamic currency conversion. Each approach has cost, complexity, and risk implications that require treasury expertise to evaluate.

The operational challenge of managing global payment operations extends beyond technology. Organizations need payment operations staff with expertise in local requirements, relationships with regional banks and payment providers, and understanding of local customer behavior. Compliance becomes significantly more complex across jurisdictions. Organizations must understand and implement PCI DSS (global), General Data Protection Regulation (GDPR; Europe), various data protection laws, anti-money laundering requirements, sanctions screening obligations, and tax reporting mandates, each varying by country. Maintaining compliance across multiple jurisdictions requires legal expertise, process controls, and audit capabilities that far exceed domestic compliance requirements.

The digital payments add-on addresses these challenges through architectural approaches that reduce the compliance scope, standardize integration across payment methods and channels, eliminate manual processing through automation, enhance security through tokenization and cloud-based processing, and provide consistent capabilities across geographic regions.

## 1.3 SAP Digital Payments Add-On Overview

The digital payments add-on was introduced in 2017, as a cloud-based software-as-a-service (SaaS) offering, hosted on SAP BTP. The solution fundamentally changes how organizations integrate payment processing with SAP applications.

Rather than building and maintaining point-to-point connections between SAP systems and individual PSPs, the digital payments add-on provides standardized middleware that handles complexity centrally while reducing the PCI compliance scope for SAP environments.

The add-on operates as an intermediary infrastructure between consumer applications and PSPs. Consumer applications including SAP S/4HANA, SAP Commerce Cloud, SAP Subscription Billing, or custom applications initiate payment operations by calling standard digital payments add-on APIs. The add-on then orchestrates communication with appropriate PSPs based on configuration rules, handles data transformation between application formats and PSP requirements, manages tokenization to protect sensitive payment data, and returns the payment results to calling applications.

The add-on's cloud-based architecture provides several advantages over traditional point-to-point integrations, as follows:

- SAP maintains and upgrades the core infrastructure, eliminating customer responsibility for server management, security patching, and capacity planning.
- The multitenant SaaS model distributes costs across subscribers rather than requiring a dedicated infrastructure for each customer.
- Elastic scalability automatically adjusts to transaction volume fluctuations without manual intervention.
- Geographic distribution across SAP BTP data centers provides high availability and disaster recovery capabilities built into the service.

In addition, the deployment model keeps sensitive payment data outside customer SAP systems entirely. When customers enter payment card information, that data flows directly to PSPs through the digital payments add-on without touching SAP systems. The add-on returns *tokens*, randomly generated identifiers with no intrinsic value, which SAP applications store instead of actual card numbers. This architectural decision fundamentally reduces the PCI DSS compliance scope for SAP environments.

The token vault is a piece of critical infrastructure within the digital payments add-on. When customers register payment cards, the digital payments add-on forwards card details to the PSP, which returns a network token from Visa Token Service or Mastercard Digital Enablement Service. The add-on stores this token in its secure vault and provides an add-on-specific reference token to calling applications. This two-tier tokenization architecture ensures that even compromising the digital payments add-on vault would not expose usable payment credentials.

Token lifecycle management handles scenarios beyond simple storage. The add-on handles these lifecycle events through APIs that consumer applications invoke as needed:

- Tokens can be updated when underlying card details change. Card networks provide token update services that propagate new expiration dates or replacement card numbers without customer interaction.
- Tokens can be suspended if fraud is suspected, preventing their use without deleting the token entirely.
- Tokens can be deleted when customers remove payment methods or close accounts.

The adapter framework isolates consumer applications from PSP specifics. Each PSP maintains proprietary APIs with different authentication mechanisms, data structures, error codes, and operational characteristics. Direct integration requires custom code for each PSP, multiplying development and maintenance effort. The digital payments add-on abstracts these differences through standardized adapters.

Each adapter implements standard interfaces defined by the digital payments add-on while handling PSP-specific communication requirements internally. When consumer applications request payment authorization, they call a generic digital payments add-on API without specifying which PSP to use. The add-on determines the appropriate PSP based on merchant mapping rules, invokes the corresponding adapter with standardized parameters, and translates responses into a uniform format for the calling application. This architecture provides substantial benefits:

- Organizations can switch PSPs by changing their configuration rather than modifying application code.
- Multiple PSPs can be supported simultaneously for different regions, business units, or payment methods.
- New PSP adapters become available through SAP and partners without requiring changes to consumer applications.
- Organizations reduce vendor lock-in because payment logic remains independent of specific PSP implementations.

If you need partners to create adapters for PSPs not yet certified by SAP, the adapter development kit allows you to do so. This extensibility ensures that organizations can integrate with regional or specialized PSPs as business requirements demand. Partners certify adapters through SAP's certification process, providing quality assurance and compatibility guarantees for customers.

PSPs communicate transaction updates through *webhooks*, HTTP callbacks triggered by payment events. When a payment status changes—an authorization expires, a chargeback is filed, a payment completes settlement—the PSP sends a webhook notification to configured endpoints. The digital payments add-on provides webhook management infrastructure that consumer applications leverage.

Webhook endpoints require several technical considerations: URLs must be publicly accessible so that PSPs can reach them, requiring appropriate firewall and network configuration. Endpoints must authenticate webhook requests to prevent spoofing—typically through signature validation using shared secrets. Processing must be idempotent because PSPs may send duplicate notifications, requiring applications to handle repeat events gracefully. High availability becomes critical because missed webhooks may require manual intervention to reconcile payment state.

The digital payments add-on handles these technical requirements centrally. Organizations configure webhook endpoints once in the add-on, which then manages signature validation, duplicate detection, and reliable delivery to consumer applications. This centralizes expertise and infrastructure rather than requiring each consumer application team to implement webhook handling independently.

When it comes to certifications and support, the add-on has PCI DSS Service Provider Level 1 certification, the highest validation level, requiring annual audits by Qualified Security Assessors. This certification covers the entire digital payments add-on infrastructure, including the token vault, API layer, PSP adapter framework, and webhook processing. Customers benefit from this certification without maintaining separate PCI environments or conducting separate assessments for payment infrastructure.

API authentication uses industry-standard OAuth 2.0 with service-to-service flows appropriate for system-to-system integration. Consumer applications first obtain bearer tokens from SAP BTP identity services, then include these tokens in API requests. The digital payments add-on validates tokens, checks authorization for requested operations, and processes requests only after successful authentication and authorization. This prevents unauthorized access even if network-level controls fail.

The digital payments add-on and certified PSP adapters support extensive payment method coverage, spanning traditional cards, digital wallets, alternative payment methods, and real-time payment rails. The specific methods available depend on PSP selection and geographic markets served. Card payment support includes major global networks: Visa and Visa Electron, Mastercard and Maestro, American Express, Discover and Diners Club, JCB (Japan), and UnionPay (China).

Digital wallet support varies by PSP but commonly includes Apple Pay, Google Pay, PayPal, Samsung Pay, Alipay, and WeChat Pay. These wallets require *network tokenization support*—a capability that certified PSP adapters provide through integration with Visa Token Service, Mastercard Digital Enablement Service, and proprietary wallet-tokenization systems.

Global companies require payment processing across multiple currencies with appropriate FX handling. The digital payments add-on supports multicurrency scenarios through PSP determination logic and merchant account configuration. Organizations can configure different merchant accounts for different currencies.

The digital payments add-on also maintains compliance with relevant payment industry standards and regulations:

- Strong Customer Authentication compliance for European transactions implements PSD2 requirements through 3D Secure 2.0 protocol integration. When processing European card payments, the add-on triggers authentication flows that collect two of three authentication factors—knowledge (password), possession (phone), or inherence (biometric).
- GDPR compliance addresses personal data handling for European customers. The add-on provides data subject access request capabilities, allowing individuals to retrieve their stored payment tokens, use data deletion APIs for right-to-be-forgotten requests, and have data processing agreements establishing SAP's role as data processor rather than controller.
- Regional certifications vary by PSP. Cybersource holds certifications across multiple regions, including North America, Europe, Asia-Pacific, and Latin America. Adyen maintains licenses in jurisdictions worldwide. Organizations selecting PSPs must verify the coverage for markets they serve, ensuring the PSP adapter combination supports required capabilities in each geographic area.
- The multi-PSP architecture provides compliance redundancy. If one PSP encounters regulatory issues in a specific market, then organizations can route transactions through alternative certified providers without system downtime. This redundancy particularly matters for global operations in which regulatory landscapes evolve differently across jurisdictions.

**Enterprise-Grade Availability, Performance, and Support Service-Level Agreements**

As a cloud service, the digital payments add-on operates within defined service-level agreements (SLAs) that cover availability, performance, and support. SAP schedules maintenance during low transaction periods and provides advance notice for planned maintenance. Unplanned outages trigger incident management processes, with priority based on business impact.

Transaction volume limits exist per merchant account and overall tenant. PSPs also impose velocity limits, preventing rapid-fire transactions that might indicate fraud or system errors. The digital payments add-on enforces configured throttling limits, rejecting requests that exceed established thresholds.

Support follows SAP's standard incident management processes: Organizations report issues through SAP Support Portal to be analyzed when problems originate in their systems and get resolution.

## 1.4 Business Case for Payment Modernization

A compelling business justification for payment modernization initiatives requires quantifying both the costs of maintaining the current approaches and the benefits achievable through transformation. Decision-makers evaluating digital payments add-on projects need a clear financial analysis showing the return-on-investment (ROI) timelines, risk-reduction value, and strategic advantages beyond immediate cost savings.

Payment modernization initiatives often compete for budget and resources against other strategic projects. You must demonstrate that payment improvements deliver a measurable value justifying the investment. The business case incorporates hard cost savings, soft cost avoidance, risk mitigation value, revenue enablement, and strategic positioning benefits. Each category contributes to the overall justification, though relative weights vary depending on organizational priorities and current pain points.

This section guides you through four key aspects of building a business case: first, establishing a comprehensive baseline by quantifying all current payment processing costs, including hidden expenses; second, articulating strategic benefits that go beyond cost savings, enabling business growth and competitive advantage; third, structuring your business case presentation to address diverse stakeholder perspectives; and fourth, positioning payment modernization within broader transformation initiatives to strengthen executive sponsorship.

### 1.4.1 Quantifying Current State Costs

Understanding current payment processing costs gives a good baseline for calculating potential savings. Many organizations lack clear visibility into total payment costs because expenses may be distributed across multiple budgets—IT infrastructure for payment systems, treasury operations for payment processing and reconciliation, compliance and audit for PCI validation, customer service for payment exceptions, and business units for transaction fees and fraud losses. Table 1.9 provides ways to calculate these payment processing costs.

| Cost Category | Cost Components | Calculation Method |
|---|---|---|
| Transaction processing | Processing labor, payment runs, exception handling | Cost per transaction × annual volume |
| PCI compliance | Qualified Security Assessor audit fees, security tools, vulnerability scanning, penetration testing | Fixed compliance costs + variable by scope |
| Integration maintenance | Developer time for PSP API updates, bug fixes, enhancements | Development hours × blended rate |
| Infrastructure | Servers, storage, network for payment systems, backup and disaster recovery | Allocated infrastructure costs |

*Table 1.9: Calculate Current Payment Processing Costs*

| Cost Category | Cost Components | Calculation Method |
|---|---|---|
| Payment failures | Failed authorization impact, manual retry efforts, lost sales | Failure rate × average transaction value |
| Fraud losses | Chargeback costs, fraudulent transaction losses, investigation time | Fraud rate × transaction volume |
| Opportunity costs | Delayed feature delivery, inability to support new payment methods | Strategic impact assessment |

*Table 1.9: Calculate Current Payment Processing Costs (Cont.)*

Manual processing costs is a key area to pay attention to as these costs scale linearly with volume. Organizations experiencing growth find these costs increasing proportionally; doubling the transaction volume doubles the manual processing costs unless workflow improvements occur. Payment modernization projects can drive automation and significantly reduce manual processing costs, creating immediate savings over a reasonable timeframe, which is something to highlight clearly in the business case.

Another key area to build a business case is to highlight how organizations storing cards on-premise face full Self-Assessment Questionnaire (SAQ) D requirements exceeding 300 controls, requiring extensive security infrastructure, quarterly vulnerability scans, annual penetration testing, and either self-assessment or external audit depending on transaction volume. Qualified Security Assessor fees for external audits also add to the cost.

Moving to tokenized architecture or cloud-based processing reduces compliance obligations substantially, eliminating many security infrastructure costs and reducing audit scope. Table 1.10 quantifies the potential savings categories with calculation approaches.

| Savings Category | Savings Mechanism | Realization Timeline |
|---|---|---|
| Manual processing elimination | Automation reduces the processing cost per transaction. | Immediate after go-live |
| PCI compliance reduction | Tokenization reduces the scope from SAQ D to SAQ A/A-EP. | First annual assessment after implementation |
| Integration maintenance | Standard adapters eliminate custom code maintenance. | Ongoing after PSP migrations complete |
| Infrastructure decommissioning | Cloud migration eliminates on-premise payment servers. | After on-premise systems are retired |

*Table 1.10: Savings Categories and Realization Timeline*

| Savings Category | Savings Mechanism | Realization Timeline |
|---|---|---|
| Payment failure reduction | Better routing and retry logic improve authorization rates. | Three to six months after optimization |
| Fraud loss reduction | Contains advanced fraud tools and real-time monitoring. | Six to 12 months as patterns establish |
| Early payment discounts | Faster processing enables capturing early payment discounts. | Immediate if payment terms support it |

*Table 1.10: Savings Categories and Realization Timeline (Cont.)*

Payment failure reduction is another key area to highlight in your business case because the impact compounds. Each failed authorization potentially represents lost revenue. If a customer's payment fails during an e-commerce checkout, that customer may abandon the purchase entirely, may switch to a competitor offering smoother checkout, or may call customer service, creating additional cost. Even small improvements in the authorization success rate create measurable revenue impact at scale. The digital payments add-on supports this improvement through intelligent routing to PSPs with higher approval rates for specific card types, automatic retry logic with different PSP accounts, and real-time optimization based on observed success patterns.

### 1.4.2 Strategic Benefits Beyond Direct Cost Savings

Financial justification extends beyond quantifiable cost savings to strategic capabilities that enable business growth, reduce risk exposure, and improve competitive positioning. Although harder to quantify precisely, these benefits often drive executive sponsorship more effectively than cost savings alone. Table 1.11 lists benefit categories that can be highlighted in a typical modernization business case.

| Benefit Category | Strategic Value | Business Impact |
|---|---|---|
| Time to market | Launch new payment methods in weeks, not months | Faster market response; competitive advantage |
| Global expansion | Enter new markets without payment infrastructure barriers | Revenue from new geographies; market share growth |
| Customer experience | Seamless checkout; preferred payment method availability | Higher conversion rates; increased customer satisfaction |

*Table 1.11: Strategic Benefits of Payment Modernization*

| Benefit Category | Strategic Value | Business Impact |
|---|---|---|
| Operational agility | Rapidly switch PSPs for pricing or feature advantages | Negotiating leverage; cost optimization |
| Risk diversification | Multiple PSP relationships reduce single vendor dependency | Resilience; business continuity; leverage |
| Compliance posture | Reduced audit scope; centralized security controls | Lower breach probability; faster audit cycles |
| Innovation enablement | Platform for emerging payment technologies and business models | New revenue streams; digital transformation progress |

*Table 1.11: Strategic Benefits of Payment Modernization (Cont.)*

The digital payments add-on helps to realize strategic benefits of payment modernization in the following ways:

- Time-to-market improvements create a competitive advantage in dynamic markets. Traditional payment integration projects requiring six to 12 months of development allow competitors to capture first-mover advantages with new payment methods. The digital payments add-on compresses this timeline to weeks for methods supported by certified PSP adapters, enabling faster response to market opportunities.
- Global expansion scenarios demonstrate significant added value. Organizations entering new geographic markets traditionally needed to establish banking relationships, integrate regional payment methods, ensure compliance with local regulations, and develop market-specific payment processing capabilities. The add-on reduces these timelines by providing preintegrated regional payment methods through global PSPs, enabling faster market entry.
- Customer experience improvements translate to measurable revenue impact. Organizations broadening their payment method support through the digital payments add-on can reduce customer abandonment, directly increasing conversion rates. Even small conversion improvements create substantial revenue at scale.
- The operational agility to switch PSPs provides negotiating leverage and cost optimization opportunities. PSP transaction fees vary by provider, card type, and merchant agreement terms. The add-on architecture allows PSP changes through configuration, creating credible alternatives that strengthen rate negotiations.

Payment processing carries multiple risk categories, including operational risk from payment system failures, compliance risk from PCI violations or data breaches, financial risk from fraud or processing errors, and reputational risk from payment-related customer dissatisfaction. Payment modernization through the digital payments add-on reduces exposure across the following risk categories:

- Operational risk reduction occurs through eliminating single points of failure. Organizations running payment processing on aging on-premise infrastructure face disaster recovery challenges, capacity constraints, and potential extended outages if critical systems fail. The cloud-based digital payments add-on architecture distributes processing across multiple data centers with automatic failover, substantially reducing outage probability and duration.
- Compliance risk represents potentially severe financial exposure. PCI violations can result in fines depending on the violation severity and transaction volume. Data breaches involving cardholder data expose organizations to notification costs, regulatory penalties, and class-action litigation. The add-on architecture systems eliminate many breach scenarios entirely, reducing this risk substantially.
- Financial risk from fraud requires ongoing management. Chargebacks filed by legitimate customers who don't recognize transactions or by fraudsters using stolen cards create direct costs. Organizations with high chargeback rates face increased transaction fees or loss of processing privileges. Advanced fraud detection tools provided by enterprise-grade PSPs integrated through the digital payments add-on help reduce fraud rates below the levels that are achievable with basic fraud screening.
- Reputational risk occurs when payment problems frustrate your customers. Reliable payment processing with strong security reduces these reputational risks.

Some modernization drivers may not be financially quantifiable but carry strategic weight in decision processes. You should address the following qualitative factors explicitly in business cases:

- Technical debt reduction provides a long-term value that is difficult to measure annually. Legacy payment integrations that were built years ago using deprecated APIs, unsupported platforms, or outdated security practices accumulate technical debt. Modernization addresses technical debt proactively rather than reacting to crisis situations that force expensive emergency remediation.
- Talent acquisition and retention benefits from modern technology stack. Developers prefer working with current technologies, cloud platforms, and contemporary integration patterns rather than maintaining legacy codebases using obsolete approaches. Organizations struggling to retain payment processing expertise find that modernization helps attract and retain skilled staff.
- Business continuity improvements enhance organizational resilience. When payment processing depends on a few individuals who understand the legacy system details, this key person risk creates vulnerability. Standardized platforms with extensive documentation and support communities reduce this dependency, making knowledge more transferable and reducing single-person risk.

- Regulatory preparedness positions organizations to respond efficiently to evolving compliance mandates. Payment regulations continue evolving, with new requirements emerging regularly. Organizations with a flexible, modern payment infrastructure can adapt to new mandates more readily.

### 1.4.3 Stakeholder Alignment and Business Case Presentation

Effective business cases should address different stakeholder perspectives with relevant metrics. Finance executives focus on ROI and payback periods. Technical leaders emphasize architectural benefits and risk reduction. Business unit sponsors care about customer experience and revenue enablement. Your business case should address all perspectives while maintaining a cohesive narrative.

Pilot programs offer risk mitigation for organizations that are uncertain about the projected benefits. Rather than implementing an enterprise-wide deployment, organizations can implement the digital payments add-on for a limited scope, such as a single business unit, specific geographic region, or sales channel. The pilot results provide concrete data on the actual cost savings, performance improvements, and implementation challenges, building confidence for a broader rollout.

### 1.4.4 Linking Payment Modernization to Broader Digital Transformation

Payment modernization initiatives gain strategic importance when positioned as components of broader digital transformation rather than isolated IT projects. Organizations pursuing digital transformation across multiple dimensions benefit from integrating payment modernization as a key workstream. The following are typical digital transformation roadmaps in which payments modernization naturally fits as a key workstream:

- **Digital commerce expansion**
  Organizations building their e-commerce capabilities, launching mobile applications, or enabling omnichannel customer experiences need payment processing that works consistently across all touchpoints. The digital payments add-on provides unified payment processing that supports web, mobile, call center, and POS channels through consistent API architecture.
- **Subscription business model transitions**
  Organizations moving from one-time sales to subscription-based revenue need infrastructure to handle recurring billing, automatic payment method updates, retry logic for failed payments, and subscription lifecycle management. The integration of SAP Subscription Billing with the digital payments add-on enables these scenarios without custom development.
- **Customer self-service initiatives**
  Organizations may need to enable their customers to manage their own payment methods, view transaction history, update billing information, and resolve payment issues without contacting customer service. The digital payments add-on APIs support building

these self-service capabilities, reducing customer service costs while improving customer satisfaction through convenient self-management.

- **Data analytics and AI initiatives**
  The digital payments add-on captures comprehensive payment event data, including authorization attempts and results, settlement timing, payment method performance, and failure reason codes. This data feeds analytics identifying payment optimization opportunities, fraud pattern detection, customer behavior analysis, and operational efficiency improvement. Organizations pursuing data-driven decision-making find that payment data provides valuable signals about customer health, market trends, and operational performance.
- **Cloud migration strategies**
  Organizations moving SAP workloads to cloud environments, whether via SAP BTP, hyperscale cloud providers, or hybrid approaches, can modernize payment processing concurrently. This consolidates migration efforts rather than requiring a separate payment modernization initiative later. The digital payments add-on, already cloud-native, integrates seamlessly with SAP S/4HANA Cloud deployments.

**Executive Messaging: Making the Strategic Case**

Although detailed financial models and technical specifications matter for project approval, concise executive messaging determines whether initiatives receive serious consideration initially. Your business case should include an executive summary suitable for board presentation or senior leadership review.

Effective executive messaging includes a problem statement framed in business terms rather than technical jargon, the quantified impact of current challenges on the business metrics executives track, a proposed solution described in terms of business capabilities rather than technical architecture, a financial justification with a clear ROI and payback period, strategic alignment with organizational initiatives and digital transformation objectives, an implementation approach showing a realistic timeline and resource requirements, and risk mitigation that addresses potential concerns proactively.

## 1.5 Summary

This chapter laid the groundwork for payment modernization in your organization by exploring the evolution of enterprise payment systems, highlighting present-day challenges, introducing the SAP digital payments add-on, and outlining frameworks for business justification. Collectively, these elements illustrate why organizations striving for operational efficiency, compliance, and competitive advantage now view payment modernization as essential. Over time, enterprise payment processing has shifted to cloud technology, instant settlements, mobile payments, and diverse global schemes.

To address these new demands, the digital payments add-on offers an integrated, cloud-hosted middleware solution via SAP BTP. Acting as a central hub, it connects your enterprise applications to various payment service providers through standardized interfaces. This design streamlines integration, removes the need for custom PSP connections, allows quick adaptation or addition of PSPs, strengthens security with PCI-compliant data management, and shifts operational support to SAP. The security architecture uses controls like tokenization, encryption, access restrictions, monitoring, and vulnerability management to protect card data. SAP's PCI DSS Service Provider Level 1 certification eases your compliance compared to on-premise storage.

Financial justification for payment modernization requires comprehensive business cases that quantify costs such as transaction processing, PCI compliance, integration maintenance, infrastructure, failures, and fraud. Modernization saves costs by eliminating manual tasks, reducing PCI scope, removing integration maintenance, decommissioning legacy infrastructure, and minimizing payment failures. Strategic benefits include faster time to market for new payment options; smoother global expansion with preintegrated regional methods; improved customer experience; greater PSP negotiation leverage; and lower operational, compliance, and fraud risks.

Modernizing payments with the SAP digital payments add-on is a substantial but achievable project. By building a thorough business case, using the right resources, and ensuring a well-planned implementation, you can certainly gain lower costs, stronger security, better customer experience, and flexible strategies. This book offers detailed guidance for turning payment modernization plans into practice.

Chapter 2

# SAP Digital Payments Add-On Architecture

*Understanding the architecture of the digital payments add-on is essential before configuring or implementing it, because every decision you make later, including SAP BTP setup, PSP integration, and SAP integration, depends on how the add-on's components, APIs, and security layers work together. Building on the business context and challenges introduced in Chapter 1, this chapter provides the technical foundation that transforms the why of payment modernization into the how, positioning you to move confidently into the hands-on implementation that follows.*

This chapter introduces the technical foundation of the SAP digital payments add-on and explains how it fits within the overall SAP ecosystem. Acting as a secure middleware service between SAP applications and external payment service providers (PSPs), it standardizes payment operations and helps reduce Payment Card Industry Data Security Standard (PCI DSS) compliance exposure. The discussion here begins with the technical architecture that positions the add-on as a centralized payment hub on SAP Business Technology Platform (SAP BTP), then details the add-on's core components and services, including the core service, routing engine, API layer, PSP adapters, and configuration UIs that make this hub work. From there, we examine the integration points with your SAP landscape and non-SAP solutions. We then explore the security architecture and token vault, showing how tokenization, PCI DSS scope reduction, and secure credential storage protect sensitive payment data. Finally, we address the high availability and scalability features built into the Cloud Foundry environment that ensure reliable performance under varying transaction volumes. By the end of this chapter, you'll have the architectural background needed to follow the configuration and implementation discussions that appear in later chapters of this book.

## 2.1 Technical Architecture

The SAP digital payments add-on is a cloud-based middleware component hosted on SAP BTP. Its primary role is to act as a centralized payment hub that decouples business applications—SAP or non-SAP—from the complexity of integrating with multiple PSPs. By serving as this central hub, the add-on unifies communication protocols and manages a broad range of payment activities, including sales order authorization, settlement, invoice and refund handling, payment advice reconciliation, and pay-by-link scenarios. The result is a single, secure gateway (see Figure 2.1) through which all digital payment transactions that touch your SAP landscape are processed and monitored.

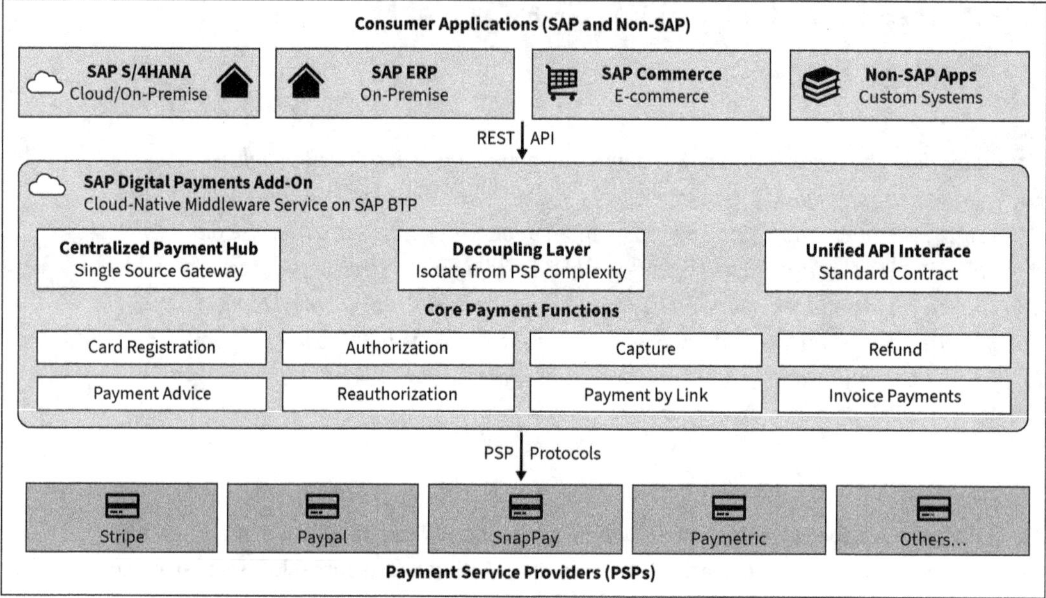


*Figure 2.1: SAP Digital Payments Add-On: Architecture*

At its core, the architecture positions the add-on service between consumer applications—such as SAP ERP, SAP S/4HANA, SAP Subscription Billing, or other, non-SAP solutions on the one hand, and external PSPs like Stripe, PayPal, SnapPay, Worldpay B2B Payments, and the like on the other. This layered arrangement delivers the key advantage of the add-on: Business applications interact through a single, consistent API interface, while the add-on handles the individual PSP integrations and routing logic behind the scenes.

Logically, the architecture has three main layers or component groups:

- **Consumer application interface**
  This layer provides secure APIs, which SAP and non-SAP applications use to initiate and manage payment transactions. Its key characteristics include the following:
  - REST-based web services
  - OpenAPI Specification compliant
  - OAuth 2.0 authentication support
  - Standardized request/response formats
  - Version-controlled API end points

  Table 2.1 lists available API operations commonly used in customer landscapes.

| Available API Functions | | |
|---|---|---|
| Card registration | Card deletion | Refund |
| Authorization | Reauthorization | Payment advice |
| Capture | Settlement | |

*Table 2.1: Consumer API Layer: Available Functions*

- **SAP digital payments add-on core service**
  This is the central engine running on SAP BTP. It receives requests from consumer applications, performs routing determination, orchestrates the end-to-end payment flow, and manages security. Table 2.2 lists the primary responsibilities.

| Routing Engine | Transaction Orchestration | State Management |
|---|---|---|
| Dynamically selects PSP and merchant account based on the following:<br>■ Payment method<br>■ Currency<br>■ Geographic region<br>■ Business rules<br>■ Configured priorities | Orchestrates the end-to-end payment flow:<br>■ Request validation<br>■ PSP adapter invocation<br>■ Response transformation<br>■ Error handling<br>■ Transaction logging | Manages transaction state and security:<br>■ Lifecycle tracking<br>■ Token-to-PSP mapping<br>■ Credential management<br>■ Configuration storage<br>■ Audit trail persistence |

*Table 2.2: Add-On Core: Primary Responsibilities*

- **PSP adapters**
  These components are deployed alongside the core service on SAP BTP. They are responsible for translating the standardized requests from the add-on core into PSP-specific protocols and message formats. Table 2.3 lists several common PSP adapters, together with their key functions and responsibilities.

| Adapter Responsibilities | Common PSP Adapters |
|---|---|
| ■ Protocol translation (REST to PSP format)<br>■ Request/response mapping<br>■ PSP-specific authentication handling<br>■ Error code normalization<br>■ Webhook event processing | ■ Stripe adapter<br>■ PayPal Braintree adapter<br>■ Cybersource adapter<br>■ Worldpay (formerly Paymetric) adapter<br>■ Additional PSP adapters |

*Table 2.3: PSP Adapters: Primary Responsibilities*

A notable concept in this architecture is the use of merchant ID aliases. Each technical PSP merchant account is linked to a user-friendly alias configured within the consumer application (e.g., a virtual bank account). These aliases connect the add-on's routing logic with the application's financial structure, enabling accurate reconciliation and simplified maintenance.

This architectural approach has significant benefits, as follows:

- **Simplified integration**
  Consumer applications can connect with the add-on's APIs without developing or maintaining multiple point-to-point connections with various PSPs. Table 2.4 compares the traditional, tightly coupled model with the add-on-based approach.

| Integration Complexity (Traditional Approach) | Simplified Integration (Add-On Approach) |
|---|---|
| ▪ Direct point-to-point integration with each PSP<br>▪ Maintenance of multiple custom interfaces<br>▪ PSP-specific code in consumer applications<br>▪ *N* applications x *M* PSPs = *N* x *M* integrations | ▪ Single integration point for all PSPs<br>▪ Standardized API eliminates PSP-specific code<br>▪ One-time integration per consumer application<br>▪ *N* applications + *M* PSPs = *N* + *M* integrations |

*Table 2.4: Integration Complexity: Traditional Versus Add-On Approach*

- **Centralized operations and maintenance**
  The add-on consolidates connectivity management, credential handling, and PSP monitoring in one place. This reduces the administrative effort required from application teams. SAP takes care of the operation and maintenance of the cloud service based on defined service-level agreements (SLAs). Table 2.5 shows how this approach minimizes operational overhead compared to decentralized models.

| Operations Burden (Traditional Approach) | Centralized Operations (Add-On Approach) |
|---|---|
| ▪ Distributed credential management<br>▪ Security patches required in multiple systems<br>▪ Separate monitoring for each integration<br>▪ Manual coordination across application teams | ▪ Single point for credential management<br>▪ SAP-managed SLA for cloud service updates<br>▪ Unified monitoring and logging<br>▪ Dedicated operations team management |

*Table 2.5: Administrative Overhead: Comparison*

- **Enhanced security and compliance**
  Through tokenization, sensitive payment data, such as cardholder information, never traverses or resides within business systems (e.g., SAP S/4HANA). This greatly reduces the PCI DSS compliance requirements for the core ERP system. The add-on insulates the application from costly audit and security requirements, serving as a protective compliance layer. Table 2.6 shows the compliance impact of both traditional and add-on architectures.

| Compliance Risk (Traditional Approach) | Enhanced Security (Add-On Approach) |
|---|---|
| ■ Broad PCI DSS scope (entire SAP landscape)<br>■ Cardholder data stored in multiple systems<br>■ Complex compliance audit requirements<br>■ High security implementation costs | ■ Tokenization isolates cardholder data<br>■ Consumer applications not in PCI audit scope<br>■ Reduced compliance costs<br>■ Simplified security control implementation |

*Table 2.6: Security and Compliance: Comparison*

- **Flexibility and scalability**
  Adding new PSPs or switching between them is much simpler. You only need to configure the add-on instead of making major changes to consumer applications, and fewer changes means less retesting of payment flows, which are typically hard to simulate in nonproduction environments. Test cards and test PSP environments do not accurately reflect the end-to-end payment flows in production, where live credit cards are used. In addition, when businesses expand into new markets, they need new payment methods and payment flows to be supported rapidly in order to get new customers. The cloud-native architecture on SAP BTP offers built-in scalability, allowing businesses to scale horizontally and vertically; that is, they can manage sudden increases in transaction volumes during the peak season or inorganic growth through a major acquisition. Table 2.7 shows a comparison of the flexibility benefits of the add-on-based approach compared to the traditional approach.

| Limited Flexibility (Traditional Approach) | Maximum Flexibility (Add-on Approach) |
|---|---|
| ■ PSP switching requires extensive redesign<br>■ Testing required across all consumer apps<br>■ Long implementation cycles for new PSPs<br>■ Difficult to adapt to changing requirements | ■ PSP switching managed easily via configuration<br>■ Minimal testing in consumer applications<br>■ Cloud-native scalability is built in<br>■ Rapid adaptation to business changes |

*Table 2.7: Flexibility and Scalability: Comparison*

## 2.2 Core Components and Services

The SAP digital payments add-on consists of a collection of interconnected components hosted on SAP BTP, primarily within the Cloud Foundry environment. Each component delivers a specific service that contributes to the add-on's role as a central payment hub. Understanding these building blocks and how they interact provides valuable insight into how the add-on handles payment flows, communicates with different systems, and maintains security, configurability, and operational transparency. This section walks through each of these building blocks in detail. We start with the core service, the central orchestrator that handles API requests, transaction processing, and state management. Next, we cover the routing engine, which dynamically determines the correct PSP and merchant account for every transaction based on configurable business rules. We then look at the API layer that defines the service contract between the add-on and consuming applications, followed by the UI integration APIs that enable embedded payment forms and hosted payment pages for e-commerce and self-service scenarios. From there, we examine PSP adapters, the microservices responsible for translating standardized requests into each provider's unique protocols and formats. We then cover the configuration UIs that administrators use to manage PSP activation, routing rules, payment advice, payment pages, and merchant mappings. Finally, we discuss the security components that protect sensitive data through tokenization and secure credential storage, plus the logging and monitoring services that provide operational visibility across the entire payment flow.

### 2.2.1 Core Service

The core service forms the backbone of the add-on. It is built as a microservices-based application running on SAP BTP, Cloud Foundry runtime. This component acts as the main orchestrator, receiving and processing all incoming requests from consumer applications through the API layer. Its main responsibilities include the following:

- **API request handling**
  This involves securely receiving incoming requests from consumer applications, validating API calls against the defined service contract, and parsing request payloads. Given that business users are waiting for results in real time, API request handling also involves conforming to the agreed-upon SLAs.
- **Transaction orchestration**
  This means managing the entire lifecycle of a payment transaction. This involves invoking necessary internal and external services in the correct order and maintaining the transaction's state. In addition, transaction orchestration involves linking capture requests to original authorizations and managing multistep authentication flows.
- **State management**
  This refers to using SAP BTP persistence services to securely store transaction data, statuses, internal token mappings, configuration details, and audit logs.
- **Routing execution**
  This is working with the routing engine component to find the correct PSP and merchant account based on the transaction context and set business rules.

- **Adapter communication**
  This means interacting with the right PSP adapter component, sending standardized request data, and receiving standardized responses for further processing.
- **Security context management**
  This involves handling authentication and authorization information for incoming requests and managing secure communication for outbound calls to adapters and other SAP BTP services.
- **Compliance and auditing**
  Beyond just logging audit trails, specific functions ensure adherence to payment industry regulations (like PCI DSS compliance) by strictly controlling where sensitive data resides and how it is accessed.
- **Error and exception handling**
  This covers comprehensive mechanisms to log errors, manage transaction timeouts, and define fallback logic or alternative PSPs in case of a failed communication attempt.
- **Monitoring and observability**
  Integration with SAP BTP's monitoring services provides visibility into transaction throughput, success rates, latency, and system health.

Figure 2.2 presents a logical view of the add-on core services' internal architecture and its various components, along with its configuration services and persistence services.

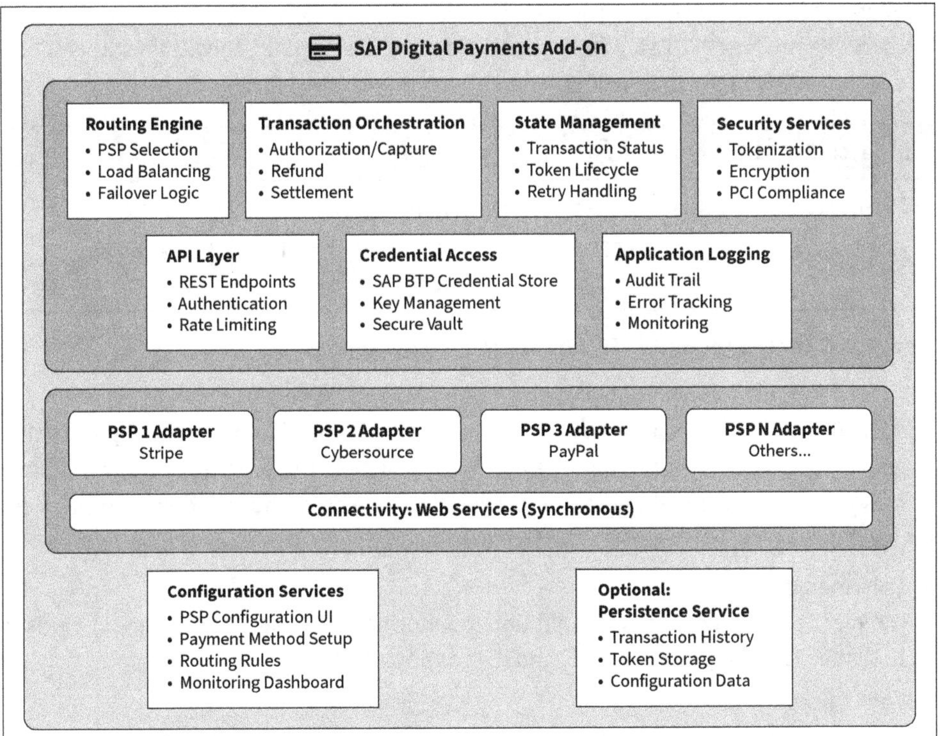


*Figure 2.2: Digital Payments Add-On: Internal Architecture*

### 2.2.2 Routing Engine

A core element of the add-on is the routing engine, which determines the target PSP and merchant account for every transaction. Administrators or key users configure routing rules through the provided UI, defining how the system evaluates each incoming API request. During runtime, the engine sequentially checks the rules until one matches all specified criteria, then selects the appropriate PSP and merchant combination. Common routing parameters include identifiers such as the following:

- Company code
- Country
- Currency
- Payment method
- Payment type

If no conditions match, then the transaction is rejected, making it essential to define a comprehensive ruleset with a default or catchall entry to prevent processing gaps.

### 2.2.3 API Layer (Consumer API)

The API layer defines the official service contract between the add-on and its consuming applications. It follows the OpenAPI Specification and is published in the SAP Business Accelerator Hub for easy discovery and testing. Table 2.8 highlights design principles aligned with modern REST standards such as versioned endpoints, consistent status codes, and secure authentication flows.

| **API Layer Characteristics** | | | |
|---|---|---|---|
| **REST-based**<br>(HTTPS/JSON) | **Language-agnostic**<br>(Any HTTP client) | **OpenAPI 3.0**<br>(Published specification) | **OAuth 2.0**<br>(Bearer token) |

*Table 2.8: Digital Payments Add-On API Layer: Key Characteristics*

The functional services exposed by this API can be categorized into six logical groups that together support enterprise-grade payment capabilities. Table 2.9 shows the first two groups—card management and token preparation—which are foundational for payment processing:

- **Card management**
  Handles registration, lookup, expiration monitoring, and deletion of payment cards in both the digital payments add-on and the connected PSP.
- **Token preparation**
  Registers cards, authorizations, or direct captures created outside the add-on by validating them with the PSP and returning secure add-on tokens for subsequent use.

| Card Management | Token Preparation |
|---|---|
| GET /cards/getregistrationurl<br>Initiate card registration | POST /tokens/getforpaymentcard<br>Register PSP card token |
| GET /cards/poll/{sessionid}<br>Poll card registration status | POST /tokens/getforauthorization<br>Register external payment authorization |
| GET /card/{token}<br>Retrieve card data by token | POST /tokens/getfordirectcapture<br>Register direct capture |
| GET /cards/expiring<br>Find expiring cards | POST /tokens/getforpaymentcardauthoriza-<br>tion<br>Register card authorization |
| POST /cards/delete<br>Delete cards at add-on and PSP | POST /tokens/getforpaymentcardwithautho-<br>rization<br>Register card plus authorization |

*Table 2.9: API Function Groups: Card Management and Token Preparation*

Table 2.10 represents the next two groups—authorization and capture, and payment by link—which cover transactional payment execution:

- **Authorization and capture**
  Supports creating, confirming, canceling, and refreshing payment authorizations as well as executing captures or direct charges against those authorizations. This functionality mirrors standard payment lifecycles used in order-to-cash scenarios.
- **Payment by link**
  Allows creation, retrieval, extension, and cancellation of secure payment request URLs that organizations can share via email, SMS, or portals so that customers can complete payments through a self-service checkout page.

| Authorization and Capture | Payment by Link |
|---|---|
| POST /authorization<br>Authorize payment amount | POST /paymentbylink/create<br>Generate payment URLs |
| POST /authorization/cancel<br>Cancel/void authorization | POST /paymentbylink/getbyid<br>Retrieve link details |
| POST /charges<br>Capture/settle payment | POST /paymentbylink/adjustvalidity<br>Extend link expiration |
| POST /reauthorizations<br>Refresh expired authorization | POST /paymentbylink/cancel<br>Invalidate payment links |

*Table 2.10: API Function Groups: Authorization and Capture, and Payment By Link*

Table 2.11 illustrates the final two functional groups—refunds, and advice and reconciliation—which focus on postpayment activities:

- **Refunds**
  Handles refunds and blind credits back to the original funding source. Requests can be triggered using either the PSP's transaction ID or the add-on's internal payment identifier, ensuring full traceability across systems.
- **Advice and reconciliation**
  Retrieves settlement or transaction advice data from connected PSPs for given date ranges and merchants. Standard and paginated response options allow efficient reconciliation and reporting of payment outcomes within financial processes.

| Refunds | Advice and Reconciliation |
|---|---|
| `POST /refunds`<br>Refund by PSP payment ID | `POST /advice`<br>Get transaction advice data |
| `POST /refunds/bypaymentbydigitalpaymentservice`<br>Refund by digital payments add-on payment ID | `POST /paginateadvice`<br>Get advice in pages |

*Table 2.11: API Function Groups: Refunds, and Advice and Reconciliation*

### 2.2.4 UI Integration APIs

In addition to backend APIs, the add-on provides a dedicated set of UI integration endpoints that enable frontend payment experiences such as e-commerce checkout or invoice payment through embedded forms and hosted pages. Table 2.12 summarizes the endpoints available for these scenarios.

| DPJSLIB Loader | Payment Page |
|---|---|
| `GET /dpjslib/loader`<br>Load javascript library URL<br>*Use case:* Embed secure payment UI elements (iFrame) with 3D Secure support | `POST /paymentpage/initiate`<br>Start payment process |
| | `POST /paymentpage/finalize`<br>Complete and get result |

*Table 2.12: API Function Group: UI Integration*

The following sections walk through representative business use cases that demonstrate how to implement these APIs within typical SAP or custom applications.

#### DPJSLIB Loader: Embedded Payment Form Scenario

In the embedded payment form scenario, the `GET /dpjslib/loader` endpoint enables secure, in-page card entry directly within a customer-facing web application or an SAP product such as SAP Commerce Cloud. The following are the main use cases supported by this approach:

- E-commerce checkout: Customers buying products without page redirects
- Self-service portal: Customers managing payment methods in their accounts
- Call center support: Agent-assisted card registration

Now, let's walk through the logical sequence of steps in a typical e-commerce checkout process. After a customer adds items to the shopping cart and proceeds to payment, the API dynamically loads the required JavaScript to display a secure iFrame for card entry. The service then tokenizes the information and returns the token to the calling application for subsequent authorization or capture. The steps are as follows:

1. **Customer reaches checkout**
   The customer adds items to their cart and proceeds to payment.
2. **Call DPJSLIB Loader API**
   `GET /dpjslib/loader` requests the library URL.
3. **Load JavaScript library**
   This creates the secure iFrame for card entry.
4. **Customer enters card securely**
   The entered data goes to the PSP via the iFrame and is handled by 3D Secure.
5. **Token returned, order complete**
   A sales order is created in SAP with a token.

This design offers several important benefits:

- The customer completes the payment without leaving the merchant's website.
- Card data entry occurs in a secure iFrame, keeping the site outside the PCI DSS scope.
- Native 3D Secure handling is included, improving fraud protection compliance.

#### Payment Page Initiate: Starting Payment Session

The Payment Page Initiate endpoint (`POST /core/v2/paymentpage/initiate`) supports asynchronous payment flows in which customers complete transactions later through links delivered by email, SMS, or hosted redirect. This capability is ideal for invoice or subscription billing scenarios in which payment occurs after order confirmation. Business applications such as SAP S/4HANA or SAP Subscription Billing use this API to generate a secure payment URL for each invoice or order. The following are the most common use cases:

- Sales order: Authorization (order create)
- Invoice payment: Payment via email link
- Subscription: Recurring billing setup
- Donations: Nonprofit payments

An SAP S/4HANA application can automatically call the Initiate API while creating an invoice, attach the resulting payment link to the document or email notification, and send it to the customer. When the customer opens the link, they are redirected to a hosted page managed by the add-on for secure checkout and payment processing. The process flow is as follows:

1. The invoice is created in SAP.
2. The Initiate API is called.
3. A link is sent to the customer.

**Payment Page Finalize: Verifying Payment Outcome**

The Payment Page Finalize endpoint (`POST /core/v2/paymentpage/finalize`) confirms the payment result generated from a previous Initiate API call. In a standard order-to-cash process, this response serves as official payment confirmation, allowing the system to release the sales order and proceed with delivery or shipment. In an invoice-to-cash workflow, the Finalize API validates the payment status so that financial documents can be cleared and a receipt issued to the customer. The following steps illustrate how the Initiate and Finalize APIs operate together to complete the end-to-end payment cycle:

1. An invoice is created in SAP, triggering the payment requirement.
2. The Initiate API is called to create a unique payment session with the PSP, returning a secure hosted page URL.
3. A secure link containing this URL is sent to the customer.
4. The customer opens the link and securely enters payment details on the PSP hosted page.
5. Once the transaction is complete, the customer is redirected back to the SAP system.
6. The Finalize API is called using the session ID to retrieve the payment details and verify the result.

### 2.2.5 PSP Adapters

PSP adapters are individual microservices that handle communication between the add-on and each specific PSP. They accommodate the unique API structure, protocols, authentication methods, and data formats of every provider—offering, for example, dedicated adapters for Stripe, PayPal, and Worldpay. Their key functions include the following:

- **Protocol and format translation**
  Transform the add-on's standard request structure into the format required by the target PSP, then convert the response back into the add-on's canonical representation.
- **Secure communication**
  Manage outbound HTTPS connections to the PSP's API endpoints with appropriate encryption and timeout controls.

- **Credential management**
  Retrieve and use PSP API keys, certificates, and secrets securely from SAP Credential Store.
- **Response normalization**
  Interpret PSP-specific status codes, transaction IDs, and error messages and map them to a common data model for the add-on.
- **PSP-specific processing**
  Address unique provider logic, such as regional API variations or support for Level 2 and Level 3 data requirements.

SAP maintains and certifies adapters for selected PSPs including Stripe, PayPal, and Worldpay. Other providers can develop their own adapters using SAP's public specifications, subject to SAP certification before deployment to production landscapes.

### 2.2.6 Configuration UI

Administrators and key users configure and operate the digital payments add-on through several intuitive UIs provided on SAP BTP. Each UI focuses on a specific configuration area:

- **PSP status management**
  This interface activates or deactivates individual PSP adapters. The URL to access the UI depends on the data center and whether the environment is a test or production subdomain. Full instructions are available in the *SAP Digital Payments Add-On Administration Guide*, under **Setup Activities • Activating PSPs** (see *http://s-prs.co/v629400*). Figure 2.3 shows an example in which the Stripe V2 adapter is active.

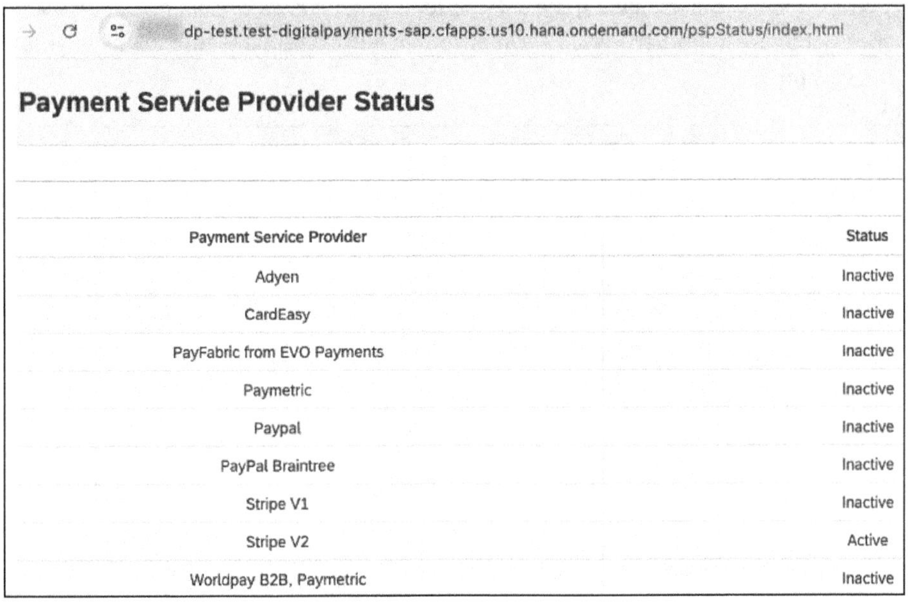

dp-test.test-digitalpayments-sap.cfapps.us10.hana.ondemand.com/pspStatus/index.html

**Payment Service Provider Status**

| Payment Service Provider | Status |
|---|---|
| Adyen | Inactive |
| CardEasy | Inactive |
| PayFabric from EVO Payments | Inactive |
| Paymetric | Inactive |
| Paypal | Inactive |
| PayPal Braintree | Inactive |
| Stripe V1 | Inactive |
| Stripe V2 | Active |
| Worldpay B2B, Paymetric | Inactive |

*Figure 2.3: Configuration UI: PSP Status Management*

- **PSP determination (routing rules)**
  This UI defines and orders the routing logic used by the engine to select the correct PSP. Its access URL varies by data center and subdomain. Refer to the *SAP Digital Payments Add-On Administration Guide* for details, under **Configuration • Setting Up the Payment Service Provider Determination** (see *http://s-prs.co/v629401*). Figure 2.4 shows an example in which Stripe acts as the payment provider across a few company codes, illustrating both mandatory and optional rule parameters.

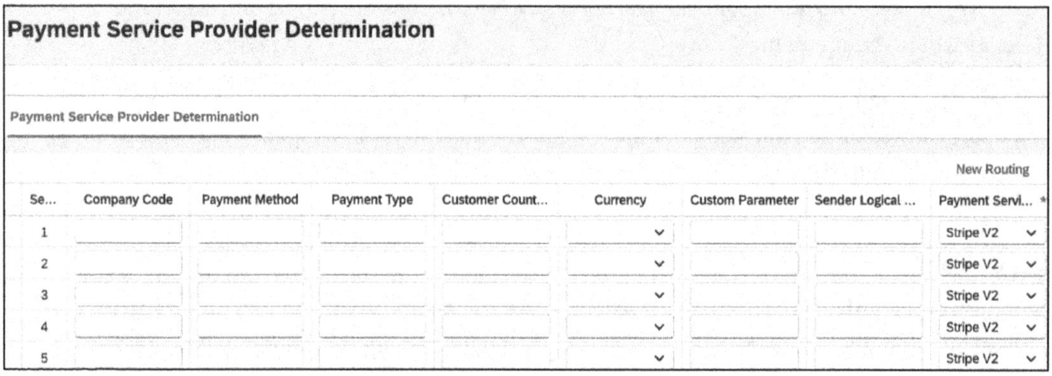

**Payment Service Provider Determination**

Payment Service Provider Determination

New Routing

| Se... | Company Code | Payment Method | Payment Type | Customer Count... | Currency | Custom Parameter | Sender Logical ... | Payment Servi... * |
|---|---|---|---|---|---|---|---|---|
| 1 | | | | | | | | Stripe V2 |
| 2 | | | | | | | | Stripe V2 |
| 3 | | | | | | | | Stripe V2 |
| 4 | | | | | | | | Stripe V2 |
| 5 | | | | | | | | Stripe V2 |

*Figure 2.4: Configuration UI: Routing Setup*

- **Digital payment advice processing setup**
  Because payment advice structures differ across PSPs, the digital payments add-on offers a setup screen for defining how and when advice data should be fetched. Currently, templates exist for Stripe and PayPal. The corresponding access URLs again vary by environment, as explained in the *SAP Digital Payments Add-On Administration Guide*, under **Configuration • Setting up Payment Advice Processing.** Figure 2.5 highlights a Stripe example.

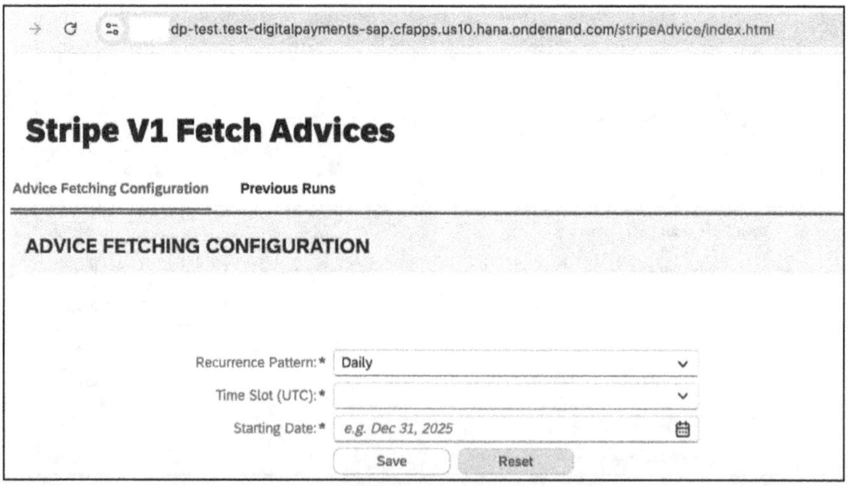

dp-test.test-digitalpayments-sap.cfapps.us10.hana.ondemand.com/stripeAdvice/index.html

**Stripe V1 Fetch Advices**

Advice Fetching Configuration    Previous Runs

**ADVICE FETCHING CONFIGURATION**

Recurrence Pattern:* Daily
Time Slot (UTC):*
Starting Date:* e.g. Dec 31, 2025
Save Reset

*Figure 2.5: Configuration UI: Payment Advice Fetch*

- **Payment page setup**
  This UI supports organizations implementing online stores or self-service portals. It allows configuration of available payment types for customer checkout. Figure 2.6 illustrates a setup in which credit card payments are enabled.

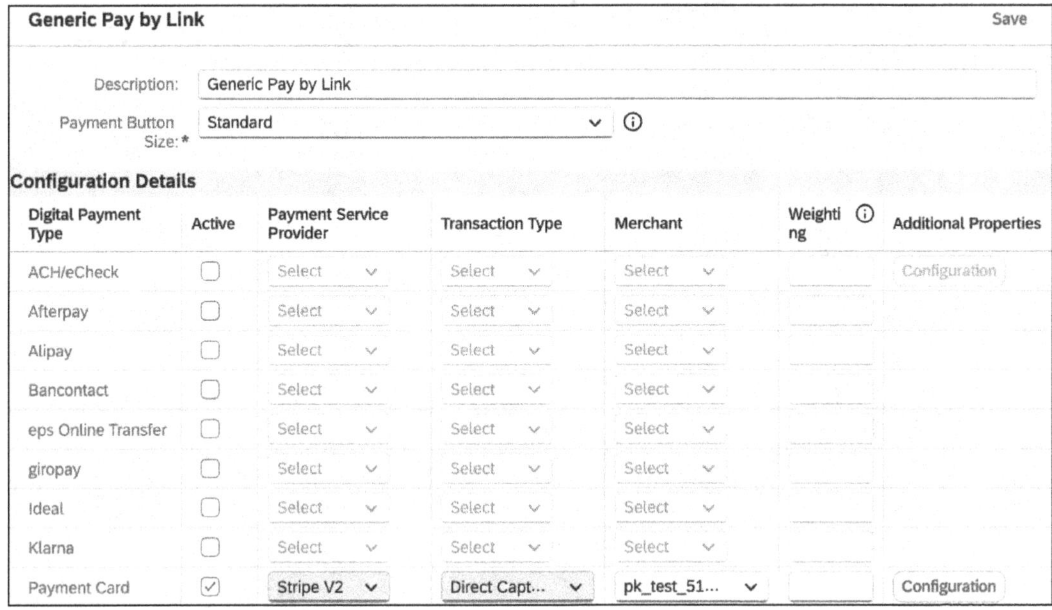


*Figure 2.6: Configuration UI: Payment Page Setup*

Figure 2.7 shows related card configuration options that appear when you click the **Configuration** button.

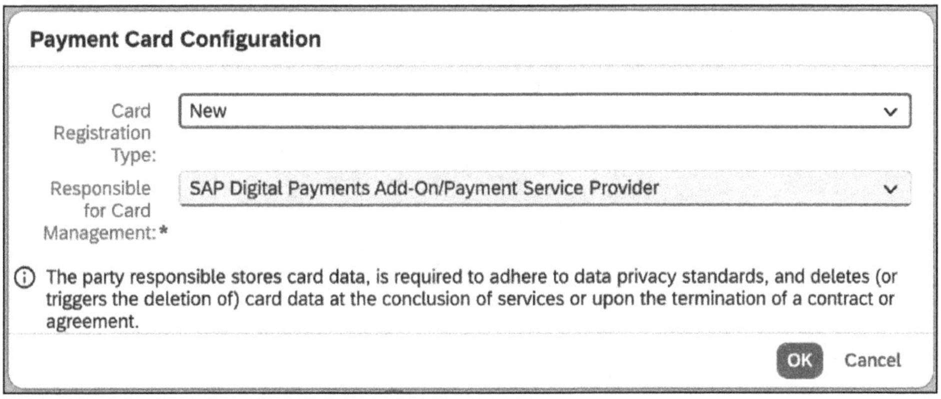

Payment Card Configuration

Card Registration Type: New

Responsible for Card Management:* SAP Digital Payments Add-On/Payment Service Provider

The party responsible stores card data, is required to adhere to data privacy standards, and deletes (or triggers the deletion of) card data at the conclusion of services or upon the termination of a contract or agreement.

OK Cancel

*Figure 2.7: Payment Card Configuration*

- **Payment by link configuration**
  This configuration UI can be used by organizations that want to generate and print a URL or a QR code in the electronic invoices they send to their customers. Customers can use

this payment link to access the checkout page (created using the payment page setup UI discussed previously), where they can choose a payment method and make a payment. Figure 2.8 shows an example that illustrates the details available in this UI. More details are available in the *SAP Digital Payments Add-On Administration Guide*, under **Configuration • Setting Up Payment by Link**.

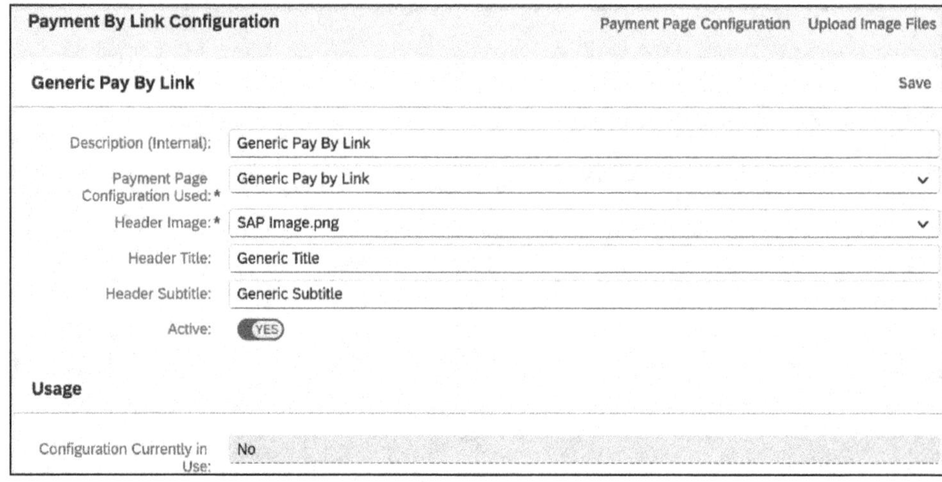


*Figure 2.8: Configuration UI: Payment by Link*

Figure 2.9 shows the API endpoints that support the payment by link feature. These API endpoints are published in SAP Business Accelerator Hub at *https://api.sap.com/api/DPCoreAPI/resource/Payment_By_Link*. You can review the request and response body and example values there as well.

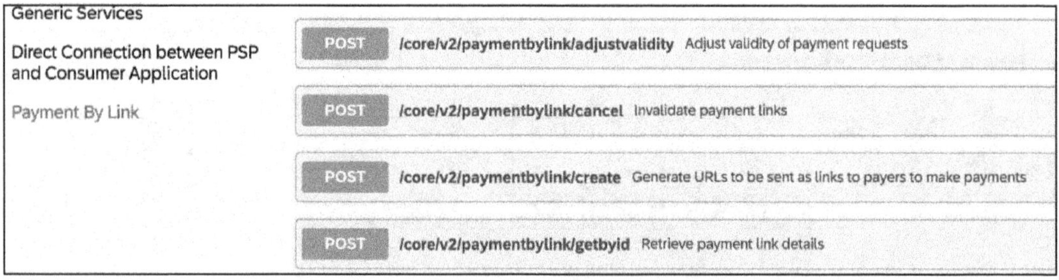


*Figure 2.9: SAP Business Accelerator Hub: Payment by Link API Endpoints*

- **Manage URLs on allowlist**
  The digital payments add-on provides this configuration UI to maintain approved redirect URLs for security reasons. Figure 2.10 shows an example URL that is configured in this UI. The add-on will allow a redirect only to this URL. You can add more URLs by selecting the **New URL** link in the top right-hand corner of the UI.

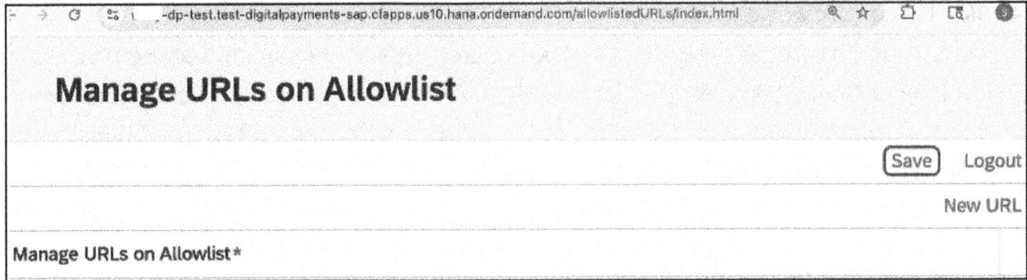


*Figure 2.10: Configuration UI: Manage Allowlist URLs*

- **Merchant mapping (merchant ID aliases)**
  This configuration UI enables grouping of technical PSP merchant IDs under logical aliases used by consuming applications (e.g., SAP virtual bank accounts) for easier configuration and reconciliation. Figure 2.11 shows the UI in which a logical alias can be provided against the PSP (e.g., Stripe) and the technical PSP merchant ID.

-dp-test.test-digitalpayments-sap.cfapps.us10.hana.ondemand.com/merchantMapping/index.html

**Merchant Mapping**

| Payment Service Provider | Merchant | Merchant Alias |
|---|---|---|
| Stripe V2 | pk_test_ | |
| Stripe V2 | pk_test_ | |
| Stripe V2 | pk_test_ | |

*Figure 2.11: Configuration UI: Manage Merchant Mapping*

### 2.2.7 Security Components

The digital payments add-on incorporates several security components designed to maintain the integrity and confidentiality of all payment operations. These elements work together to protect data and simplify compliance management:

- **Token mapping and abstraction**
  The add-on securely manages its own internal payment tokens while maintaining a confidential mapping to each PSP's token. This abstraction layer provides PSP independence—allowing merchants to switch providers or update configurations without exposing or migrating sensitive data.
- **Secure credential storage**
  Sensitive PSP credentials such as API keys or certificates are stored within SAP Credential Store. Keeping credentials isolated from application logic strengthens security and ensures controlled access during runtime.

- **SAP BTP security integration**
  The solution leverages the SAP Authorization and Trust Management service (XSUAA) to implement OAuth 2.0 protection for its APIs. It also integrates seamlessly with customer-managed identity providers (IdPs) to enable secure, role-based access for administrators and business users across configuration UIs.

### 2.2.8 Logging and Monitoring Services

The digital payments add-on provides comprehensive logging and monitoring capabilities to support secure operations and efficient troubleshooting:

- **Application logging**
  All add-on components generate detailed records of transaction processing, configuration changes, and error conditions through the SAP Application Logging service for SAP BTP. Administrators can view these logs via tools such as Kibana to trace issues or audit changes. To maintain compliance with data-protection standards, any sensitive information within the logs is automatically masked.
- **Monitoring integration**
  Within the Cloud Foundry environment, built-in monitors track application health, instance availability, and resource usage. Administrators can extend this visibility by forwarding metrics and alerts to enterprise tools such as SAP Cloud ALM or third-party monitoring solutions for centralized operations control.

## 2.3 Integration Points with SAP Landscape

The digital payments add-on functions as a core integration service through its standardized Consumer API layer. Integration touchpoints vary across the SAP landscape depending on the business process and application in use. The tightest and most comprehensive integration exists with SAP S/4HANA, where digital payment features are embedded into the sales, distribution, and finance functionalities. The add-on also supports other SAP cloud applications, earlier ERP releases such as SAP ERP, and even non-SAP systems through API integration and SAP Notes–based extensions or custom logic.

### 2.3.1 SAP S/4HANA Integration

SAP S/4HANA integrates natively with the digital payments add-on, embedding digital payment functions into business partner, sales, and finance processes.

#### Business Partner Master Data

The business partner master record is the foundation for card-based payments. The most important elements are as follows:

- **Card registration**
  Within the Manage Business Partner app or Transaction BP, users can add a payment card to a customer profile. Figure 2.12 shows the logical data flow.

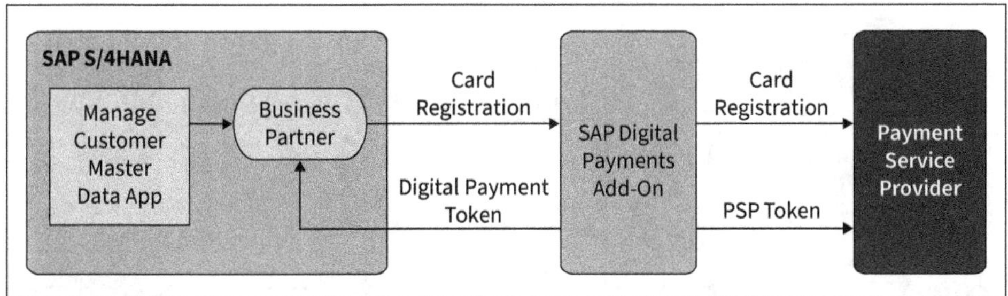


*Figure 2.12: SAP S/4HANA: Business Partner Card Registration Flow*

  Instead of entering card numbers directly in SAP S/4HANA fields, the system invokes a secure UI call to the digital payments add-on. This opens the PSP's encrypted card entry iFrame for the customer or agent to capture details securely. In the Manage Customer Master Data app, navigate to the **Payment Cards** section and select **Create** as shown in Figure 2.13.

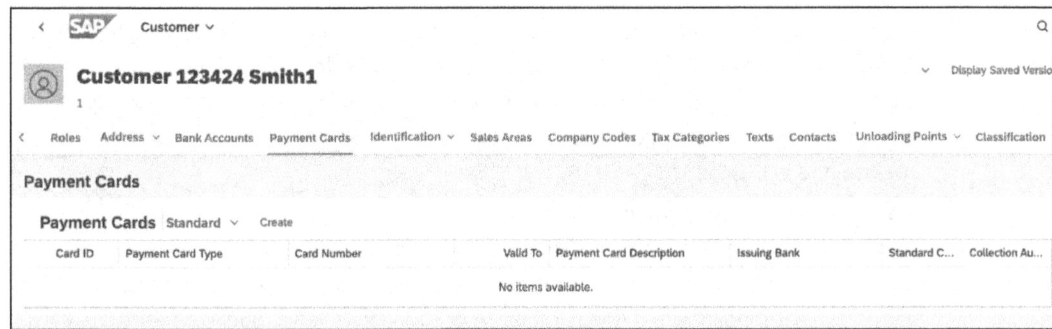


*Figure 2.13: Manage Customer Master Data: Create Payment Card*

- **Token storage**
  After processing the card, the PSP stores the primary account number (PAN) and returns a PSP-specific token. The add-on then creates its own internal digital payment token and sends it back to SAP S/4HANA along with nonsensitive metadata such as the masked card number, type, and expiration date.
- **Data model**
  Tokenization minimizes the PCI DSS scope for the SAP S/4HANA system. Payment card fields are structured to store only digital tokens, never the actual PAN. For display purposes, SAP S/4HANA uses a masked format (e.g., xxxxxxxxxxxx1111) to protect sensitive information while still enabling customer support to identify the card.

### Sales and Distribution

Figure 2.14 illustrates how the digital payments add-on integrates into the SAP S/4HANA order-to-cash process, supporting two-step (authorization and capture) payment handling.

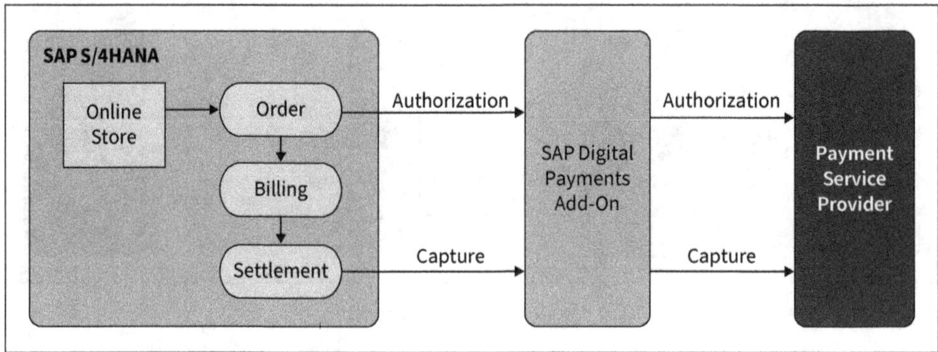


*Figure 2.14: SAP S/4HANA: Two-Step Payment Flow*

The following steps outline how the digital payments add-on participates at each stage of the order-to-cash process, from sales order creation through billing and settlement:

1. **Sales order**
   When a user creates a sales order using the Manage Sales Orders app, they can assign a tokenized payment card from the business partner record by selecting **Assign Payment Card** in the **Electronic Payments** tab. Upon saving the order, SAP S/4HANA automatically triggers an authorization request through the add-on, guided by configuration parameters such as the payment guarantee procedure (e.g., 0002) and the authorization horizon maintained in Customizing.
2. **Authorization versus preauthorization**
   If the delivery date falls within the defined authorization window (e.g., one day), then the system requests a full authorization for the order amount. For later deliveries, it performs a *preauthorization*—typically a zero- or low-value validation—to verify the card's status.
3. **Failed authorization**
   When an authorization fails (due to being declined, a card being expired, or incorrect data), the system automatically applies a credit block to the sales order, preventing downstream delivery processing.
4. **Delivery**
   During delivery creation, the system validates whether an active authorization exists. If the prior authorization has expired or is insufficient, the delivery remains blocked until reauthorization is completed—for example, using the Resolve Payment Card Issues app.
5. **Billing**
   Once the billing document (invoice) is generated, the payment token and authorization details are inherited from the sales order. When released to financial accounting, the document generates customer open items and general ledger postings, which initiate the settlement (capture) process in finance.

**Accounts Receivable**

Accounts receivable in SAP S/4HANA manages the financial side of payment transactions, covering settlement, reconciliation, and refund processing. The primary integration points with the digital payments add-on are as follows:

- **Payment settlement (capture)**
  Settlement is executed via the Payment Card Settlement background job, available in the Schedule Accounts Receivable Jobs app. The job identifies all open, authorized customer items from billing that are ready for settlement and sends batch capture requests to the add-on. Figure 2.15 shows the SAP Fiori app used to configure job scheduling.

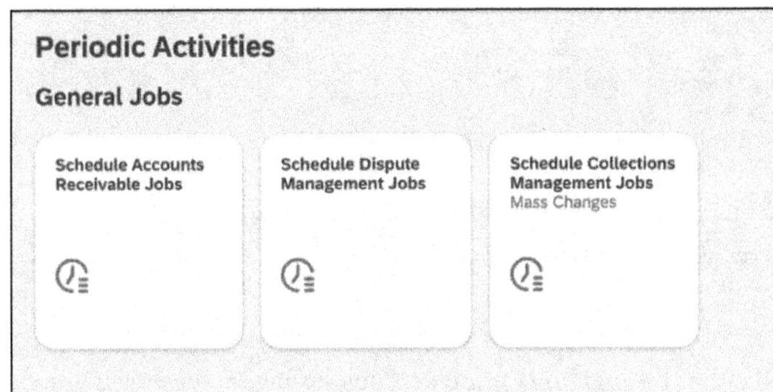


*Figure 2.15: SAP Fiori App: Schedule Accounts Receivable Jobs*

Launch the SAP Fiori app and create a new job. In the **Template Selection** screen that appears, search for the predefined "payment card settlement" job template in the **Job Template** selection box shown in Figure 2.16.

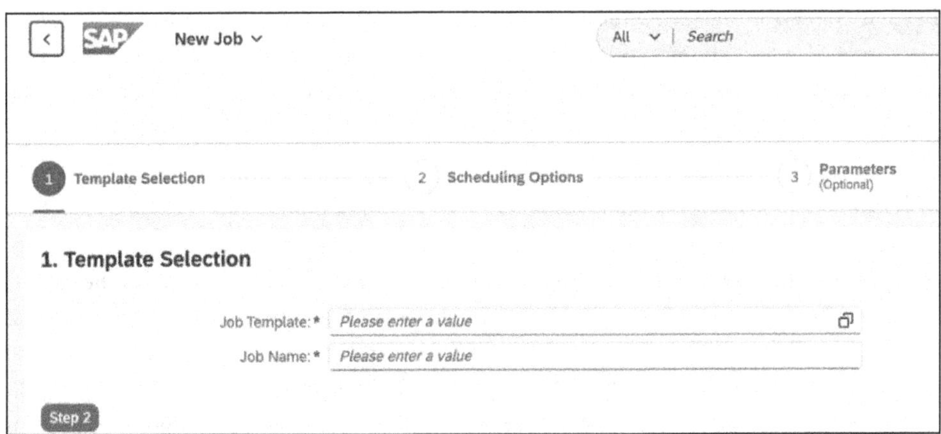


*Figure 2.16: Template Selection: Payment Card Settlement*

- **Payment advice processing**
  For automated reconciliation, the digital payments advice processing job retrieves advice data from the add-on's APIs. Each advice file includes records of successfully

captured payments, PSP fees, and processed refunds, allowing the system to automatically match payments against open items. Figure 2.17 shows the standard job template used for scheduling advice processing.

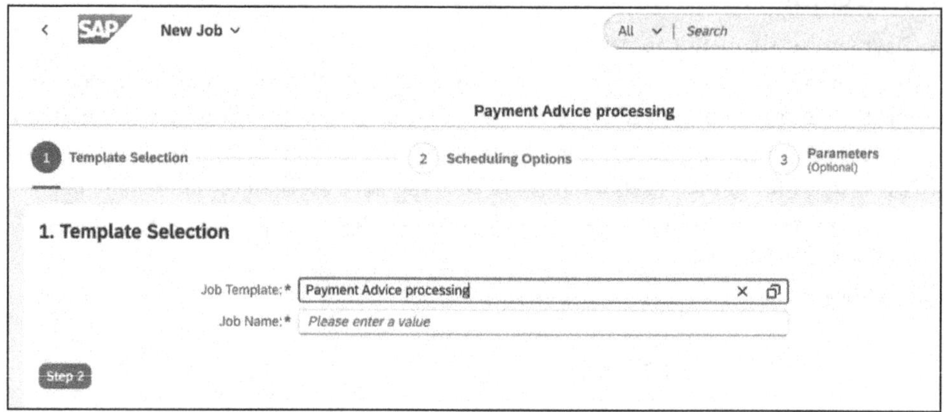


*Figure 2.17: Job Template Search: Payment Advice Processing*

- **Refunds**
  Refunds can be initiated either through the Create Refunds for Digital Payments app or as part of a standard credit memo workflow. In both cases, the add-on triggers a refund request referencing the original capture transaction, ensuring traceability between the refund and the initial payment.

### 2.3.2 Integration with Other SAP Solutions and Third-Party Solutions

Beyond SAP S/4HANA, several other SAP solutions can leverage the digital payments add-on to enable secure, tokenized card payments. In the following sections, we'll look at integration with SAP ERP, SAP Subscription Billing, SAP S/4HANA Cloud for customer payments, and non-SAP solutions.

#### SAP ERP

Integration with SAP ERP is possible, but it typically requires custom development. Common steps include applying the relevant SAP Notes, creating necessary DDIC objects and fields (e.g., adding fields to table BSEG), and implementing custom classes and programs (Z objects) to call the add-on's consumer APIs. This approach reproduces the settlement and advice logic available in SAP S/4HANA so that SAP ERP landscapes can benefit from tokenization and secure payment processing without a full system migration. A full, step-by-step SAP ERP integration approach is provided in Chapter 5.

#### SAP Subscription Billing

In a typical deployment, SAP Subscription Billing manages customer lifecycles and billing cycles while a backend system (often SAP S/4HANA) handles authorization and settlement via the digital payments add-on. Figure 2.18 shows the order of events for SAP Subscription Billing integration: card registration on sign-up, tokenization by the PSP via the add-on, storage of the returned token on the subscription record, the transfer of tokenized billing data to the backend for invoicing, and final settlement and reconciliation through the add-on.

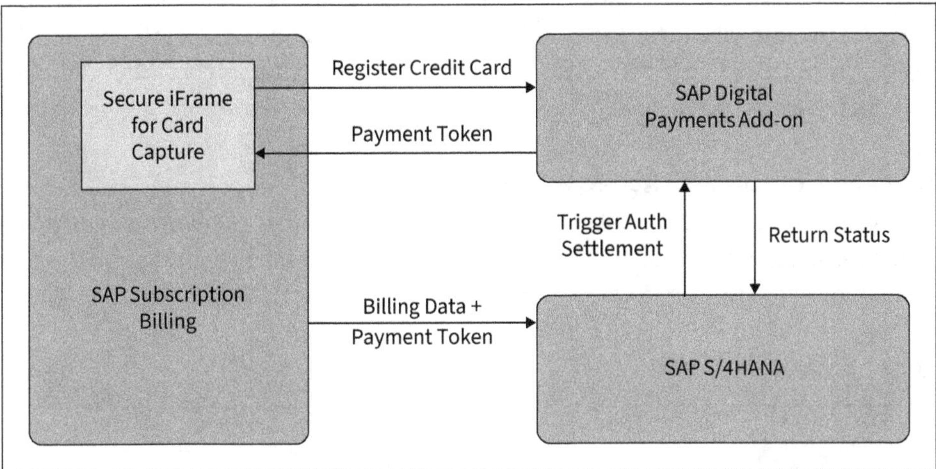


*Figure 2.18: SAP Subscription Billing: Integration Points for Digital Payments Add-On*

The integration between SAP Subscription Billing and the digital payments add-on follows a defined sequence of steps, from initial card registration through to final reconciliation:

1. **Payment card registration**
   When a subscriber signs up or updates their payment method, SAP Subscription Billing securely sends the credit card details to the SAP digital payments add-on for registration.
2. **Tokenization**
   The add-on, through the connected PSP, checks the card details and sends back a unique, nonsensitive payment token to SAP Subscription Billing. This token is saved with the customer's subscription record for future use.
3. **Billing data transfer**
   As billing cycles progress, SAP Subscription Billing creates billing data. This data includes the payment token instead of raw card details and is sent to the backend system (e.g., SAP S/4HANA) for invoicing and financial processing.
4. **Financial settlement**
   The backend system uses the payment token to initiate authorization and settlement processes directly with the add-on.
5. **Reconciliation**
   The add-on works with the PSP and sends back authorization statuses, payment confirmations, or refund statuses to the SAP S/4HANA backend for final reconciliation and financial reporting.

#### SAP S/4HANA Cloud for Customer Payments

SAP S/4HANA Cloud for customer payments, which provides a customer self-service portal for viewing and paying invoices, uses the add-on as its engine to process the payments made by customers. This product is being discontinued by SAP; you can contact your SAP account executive to discuss alternate options in light of the recent announcement from SAP. We therefore will not be discussing the integration points of this SAP solution with the digital payments add-on in detail. More details are available in the product's SAP Help page at *https://help.sap.com/docs/SAP_S4HANA_Cloud_for_Customer_Payments*.

#### Other Integrations with Non-SAP Solutions

Most global companies have online stores where their customers can browse through the product catalog and place an order directly in the online web shop application. In this scenario, the web shop payments process can be integrated with the add-on to obtain an authorization up front when the customer is ready to check out. Figure 2.19 shows the integration points in this flow.

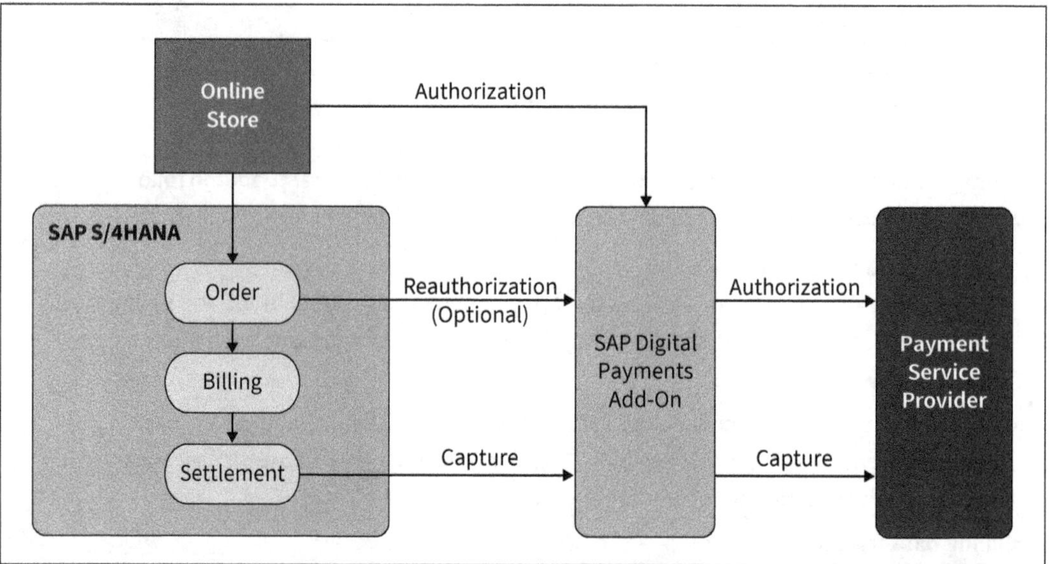


*Figure 2.19: Online Store: Integration Points with Add-On*

A variation of this flow is also possible in which the online store integrates directly with the PSP and obtains the PSP token, then integrates with digital payments add-on to obtain an authorization or both authorization and capture. The digital payments add-on provides lot of flexibility, so online stores connected directly to a PSP are also supported. Figure 2.20 shows the logical integration points in this flow.

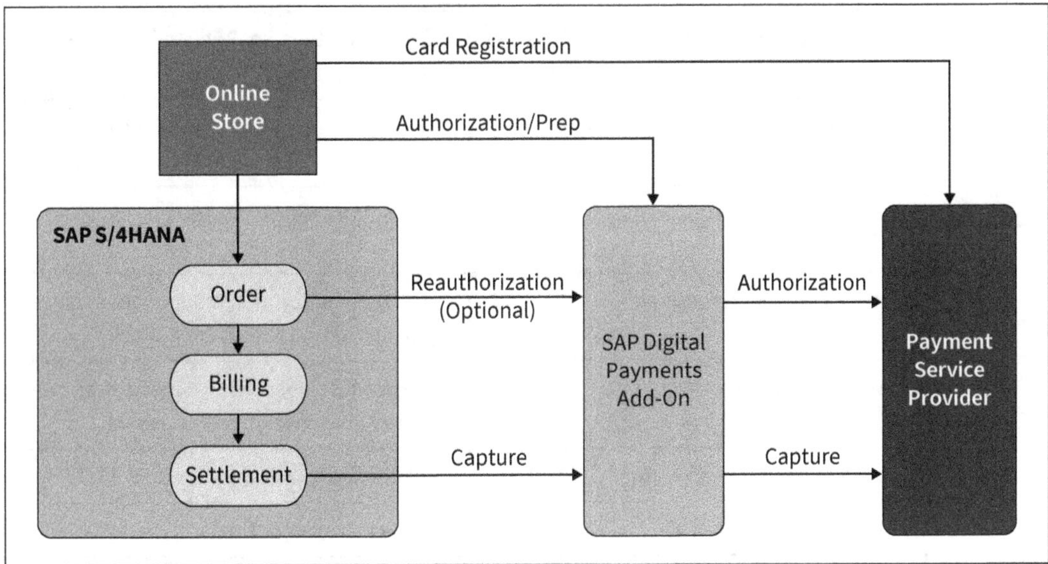


*Figure 2.20: Online Store: Integration Points with PSP and Add-On*

## 2.4 Security Architecture and Token Vault

Security is a central concern in any payment architecture, and the SAP digital payments add-on addresses it through a layered approach designed to protect sensitive data at every stage of the payment lifecycle. This section covers three key aspects of that approach. First, we examine PCI DSS scope reduction, showing how the add-on's architecture keeps sensitive card data out of your SAP systems entirely.

Next, we explore PSP abstraction, data storage, and token lifecycle management, explaining how dual-layer tokenization enables PSP independence while keeping your data model free of actual payment credentials. Finally, we cover authentication, authorization, and secure credential storage, detailing how OAuth 2.0, role-based access, and SAP Credential Store secure both user and system interactions.

### 2.4.1 PCI DSS Scope Reduction

One of the main reasons to adopt the add-on is its significant reduction of the PCI DSS compliance scope for your consumer applications. Table 2.13 shows the strict and costly security measures for systems that store, process, or transmit cardholder data, especially the PAN.

| Traditional Integration (Full PCI DSS Scope) | Add-On Architecture (Reduced PCI DSS Scope) |
|---|---|
| PAN storage and direct PSP connection using cardholder data | No PAN storage and no direct PSP integrations |
| Entire SAP system in scope | SAP is out of scope; only add-on and PSP are in PCI DSS scope |
| Database encryption required | No database encryption required as digital payments add-on tokens are not sensitive data |
| Network segmentation | No network segmentation necessary as plain card data does not traverse network |
| Annual audit of full landscape | Audit needed only for add-on and PSP |
| Extensive security controls | Minimal security controls |

*Table 2.13: PCI DSS Scope Comparison*

By ensuring that the main SAP S/4HANA system does not store, process, or transmit the PAN, an organization's PCI DSS assessment scope for its ERP environment is greatly reduced. The audit focus shifts from the entire ERP system to checking the security of the integration points. The residual scope now depends on the compliance statements provided by SAP for the add-on service and by the customer's chosen PSP.

### 2.4.2 PSP Abstraction, Data Storage, and Token Lifecycle Management

Table 2.14 shows how a consumer application, like SAP S/4HANA, stores and manages only the digital payments add-on token. The application need not store the specific token used by the PSP. This setup allows your company to change PSPs without affecting the master data in the SAP system. The add-on's internal mapping can be updated to link the existing digital payment token to a new PSP token from a different provider.

| SAP S/4HANA Database: Data Stored | SAP S/4HANA Database: Data Not Stored |
|---|---|
| Digital payments add-on token is stored | Actual PAN is *not* stored |
| Masked PAN is stored | PSP token is *not* stored |

*Table 2.14: Token Management*

Benefits of the abstraction layer are as follows:

- Consumer applications only store digital payments add-on tokens.
- Consumer applications are unaware of PSP tokens.

- PSP switching is possible without SAP data model changes.
- Security is enhanced through dual-layer tokenization.

The digital payments add-on facilitates better token lifecycle management. Table 2.15 shows example scenarios like the following:

- **Token update**
  This occurs when customers switch PSPs or when a card update happens (e.g., a card has a new expiration date).
- **Token deletion**
  When a card is removed from the business partner master data in SAP S/4HANA, a request goes to the add-on to remove its internal token mapping, and a deletion request is sent to the corresponding PSP to cancel or delete the PSP token.
- **Masked PAN storage**
  The masked PAN is kept only for reference to assist conversations between customer service agents and their company's customers. It is *not* used in the payment flows.

| Token Update | Token Deletion | Display Purpose |
|---|---|---|
| Scenario: PSP switch or card update required | Trigger: Card deleted from customer master data | Masked PAN storage: xxxxxxxxxxxx3456 |
| Process:<br>▪ Add-on updates internal mapping<br>▪ SAP digital payments add-on token remains unchanged in SAP<br>▪ New PSP token mapped | Cascade Process:<br>▪ Add-on deletes internal mapping<br>▪ PSP invalidates PSP token<br>▪ Add-on checks for in-flight transaction | Usage:<br>▪ User interface display only<br>▪ Not used for payment processing<br>▪ Human-readable reference |

*Table 2.15: Add-On: Support for Efficient Token Lifecycle Management*

### 2.4.3 Authentication, Authorization, and Secure Credential Storage

The digital payments add-on provides secure authentication and authorization mechanisms for its services by using the latest industry standard protocols, whether for a user interaction or a system interaction.

When an administrative or operational user opens the add-on configuration UI, the IdP is consulted to verify the user. Role collections defined in the SAP BTP subaccount determine permissions and enforce separation of duties between administrators and key users. Table 2.16 depicts the typical user authentication flow.

| Step 1 | Step 2 | Step 3 | Step 4 |
|---|---|---|---|
| User authentication | Authorization check | Role templates | Access granted |
| ■ User logs into IdP<br>■ Authenticates via SAP Cloud Identity Services<br>■ Federated with SAP BTP | ■ Permissions verified<br>■ Role collections checked by SAP BTP subaccount | ■ DigitalPaymentsAdmin<br>■ DigitalPaymentsKey-User | ■ User groups assigned<br>■ Separation of duties<br>■ SAP Fiori UI access |

*Table 2.16: Add-On: User Authentication Mechanisms*

Backend systems such as SAP S/4HANA authenticate to the Consumer API using OAuth 2.0 client credentials. The SAP S/4HANA system holds a client ID and client secret (set in an RFC destination) sourced from the add-on's service key in SAP BTP. This ensures that only trusted, registered systems can call payment APIs. Table 2.17 shows the system authentication sequence.

| Step | SAP ERP/SAP S/4HANA | Authentication Process | Digital Payments Add-On |
|---|---|---|---|
| 1 | Requests service key credentials | OAuth 2.0 handshake | Provides client ID and client secret |
| 2 | Exchange credentials for token | Client credential flow | Issues and validates access token |
| 3 | Executes payment APIs | Bearer token in header | Authorizes trusted system call |

*Table 2.17: Add-On: System Authentication Mechanism*

The add-on manages sensitive PSP credentials using SAP Credential Store. At runtime, service bindings provide ephemeral access to secrets; administrators never embed raw keys in application code. Table 2.18 shows best practices for secret management and the service-binding patterns recommended for production landscapes.

| Not Stored in Application | SAP Credential Store |
|---|---|
| No hardcoded credentials in application code | Secure service binding (isolated credentials) |
| No plaintext files (configuration files) | Encryption at rest (all credentials encrypted) |
| No plain environment variables | Access control (authorized services only) |

*Table 2.18: SAP Credential Store: Secret Management*

## 2.5 High Availability and Scalability

Payment processing is an important part of any business. Service interruptions or slow performance, especially during busy times, can hurt revenue and customer trust. The add-on architecture focuses on high availability and scalability using SAP BTP—particularly its Cloud Foundry environment features.

### 2.5.1 High Availability

High availability ensures that the payment service stays operational and accessible to consumer applications even while experiencing hardware failures, software issues, or infrastructure problems.

SAP BTP regions are built across several physically separate data centers called *availability zones* (AZs). The platform's infrastructure, including computing resources, networking, and storage, has redundancy across these availability zones. This setup means that if one availability zone fails, it usually does not cause a total service outage in the region.

The key characteristics of fault tolerance are as follows:

- Single AZ failure does not cause outage
- Geographically distributed data centers
- Automatic failover between zones
- No single point of failure

Figure 2.21 shows how computing, network, and storage resources are grouped within each AZ to ensure fault tolerance.

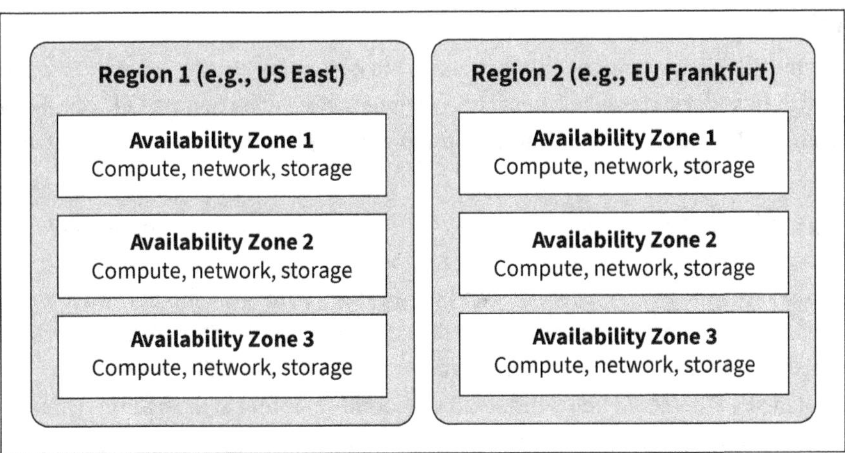


*Figure 2.21: SAP BTP: Regional Redundancy*

In addition, the Cloud Foundry environment has built-in features for application high availability (see Table 2.19). The digital payments add-on microservices run in multiple instances. These instances are spread across virtual machines and across different AZs for better fault tolerance. Also, the Cloud Foundry router acts as a smart load balancer. It distributes incoming API requests to all available healthy instances of the add-on service. This setup prevents any single instance from becoming overloaded.

<table>
<tr><th>Multiple Instances</th><th>Intelligent Load Balancing</th></tr>
<tr><td>SAP digital payments add-on service</td><td rowspan="3">Cloud Foundry router<br>This acts as a load balancer, distributing incoming API requests across all available healthy instances.<br>■ Prevent overhead<br>If an instance becomes overloaded, requests are automatically routed away.<br>■ Even distribution<br>Traffic is evenly distributed across all healthy instances.</td></tr>
<tr><td>■ Instance 1: AZ1<br>■ Instance 2: AZ2<br>■ Instance 3: AZ3<br>■ Instance N: Distributed</td></tr>
<tr><td>Auto-distribution: The platform automatically distributes instances across different AZs for maximum fault-tolerance.</td></tr>
</table>

*Table 2.19: Cloud Foundry: Load Balancing*

SAP is responsible for the high availability of the SAP BTP infrastructure and the add-on service according to specific SLAs, but end-to-end availability also relies on a few factors outside SAP's control. This includes the availability of your business applications, such as SAP S/4HANA; your network connection to SAP BTP; and the uptime of external payment service providers.

### 2.5.2 Scalability

*Scalability* refers to your system's ability to manage increasing workloads, whether through gradual growth or sudden spikes in transaction volume. This can be handled efficiently by adding resources. The add-on's cloud-native design allows for both horizontal and vertical scaling.

The main option is *horizontal scaling*. This approach involves increasing the number of running instances for the add-on's microservices. In SAP BTP, Cloud Foundry runtime, the number of instances for the core service can be adjusted, usually by the SAP operations team. As transaction volume grows, adding more instances helps distribute the load across a larger pool of resources. The Cloud Foundry router automatically includes new instances in its load balancing. This setup helps the service respond smoothly to changes in demand, like during holiday sales or marketing campaigns. Figure 2.22 shows how increasing the number of instances can help.

Although used less often for stateless applications than horizontal scaling, you can also scale individual instances *vertically* (see Figure 2.23). This involves allocating more resources, like CPU or memory, if specific processing tasks need it.

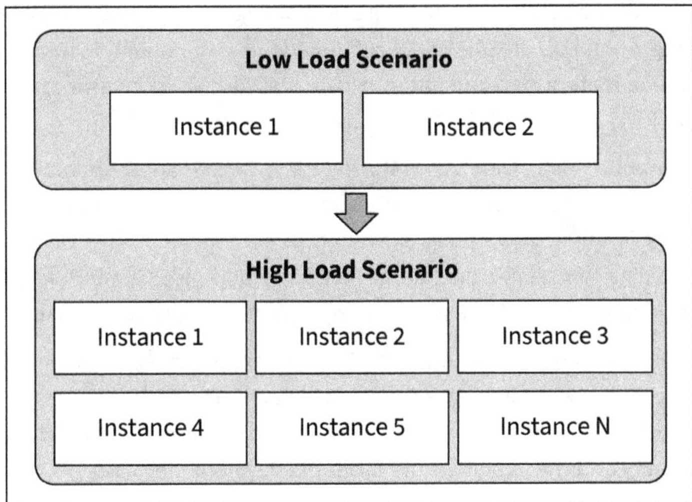


*Figure 2.22: Horizontal Scaling Example*

It's important to understand that the overall end-to-end transaction performance combines the performance of the consumer application generating the request, network latency, the SAP digital payments add-on processing time, and the response time of the external PSP. Although the add-on is designed for scalability, performance testing throughout the entire chain is crucial to find and fix potential bottlenecks outside the add-on service.

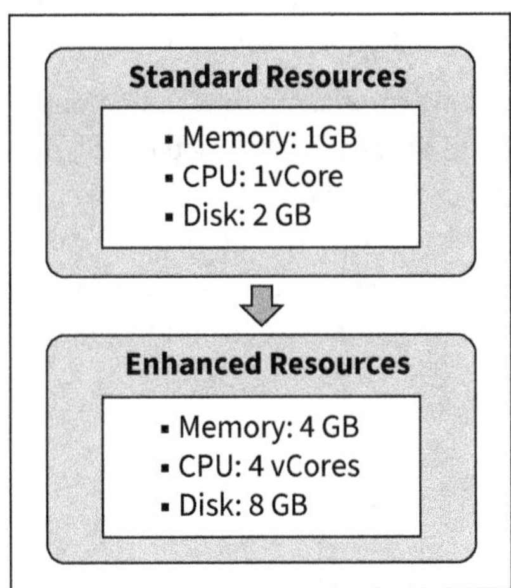


*Figure 2.23: Vertical Scaling Example*

## 2.6 Summary

This chapter introduced the SAP digital payments add-on as a vital service that links SAP applications with PSPs in a standard, secure, and scalable manner.

We started with an overview of the solution architecture, explaining how the add-on functions as a cloud-based integration hub on SAP BTP. Each functional component, including the Consumer API layer, PSP adapters, and configuration UIs, was discussed in terms of its role in managing payment flows across both SAP and non-SAP systems.

Next, we looked at the integration points with SAP S/4HANA, SAP ERP, SAP Subscription Billing, and other, non-SAP solutions. We showed how the add-on supports synchronous and asynchronous payment models, guiding processes like real-time authorizations in order-to-cash scenarios, recurring billing, and link-based collections. The add-on's OpenAPI layer guarantees consistent processing logic and PSP abstraction, regardless of the system in use.

We then examined the security architecture and token-vault design, emphasizing how the add-on reduces PCI DSS exposure by separating sensitive card data from core ERP systems. Tokenization, encrypted storage, SAP Credential Store, and the authorization framework together ensure confidentiality and traceability throughout the payment lifecycle.

The chapter also highlighted the high-availability and scalability features, showing how the add-on uses SAP BTP's multi-instance deployment, database replication, and autoscaling to provide reliable performance during varying transaction volumes. These platform features assist in ensuring business continuity and predictable response times during important financial operations.

In conclusion, the SAP digital payments add-on offers a strong foundation for updating digital payments within the SAP landscape. Its modular design, standardized APIs, and built-in compliance controls enable organizations to quickly enhance payment capabilities while upholding the security, governance, and reliability expected from SAP solutions.

This thorough understanding of the architecture, including component roles, communication flows, security measures, integration methods, and operational characteristics, provides a strong foundation for implementation. The next chapters will transition from architectural concepts to practical configuration, starting with SAP BTP setup (Chapter 3); then PSP integration, with detailed Stripe implementation examples (Chapter 4); followed by activation within the SAP ERP (Chapter 5) and SAP S/4HANA (Chapter 6) environments.

Chapter 3

# Setup in SAP BTP

*Before you integrate the SAP digital payments add-on in your SAP S/4HANA or SAP ERP system, you must establish the technical foundation of the add-on and its components in SAP BTP, a prerequisite that is critical to a successful implementation. This chapter guides you through the foundational setup that transforms a theoretical understanding of the add-on into a tangible, secure, and operational environment in which enterprise payments can be efficiently and effectively processed and managed.*

This chapter provides step-by-step instructions for provisioning, configuring, and securing the SAP digital payments add-on service in your company's SAP Business Technology Platform (SAP BTP). It covers the complete setup process from initial subscription through security configuration to validation testing. The chapter also includes troubleshooting tips for common setup issues.

This setup is a prerequisite for all subsequent implementation work. It builds the technical foundation upon which you will connect your payment service providers (PSPs) and integrate your consumer application (e.g., SAP ERP, SAP S/4HANA, etc.). We will walk through the complete technical setup, from creating the subaccount to validating the final configuration, after which application-level integration can be performed. We will begin by confirming that all necessary technical and commercial prerequisites are in place at the global account level, including entitlements, Cloud Foundry runtime availability, and region selection. We'll then explore the necessary steps to provision dedicated subaccounts for test and production environments, subscribe to the add-on application, and establish the secure machine-to-machine (M2M) communication channels that allow your ERP systems to authenticate and call the add-on's APIs. Next, we'll configure user authentication by establishing trust relationships with an identity provider (IdP) and then assign role-based access control to your implementation team. Finally, we'll validate the entire configuration to identify and resolve any setup issues before moving forward with PSP activation. By the end of this chapter, your SAP BTP environment will be fully operational, secure, and ready for integration with your SAP S/4HANA or SAP ERP system.

## 3.1 Prerequisites and Entitlements

Before you create your first subaccount or service instance, it is important to confirm that all necessary technical, commercial, and access-related prerequisites are in place. This section serves as a checklist for the SAP BTP administrator. In the following sections, we will verify that your global account has administrative access, Cloud Foundry runtime availability, and the correct region selected for your deployment. We will then confirm that your global account holds entitlements for all four required service plans, the demo and production user interface plans, and the test and production M2M API plans. We will also identify the implementation team members who will serve as SAP BTP administrators and functional digital payments key users, and determine which IdP will authenticate these users. Finally, we will verify that your SAP S/4HANA or SAP ERP systems meet the minimum version requirements, outbound network connectivity is in place, TLS 1.2 security is configured, and you have valid API credentials from at least one supported PSP.

### 3.1.1 Global Account Requirements

Your SAP BTP global account is the starting point for all services. Ensure the following items are confirmed:

- **Administrative access**
  You must have administrative privileges for the SAP BTP global account to create new subaccounts and manage service entitlements. SAP Help provides comprehensive information about SAP BTP, and details for account administration are available in the following section: **SAP Business Technology Platform (SAP BTP) • SAP Business Technology Platform • Administration and Operations • Account Administration** (*https://help.sap.com/docs/btp/sap-business-technology-platform/account-administration*).
- **Cloud Foundry runtime**
  The global account must be provisioned for SAP BTP, Cloud Foundry runtime, which is the runtime in which the add-on's microservices are deployed.
- **Region selection**
  You must have selected a target SAP BTP region that supports the SAP digital payments add-on and the specific PSP adapters you intend to use (e.g., cf-us10:US East (VA) or cf-eu10: Europe (Frankfurt)). This decision impacts data residency and network latency. More information about the data centers (regions) that support the add-on is available in the *SAP Digital Payments Add-On Administration Guide* under **Prerequisites • Data Centers (Regions)** (*http://s-prs.co/v629402*). SAP keeps this section updated as it expands the regions in which the add-on is supported.

### 3.1.2 Subscription and Service Plans

Commercially, the add-on is licensed via distinct service plans. Your SAP BTP global account must be entitled for the quota of these plans before you can subscribe a subaccount to them. Verify that your global account has an entitled quota (of at least 1) for the following four service plans:

- Application (user UI) plans: SAP digital payments add-on demo (test), and SAP digital payments add-on (production)
- M2M (API) plans: SAP digital payments add-on (test/demo M2M), and SAP digital payments add-on (productive M2M)

You will assign the *demo* plans to your nonproductive subaccount and the *production* plans to your productive subaccount.

Follow these step-by-step instructions to verify your entitlements:

1. Log into your SAP BTP global account cockpit and select **Entitlements • Service Assignments** in the left-hand panel, as shown in Figure 3.1.

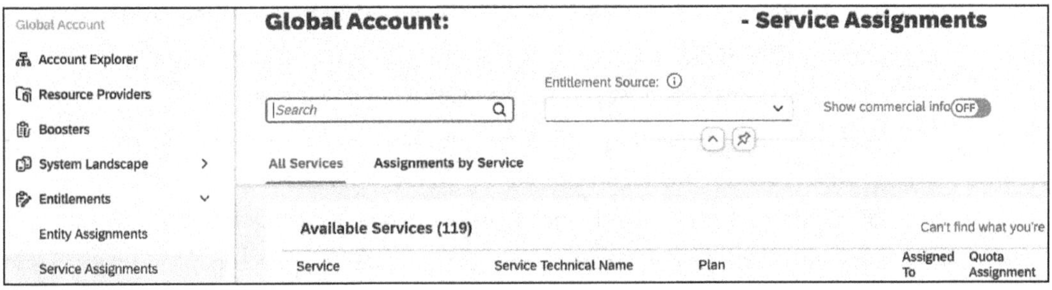

*Figure 3.1: Service Assignments View: Navigation*

2. Search for "SAP Digital Payments"; you will typically see the four service plans shown in Figure 3.2, with the **Global Quota** value set to **1** for each of them.

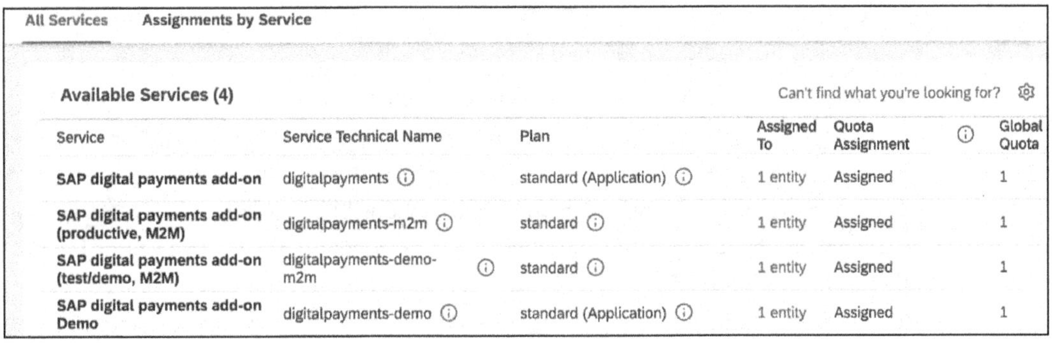

| Service | Service Technical Name | Plan | Assigned To | Quota Assignment | Global Quota |
|---|---|---|---|---|---|
| SAP digital payments add-on | digitalpayments | standard (Application) | 1 entity | Assigned | 1 |
| SAP digital payments add-on (productive, M2M) | digitalpayments-m2m | standard | 1 entity | Assigned | 1 |
| SAP digital payments add-on (test/demo, M2M) | digitalpayments-demo-m2m | standard | 1 entity | Assigned | 1 |
| SAP digital payments add-on Demo | digitalpayments-demo | standard (Application) | 1 entity | Assigned | 1 |

*Figure 3.2: Add-On Service Plans and Quotas*

### 3.1.3 User Types and Access Requirements

The add-on requires two distinct user roles for setup and configuration. Ensure the following preparations have been made:

- **Personnel identified**
  You must have identified the implementation team members who will act as SAP BTP administrators and functional digital payments key users.
- **Identity provider**
  SAP digital payments add-on provides several options to integrate with an IdP. This is needed to authenticate the implementation team members to the application. SAP ID is the default IdP for the subaccount. SAP recommends that you use an SAP Cloud Identity Services tenant as the IdP for the add-on. More details about configuring an IdP can be

found in the *SAP Digital Payments Add-On Administration Guide* under **Setup Activities • Account Setup • Configuring the Identity Provider and Setting Up Authentication** (*http://s-prs.co/v629403*). For purposes of this book, we will proceed with SAP ID as the IdP.

### 3.1.4 Add-On Prerequisites

This final checklist covers the consumer systems and external dependencies, as follows:

- **Consumer system version**
  Confirm your SAP system release meets these minimum requirements:
  - SAP S/4HANA (on-premise/SAP S/4HANA Cloud Private Edition): Release 1709 or higher.
  - SAP S/4HANA Cloud Public Edition: Relevant scope items must be active.
  - SAP ERP: Release 6.0 EhP5 or higher (this is a custom project that we will discuss in detail in Chapter 5).
- **Network connectivity**
  Confirm outbound HTTPS (port 443) connectivity from your consumer application servers to the target SAP BTP region's API domains. For on-premise/private cloud systems, the cloud connector should be installed, configured, and connected to your SAP BTP subaccount.
- **System security (ABAP)**
  Verify that your ABAP system (SAP S/4HANA or SAP ERP) is configured to support TLS 1.2 or higher for outbound RFC calls (Type G). For all SAP applications hosted on the cloud and managed by SAP, SAP is responsible for having TLS 1.2 enabled. However, for SAP applications managed on-premise, it is your responsibility to enable TLS 1.2. This typically includes checking profile parameters (via Transaction RZ10) and ensuring the root CA certificates for SAP BTP (e.g., DigiCert) are loaded in the default SSL Client (Standard) personal security environment (PSE) in Transaction STRUST. Details about this setup are available in SAP Note 510007.
- **PSP contract**
  You must have a commercial contract and active API test credentials (e.g., API keys, secrets, merchant IDs) from at least one supported PSP. The add-on is a connector, not a payment processor, so these external credentials are a prerequisite.

## 3.2 Subscription and Tenant Configuration

With the necessary entitlements confirmed at the global account level, the next step is to provision the dedicated environments, or subaccounts, where the SAP digital payments add-on will operate. In SAP BTP, each subaccount provisioned with the add-on service effectively becomes a tenant.

Following SAP Best Practices for a standard three-tier (development, quality, and production) landscape, you must create at least two subaccounts:

- **A nonproductive (test/demo) subaccount**
  This tenant will be subscribed to the demo or test service plan. It will be used for all implementation, configuration testing, and quality assurance activities. Your development and test systems will connect to this tenant.
- **A productive (prod) subaccount**
  This tenant will be subscribed to the production service plan. It is used exclusively for live, productive transactions. Your production systems will connect to this tenant.

This separation is mandatory not only for a safe development lifecycle but also because the service plans themselves are distinct. You cannot subscribe a single subaccount to both the demo and prod plans.

With the prerequisites confirmed, you are now ready to provision the dedicated environments for the SAP digital payments add-on. Then, we will discuss accessing the SAP BTP cockpit to create two subaccounts following SAP Best Practices: one for nonproductive (test/demo) use and one for productive transactions. After the subaccounts are created, we will discuss assigning the appropriate service entitlements to each subaccount. Finally, we will subscribe each subaccount to the add-on application using the correct service plan (demo/test plans for the test subaccount and production plans for the production subaccount), which will provision the add-on's core services and administrative UIs within each tenant.

### 3.2.1 Accessing the SAP BTP Cockpit

All setup activities are performed within the SAP BTP cockpit. Ensure you have a user ID with global account administrator privileges (or at least the necessary subaccount creation permissions).

The following are step-by-step instructions for accessing the SAP BTP cockpit:

1. The URL for accessing the SAP BTP cockpit is *https://cockpit.btp.cloud.sap*. Based on the location from which you are accessing the cockpit, you will be directed to the nearest regional cockpit URL. Figure 3.3 shows the regional cockpit URL for the Americas (abbreviated here as *amer*) region.

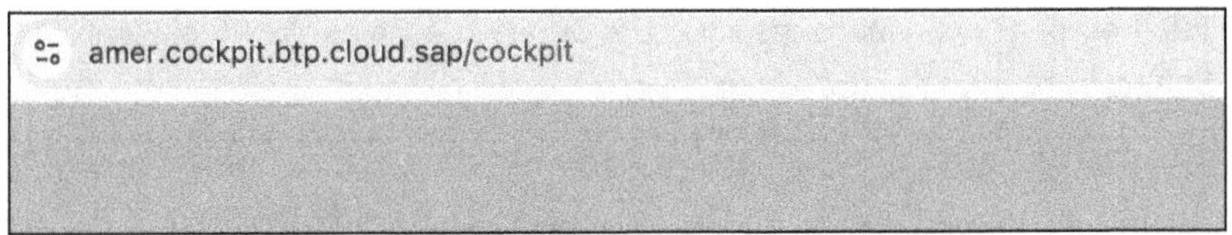


*Figure 3.3: Americas Region Cockpit: URL Redirect in Browser*

2. If your organization has more than one SAP BTP global account, you will be asked to choose which global account to log into. Figure 3.4 shows an example organization that has two SAP BTP global accounts.

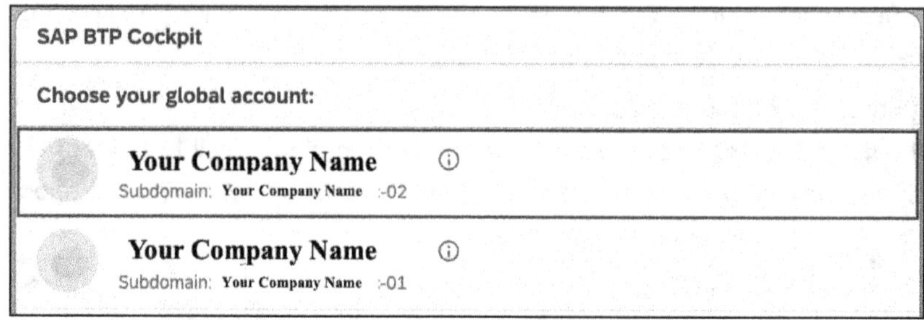


*Figure 3.4: SAP BTP Cockpit Login: Choosing Your Global Account*

3. Once you select the correct global account to log into, you will be in the account explorer view, where you can view and manage directories, subaccounts, entitlements, and security administration depending on the roles assigned to your user ID or user group. Figure 3.5 shows an example account explorer view.

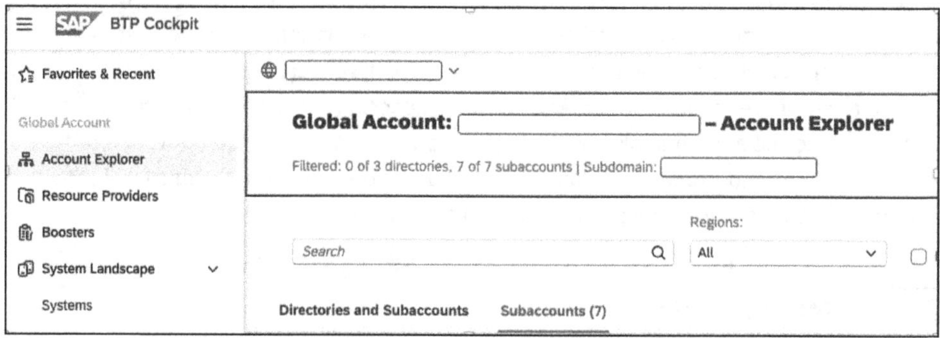


*Figure 3.5: SAP BTP Cockpit: Account Explorer View*

### 3.2.2 Creating Subaccounts

In the SAP BTP cockpit account explorer, you can create the subaccounts that will house the add-on tenants. This process must be repeated for both your test and production add-on environments.

To create a test subaccount, follow these steps:

1. From the global account overview page, click the **Create** button as shown in Figure 3.6 and select **Subaccount** from the dropdown.

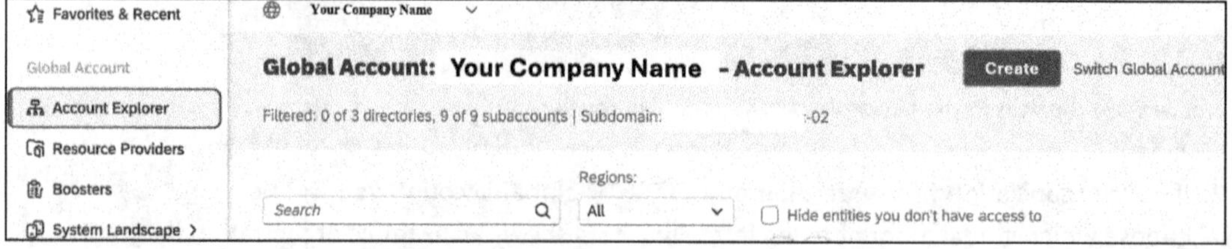


*Figure 3.6: SAP BTP Cockpit: Subaccount Creation*

2. The **New Subaccount** dialog will appear. Fill in the required fields as follows:
   - **Display Name**: Enter a clear, human-readable name that identifies the purpose and environment of the subaccount.
     - Test example: DP Test
     - Prod example: DP Production
   - **Region**: Select the region (e.g., US East (VA)). This choice is permanent for the subaccount and determines the physical data center where your add-on tenant will reside and process data.
   - **Subdomain**: This is a critical, unique identifier that will become part of your tenant's URL for both API calls and administrative UIs. It must be unique across all of SAP BTP. Choose a clear and consistent naming convention.
     - Test example: your company-dp-test
     - Prod example: your company-dp-prod
   - **Other Fields**: You can assign labels or add descriptions as needed for your internal governance.
3. Expand **Advanced**, check **Enable beta features**, then click **Create.**
4. After a few moments, the subaccount will be provisioned. You will now have a new subaccount tile visible in your global account overview, as shown in Figure 3.7.

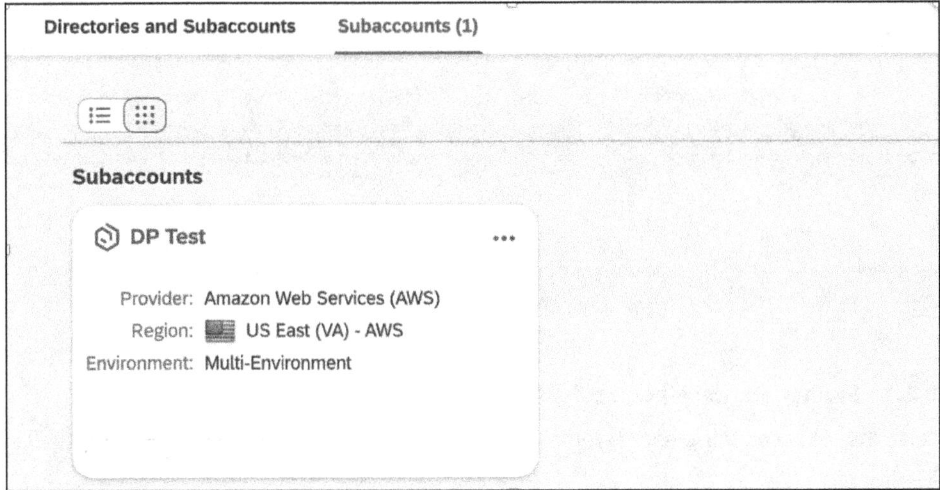


*Figure 3.7: New Subaccount Tile*

To create a production subaccount, follow these steps:

1. From the global account overview page, click the **Create** button (see Figure 3.8) and select **Subaccount** from the dropdown.

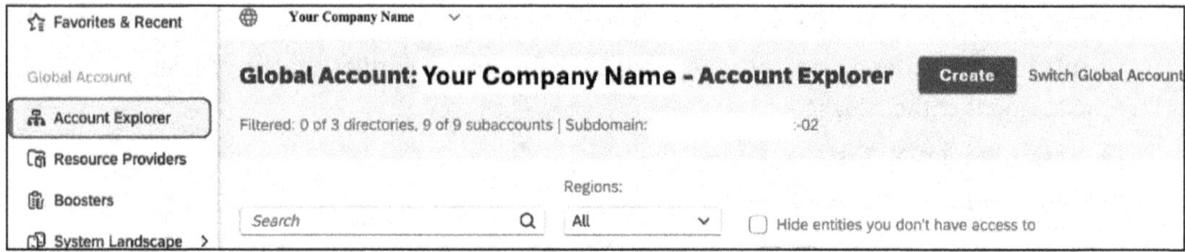


*Figure 3.8: Add-On Production Subaccount Creation*

2. The **New subaccount** dialog will appear. Fill in the required fields as follows:
   - **Display Name** = DP Production
   - **Description** = Production DP subaccount.
   - **Region** = US East (VA)
   - **Subdomain** = {your company}-dp
3. Expand **Advanced**, check **Used for production**, then click **Create**.
4. After a few moments, the subaccount will be provisioned. You will now have a new **DP Production** subaccount tile visible in your global account overview, as shown in Figure 3.9.

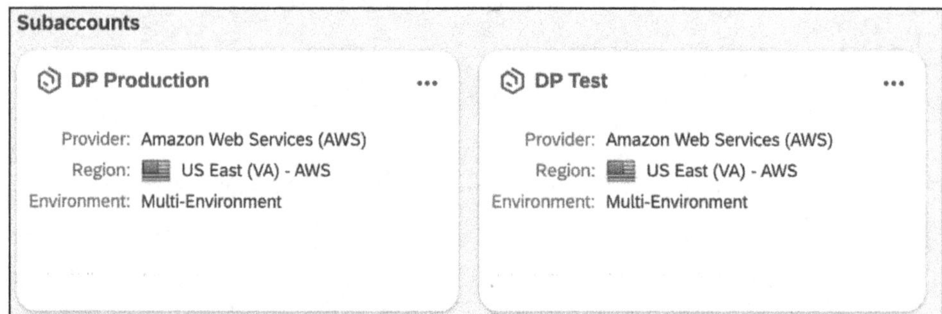


*Figure 3.9: SAP BTP Cockpit: Production Subaccount Tile*

### 3.2.3 Service Marketplace Navigation

Once the subaccount is created, you can navigate into it to subscribe to the add-on service, as follows:

1. From your global account overview, click the tile for the subaccount you created (e.g., **DP Test**). It will take you to the subaccount overview page, as shown in Figure 3.10.
2. In the **DP Test** subaccount's left-hand navigation pane, select **Services • Service Marketplace**. You will see all services and applications to which your global account is entitled, available for subscription within this specific subaccount. The ones we are interested in are shown in Figure 3.11.

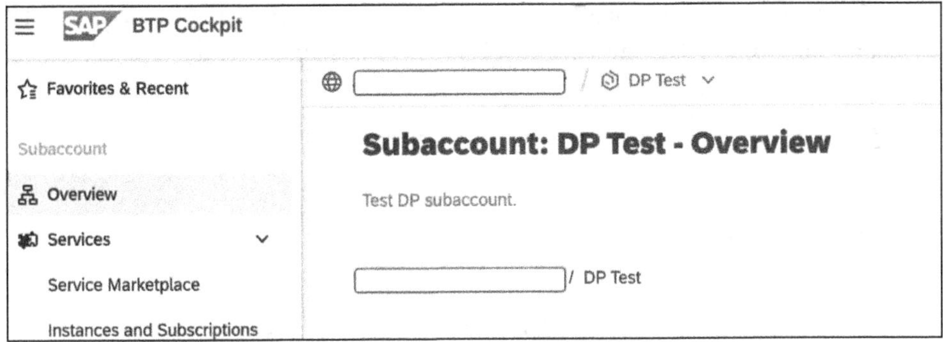


*Figure 3.10: Subaccount Overview Page*

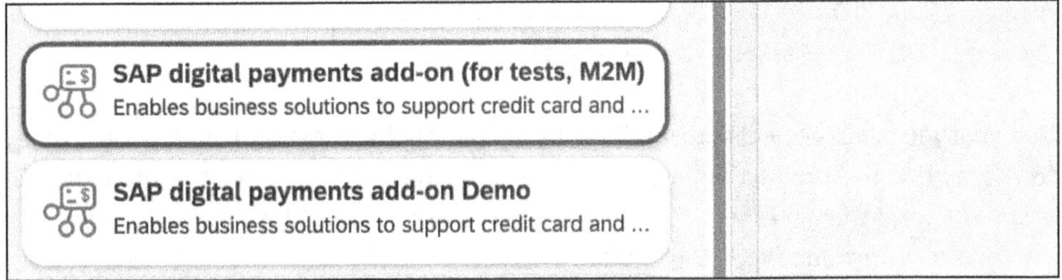


*Figure 3.11: Add-On Test Subaccount: Expected Entitlements*

3. If the expected entitlements are not visible in the subaccount, then they must be assigned first. Entitlements are managed at the global account level, so go to the global account view and choose **Entitlements • Entity Assignments**, as shown in Figure 3.12.

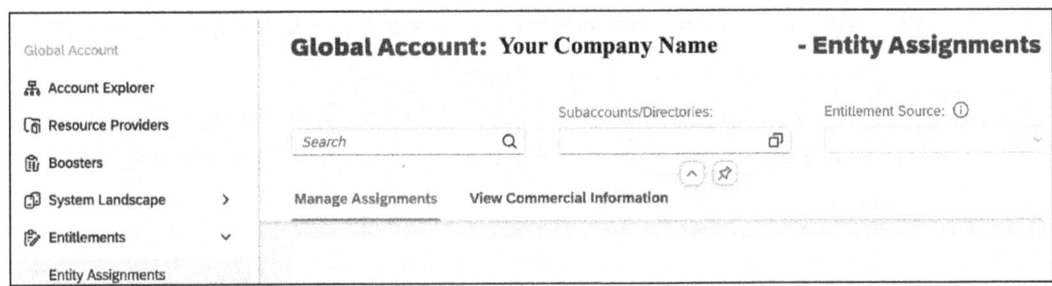


*Figure 3.12: Global Account: Entity Assignment to Subaccount*

4. In **Manage Assignments**, you should see the **DP Test** subaccount; select **Edit**. You will now be able to add a service plan to this subaccount.
5. Choose the **SAP digital payments add-on Demo** service, and in the **Service Details** pane, select the available plan (**standard**).
6. The next step is to add the add-on M2M service plan.

7. Verify the entitlement assignments by going to **Test DP subaccount • Overview • Entitlements**. You will see the entitlements reflected in the subaccount along with the subaccount quota set to **1**, as shown in Figure 3.13.

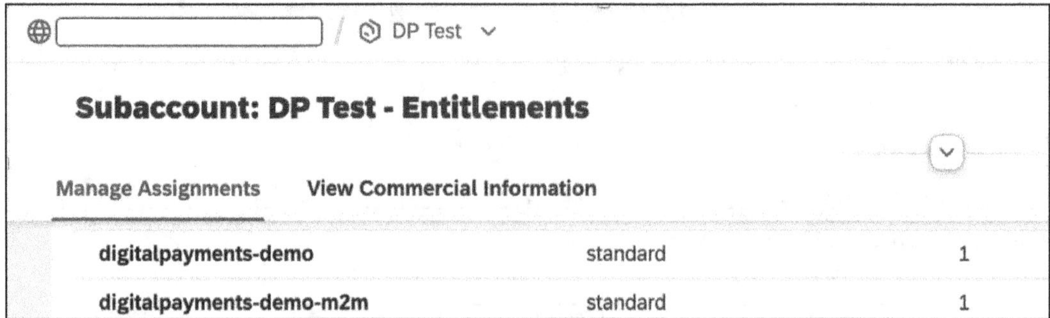


*Figure 3.13: DP Test Subaccount: Entitlements View*

Now that you have the necessary entitlements added to the test subaccount, you are ready to proceed to the production subaccount. Repeat the steps used for the **DP Test** subaccount, except that you should use the **SAP digital payments add-on** and **SAP digital payments add-on (productive, M2M)** entitlements. Note that the difference here is that you are not assigning any service plan that has **demo** in its name.

### 3.2.4 Add-On Subscription Process

In this section, we will discuss the steps to subscribe to the SAP digital payments add-on application in your subaccount in the SAP BTP cockpit. Subscribing to the application activates the add-on's core services and administrative UIs within your tenant.

To subscribe to the add-on application, follow these steps:

1. In your SAP BTP cockpit, navigate to the **DP Test** subaccount. In the subaccount explorer view, navigate to **Services • Instances and Subscriptions**, as shown in Figure 3.14.

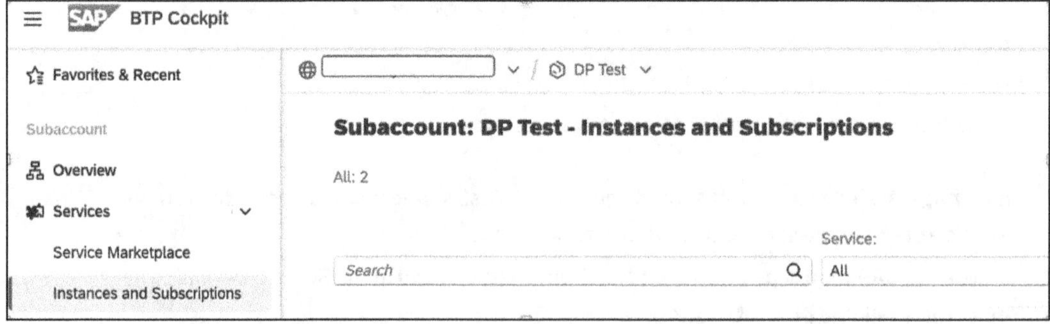


*Figure 3.14: DP Test Subaccount: Instances and Subscriptions View*

2. Select the **Create** button in the top right-hand corner, as shown in Figure 3.15.

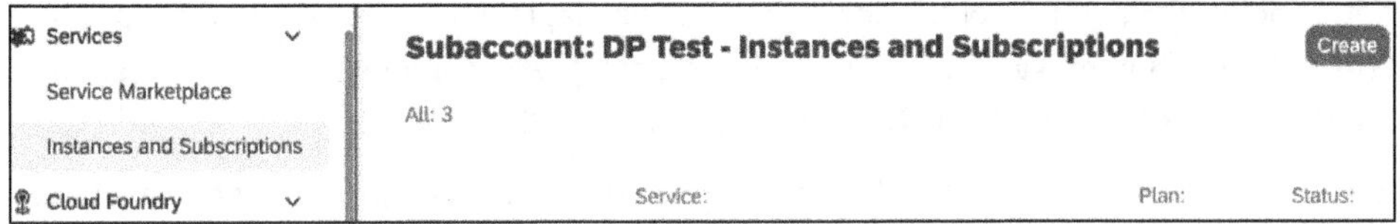


*Figure 3.15: DP Test Subaccount: Create Subscription*

3. A wizard will open to create the subscription. Choose the following settings:
   - Set **SAP digital payments add-on Demo** as the **Service**.
   - Select standard as the **Plan** (see Figure 3.16).

   Click **Create**. The system will begin provisioning the application, which may take several minutes.

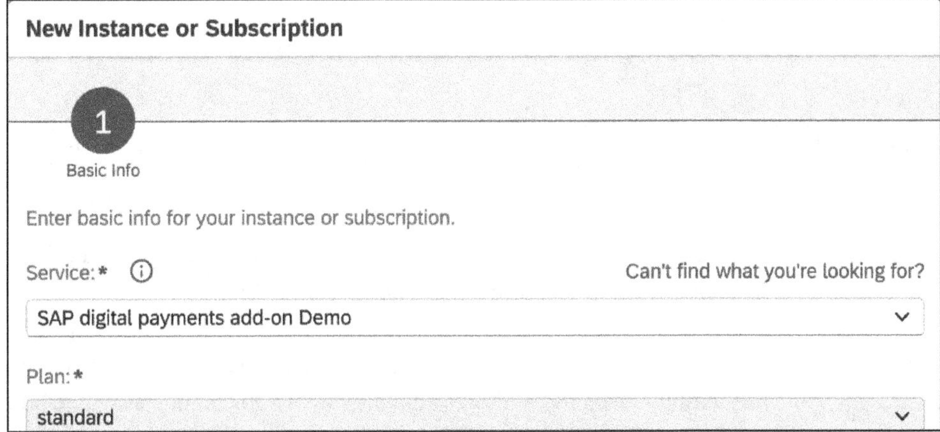


*Figure 3.16: DP Test Subaccount: Creating Subscription for Add-On*

4. Once completed, you will see **SAP digital payments add-on Demo** listed in the **Subscriptions** table with a status of **Subscribed** (see Figure 3.17).

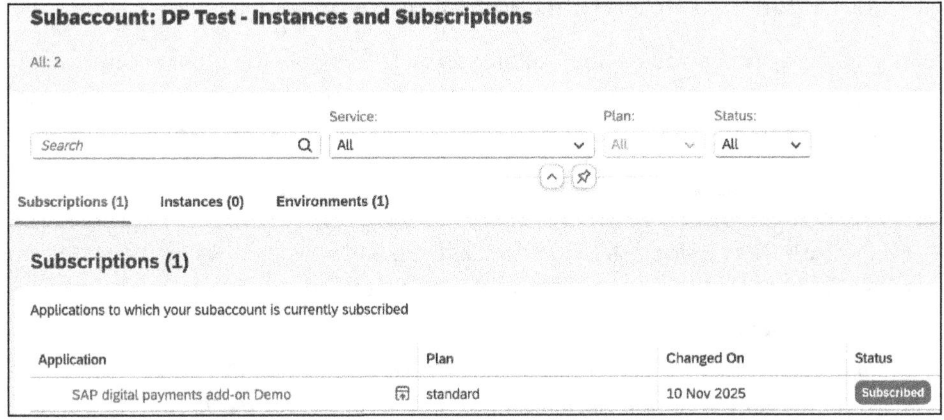


*Figure 3.17: DP Test Add-On Subscription: Successful Creation*

At this point, the add-on application is successfully deployed to your **DP Test** subaccount tenant. This action makes the add-on's administrative applications available and enables the M2M service broker, which is required for creating API credentials.

Repeat the same steps for the **DP Production** subaccount, except that when the wizard opens to create the subscription, set **SAP digital payments add-on** (*not* **SAP digital payments add-on Demo**) as the **Service** (see Figure 3.18), select **standard** for the **Plan**, and when you're done, click **Create.**

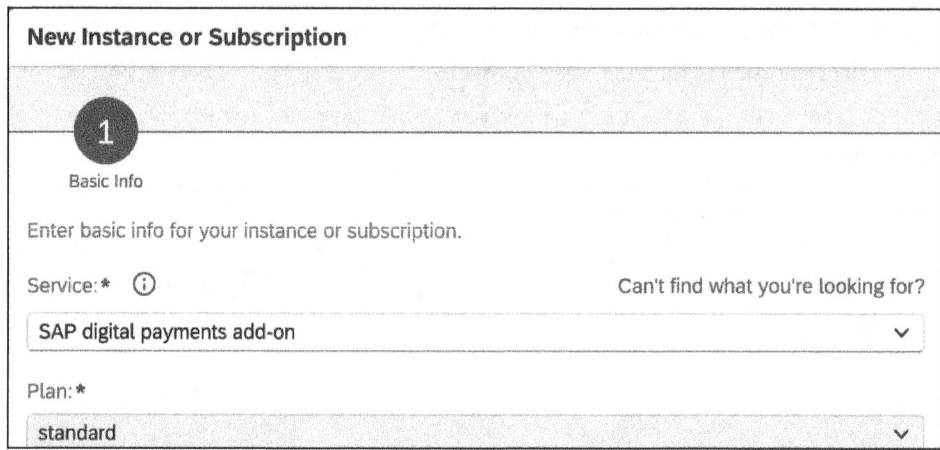


*Figure 3.18: DP Production Subaccount: Creating Subscription for Add-On*

At this point, the add-on application is successfully deployed to your DP Production subaccount tenant. This action makes the add-on's administrative applications available and enables the M2M service broker, which is required for creating API credentials. However, no users can yet access the applications, and no systems can call the APIs. The following sections will address these communication and authorization steps.

## 3.3 Communication Systems and Arrangements

After subscribing to the add-on application, you must configure the primary communication channels. This section focuses on the M2M channel, which allows the consumer applications to securely call the add-on's APIs. Configuring the M2M channel is a two part process:

1. **In SAP BTP**
   First, you create a service instance and a service key within your BTP subaccount to generate the required API credentials (client ID, client secret, and endpoints).
2. **In SAP S/4HANA (or SAP ERP)**
   Second, you use these credentials to configure HTTP destinations (RFC destinations) in your ABAP system, to securely connect and authenticate with the add-on.

### 3.3.1 Destination Configuration Basics

In the SAP environment, outbound communication to an external HTTP-based service is managed via RFC destinations (Transaction SM59). For the SAP digital payments add-on integration, you must configure two distinct RFC destinations of Type G (HTTP connection to external server). We will go through the step-by-step instructions for creating these destinations in Chapter 5 (for SAP ERP) and Chapter 6 (for SAP S/4HANA):

1. **Token endpoint destination**
   The first destination is configured to point to the SAP BTP authentication server's token endpoint (*uaa.url* from the service key of the subaccount). Its purpose is to exchange the static client ID and client secret for a temporary OAuth 2.0 bearer token.
2. **Service endpoint destination**
   The second destination points to the add-on's main application API. This destination is configured to use OAuth 2.0, and it leverages the first RFC destination to automatically fetch the bearer token before sending any business-related API calls (like authorize payment or settle payment).

The authentication methods are different for the two RFC destinations:

- **`DIGITALPAYMENTS_OAUTH` (token destination)**
  This destination uses basic authentication. The SAP system sends its client ID as the username and the client secret as the password to the SAP BTP authentication server.
- **`DIGITALPAYMENTS` (service destination)**
  This destination uses OAuth 2.0. It uses the `DIGITALPAYMENTS_OAUTH` destination to fetch the access token. It then includes this bearer token in the HTTP authorization header of the actual business call to the add-on's API.

### 3.3.2 Add-On-Specific Communication Setup

This section provides complete step-by-step instructions for configuring the M2M communication: enabling Cloud Foundry, creating a Cloud Foundry space, creating the M2M service instance, and creating digital payments service keys.

#### Enable Cloud Foundry

To enable Cloud Foundry, follow these steps:

1. Navigate to your **DP Test** subaccount in the SAP BTP cockpit and select **Overview**, as shown in Figure 3.19.

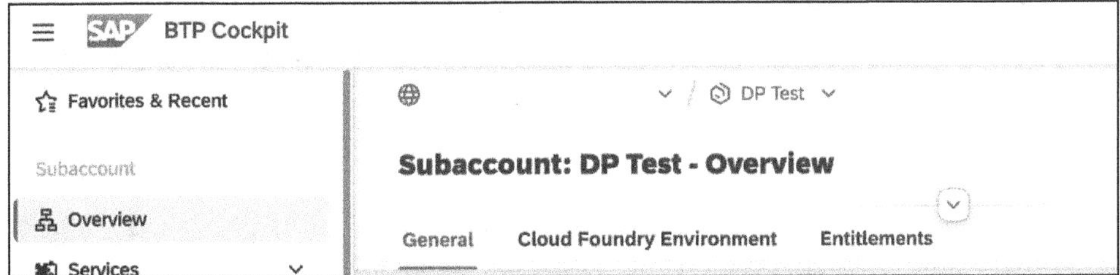


*Figure 3.19: SAP BTP Cockpit Subaccount: Overview*

2. On the **Overview** page, if not already enabled, find the **Cloud Foundry Environment** section and click **Enable Cloud Foundry**. This will create a Cloud Foundry organization, as shown in Figure 3.20.

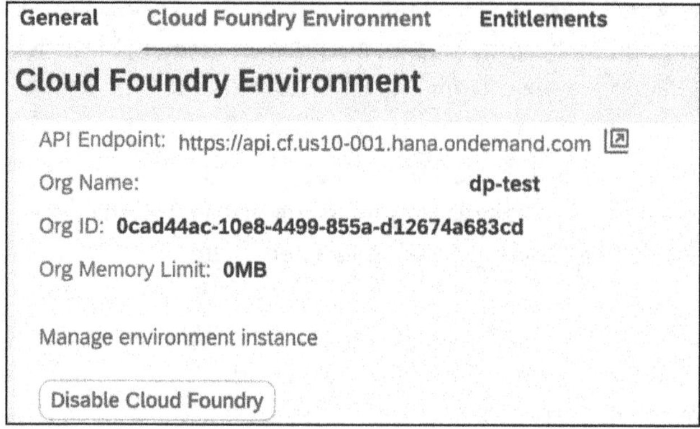


*Figure 3.20: SAP BTP Cockpit: Enable Cloud Foundry*

3. In the wizard that opens, select **standard** for the **Plan**, use the proposed instance and org names, and click **Create**. You will see a new Cloud Foundry space created as shown in Figure 3.21. Now you can use the **Add Members** button to add org members who can create service keys.

*Figure 3.21: Cloud Foundry Space: Adding Org Members*

4. Users who need to create service keys can be assigned **Org Manager** and **Org Auditor** security roles, whereas read-only users should only get the **Org Auditor** role.
5. You can repeat these steps for the **DP Production** subaccount to enable Cloud Foundry in the production environment as you near the deployment time for your implementation project.

### Create a Cloud Foundry Space

To create your Cloud Foundry space, follow these steps:

1. Navigate to your **DP Test** subaccount and select **Cloud Foundry • Spaces • Create Space**, as shown in Figure 3.22.

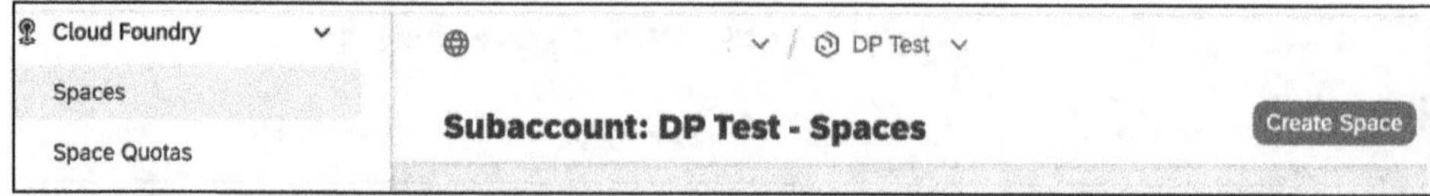


*Figure 3.22: SAP BTP Cockpit: Create Cloud Foundry Space*

2. In the dialog window that opens, enter "digitalpayments" in the **Space Name** field, as shown in Figure 3.23.

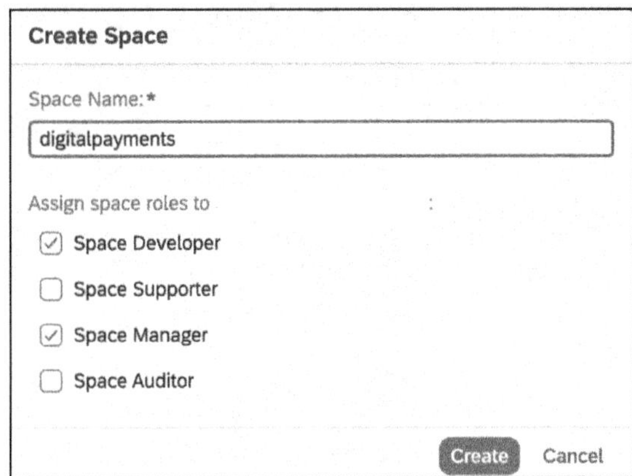


*Figure 3.23: SAP BTP Cockpit: Space Name and Space Roles Setup: Part 1*

3. Select both the **Space Developer** and **Space Manager** role checkboxes, then click **Create**.
4. You can now see the **digitalpayments** Cloud Foundry space successfully created, as shown in Figure 3.24.

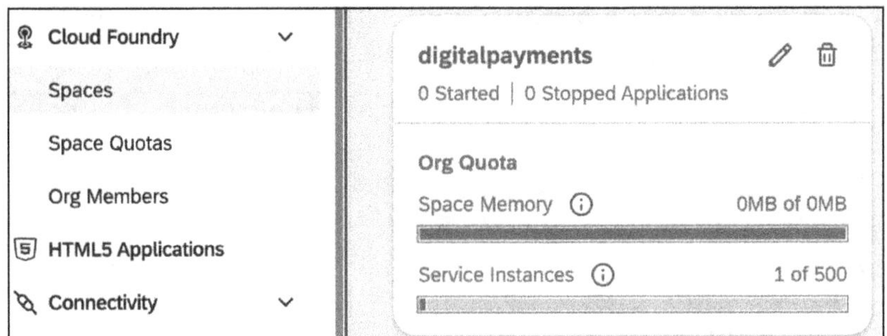


*Figure 3.24: SAP BTP Cockpit: Space Name and Space Roles Setup: Part 2*

5. Now that the new Cloud Foundry space has been successfully created, new users can be added as necessary. Navigate to **Space Members** as shown in Figure 3.25.

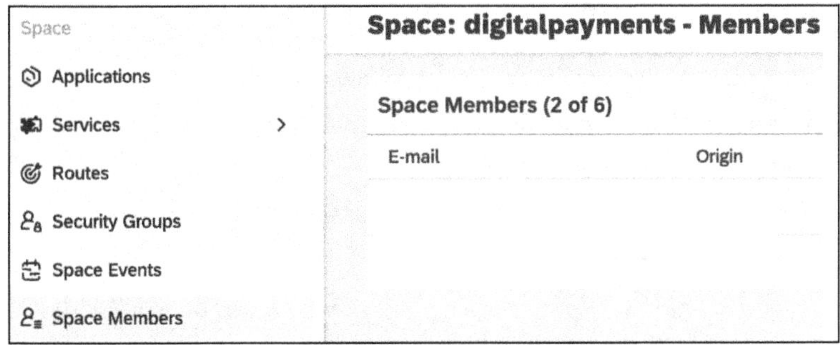


*Figure 3.25: SAP BTP Cockpit: Space Name and Space Roles Setup: Part 3*

6. Here, to add users who need to generate service keys, select the **Add Members** button as shown in Figure 3.26 and assign the users both the **Space Developer** and **Space Manager** roles. To add users who can just view service keys, assign them the **Space Supporter** and **Space Auditor** roles.

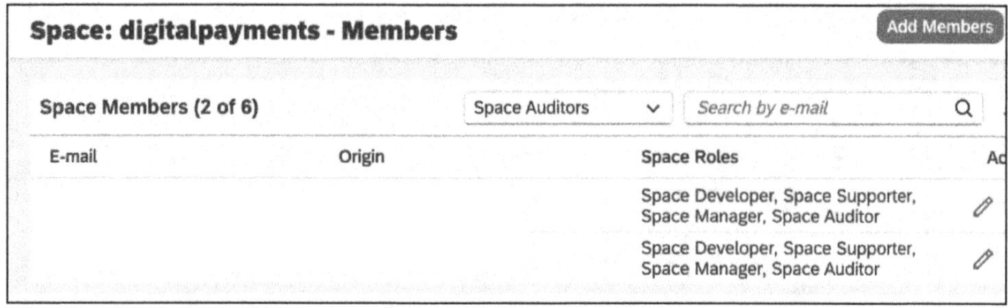


*Figure 3.26: Cloud Foundry Space: Add Members*

### Create the M2M Service Instance

An important prerequisite for creating an M2M service instance is to ensure that the M2M communication for the subaccount is enabled as we discussed in Section 3.2.3. Remember to complete those steps before you follow these steps:

1. Go to the SAP BTP cockpit and choose the **DP Test** subaccount. Now, go to **Services • Instances and Subscriptions** and click **Create**, as shown in Figure 3.27.

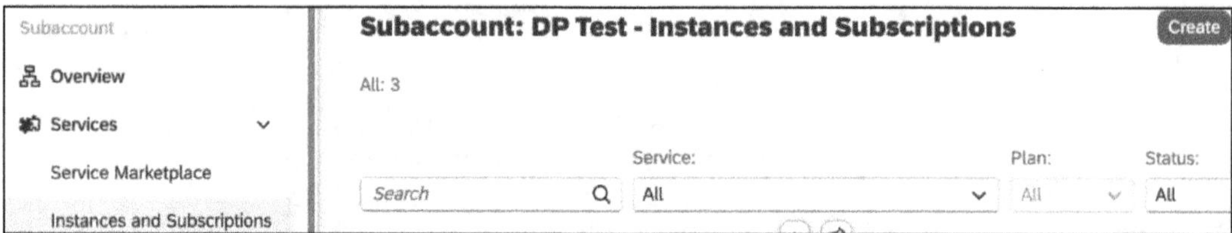


*Figure 3.27: SAP BTP Cockpit: Instances and Subscriptions*

2. In the wizard that opens, do the following (see Figure 3.28):
   - Select **SAP digital payments add-on (for tests, M2M)** under **Service**.
   - Enter “dp-m2m” in the **Instance Name** field.
   - Click **Create**.
3. The test dp-m2m instance should now be created, as shown in Figure 3.29.
4. Repeat these steps for the DP Production subaccount. In the wizard that opens, do the following:
   - Select **SAP digital payments add-on (for productive usage, M2M)** under **Service**.
   - Enter “dp-m2m” in the **Instance Name** field.
   - Click **Create**.

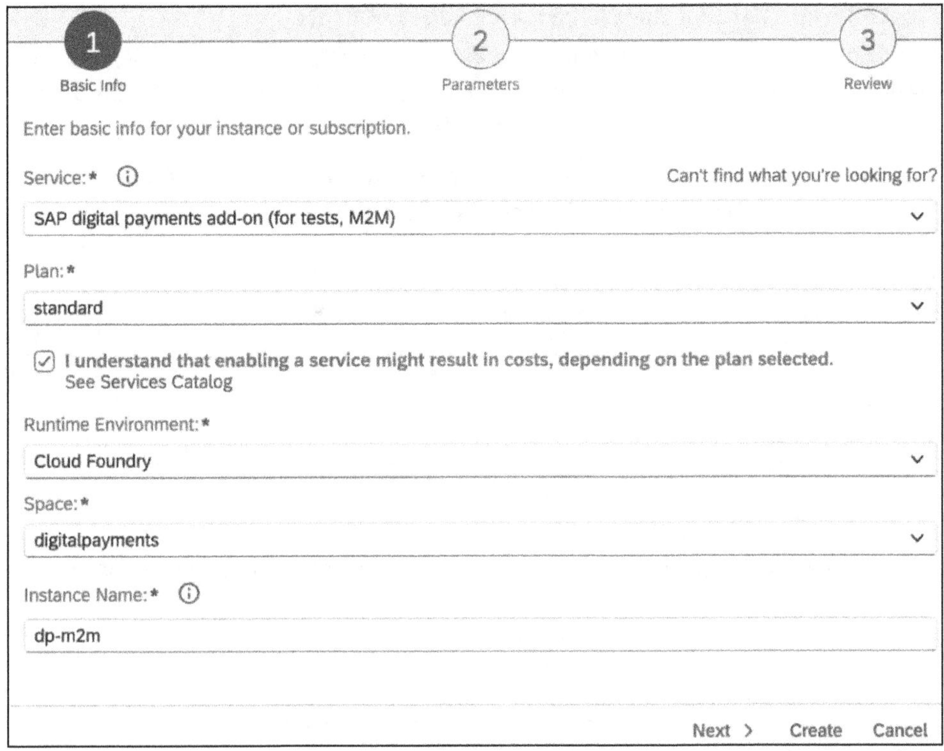


*Figure 3.28: Instances and Subscriptions: Wizard*

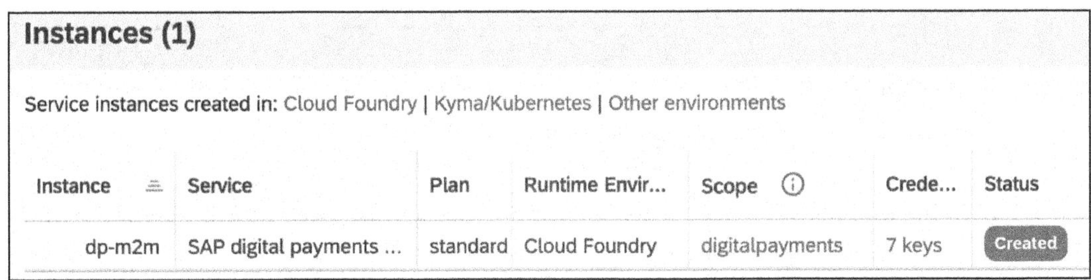


*Figure 3.29: M2M Test Instance: Successful Creation*

5. The dp-m2m production instance should now be successfully created, as shown in Figure 3.30.

Service: SAP digital payments add-on (for productive usage, M2M
Service Technical Name: **digitalpayments-m2m**
Status: Created

*Figure 3.30: Production M2M Instance: Successful Creation*

### Create Service Keys for Digital Payments Add-On Clients

A service key needs to be created for each digital payments add-on client (SAP ERP, SAP S/4HANA, etc.) to enable the client applications to securely communicate with the add-on. Let's walk through how to create a service key step by step:

1. In the SAP BTP cockpit, navigate to **Services • Instances and Subscriptions**, click the "..." button, and choose **Create Service Key**, as shown in Figure 3.31.

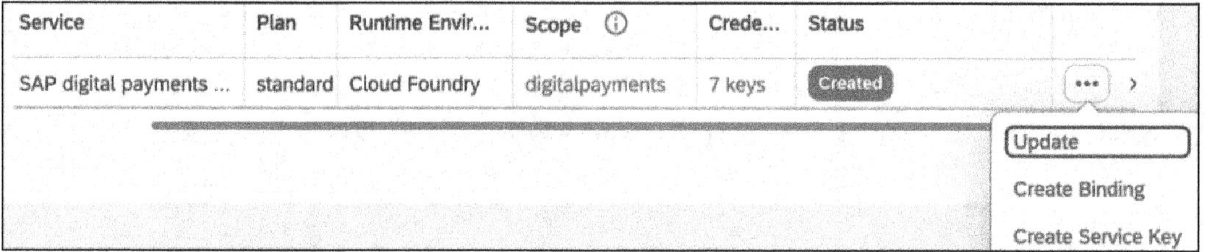


*Figure 3.31: SAP BTP Cockpit: Create Service Key*

2. Enter a service key name that indicates the client application—for example, "SAP ERP"—then click **Create**, as shown in Figure 3.32.

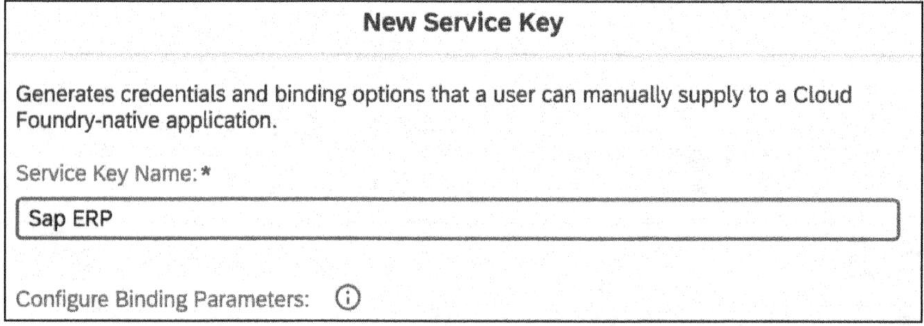


*Figure 3.32: SAP BTP Cockpit: Service Key Name*

3. The client ID and client secret will be used by the digital payments add-on client for secure communication with the add-on, as shown in Figure 3.33.

**Credentials**

SAP ERP

Form | JSON


```
{
    "uaa": {
        "clientid": "                    digitalpayments
            -test-m2m!b21537",
        "clientsecret":
            -                                  ,
        "url": "https://       dp-test.authentication.us10.hana.ondemand.com",
        "identityzone":        -dp-test",
```


*Figure 3.33: Client ID and Client Secret*

## 3.4 Security Configuration and Authentication

This section focuses on configuring user authentication, which allows administrators and key users to access the configuration screens used to configure the SAP digital payments add-on. User authentication is different from the M2M authentication used for API calls. In the sections that follow, we will first discuss the Identity Authentication architecture and the role of an IdP in establishing trust between your SAP BTP subaccount and your user directory. We then follow the step-by-step process to configure trust relationships with your chosen IdP. Finally, we explore certificate and metadata management, which establishes the cryptographic foundation for secure authentication exchanges between your subaccount and the IdP.

### 3.4.1 Identity Authentication Overview

The digital payments add-on applications are protected by the SAP Authorization and Trust Management service (XSUAA). By default, this service is not connected to your company's user base. To grant access to your project team, you must establish a trust relationship between your SAP BTP subaccount and an IdP.

SAP ID is the default provider used for logging into the SAP BTP cockpit, except for China (Shanghai region) or Government Cloud (US). You can start using it without any additional configuration. SAP ID provides a central user store and a single sign-on (SSO) service so that users can log on once and get access to all their applications, as shown in Figure 3.34.

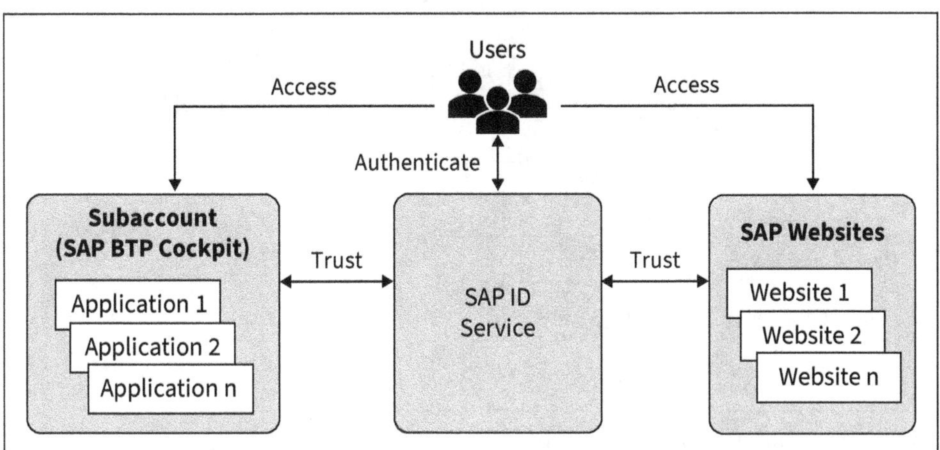


*Figure 3.34: SAP ID Authentication Flow*

New users can use the self-service registration option at the SAP website, *https://www.sap.com/index.html*. In the top-right corner, select the person icon; a dialog window will open. When you select **Create your SAP account**, as shown in Figure 3.35, and follow the steps, the SAP website registers you for an SAP Universal ID. During this process, you are also registered with SAP ID (via SAP Cloud Identity Services).

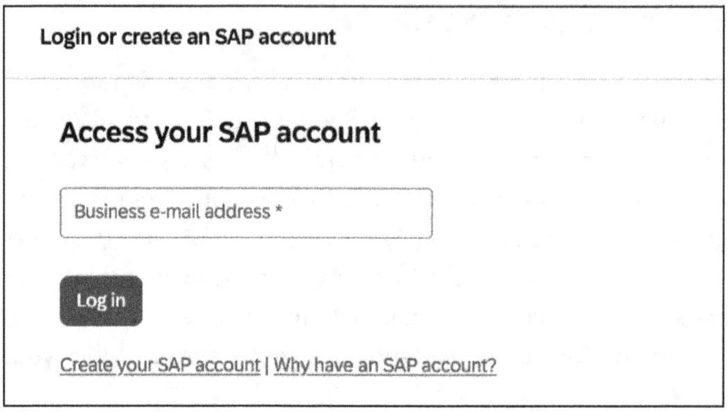

Login or create an SAP account

Access your SAP account

Business e-mail address *

Log in

Create your SAP account | Why have an SAP account?

*Figure 3.35: Create Your SAP Account*

SAP ID acts as a proxy for SAP Universal IDs when users log in with their email addresses. Users can log in and manage all their user accounts with an SAP Universal ID.

You can use your Identity Authentication tenant as the primary IdP for your SAP BTP subaccount. You can also federate your Identity Authentication tenant with your corporate IdP (see Figure 3.36), such as Microsoft Azure AD, to enable SSO for your corporate users.

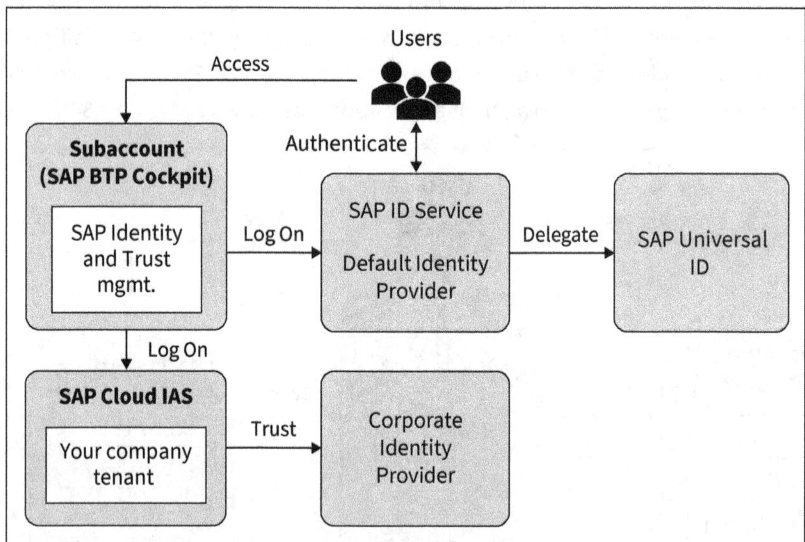


*Figure 3.36: Identity Providers and Federation*

You can also configure a direct SAML 2.0 trust relationship between your subaccount and your existing corporate IdP.

### 3.4.2 Trust Configuration Steps

The most efficient way to establish trust with Identity Authentication is by using the automated OpenID Connect (OIDC) setup available in the SAP BTP cockpit.

In your SAP BTP cockpit, select the add-on subaccount, then, in the left pane, expand the **Security** section and select **Trust Configuration**. Click the **Establish Trust** button in the top right-hand corner, as shown in Figure 3.37. In the wizard that appears, select **SAP Cloud Identity Services** from the IdP list; the system will automatically detect your available Identity Authentication tenants. Select the tenant you wish to use. The system will prefill the necessary details. Click **Establish Trust** to complete the process. Once the trust is established, you will see your Identity Authentication tenant listed on the **Trust Configuration** page.

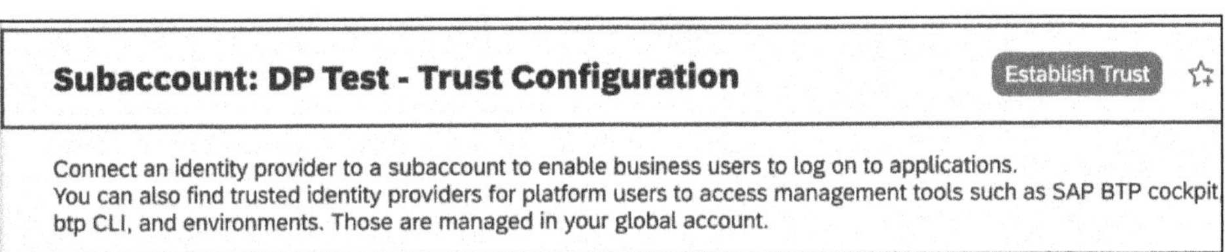


*Figure 3.37: SAP BTP Cockpit: Trust Configuration*

### 3.4.3 Certificate Management

If you are using a manual SAML 2.0 trust configuration instead of the automated OIDC setup, you must manually exchange metadata files. This typically involves downloading the SAML metadata from your SAP BTP subaccount and uploading it to your IdP to register the subaccount as a trusted service provider. You then export the metadata from your IdP and upload it back into SAP BTP's **Trust Configuration** screen. This manual exchange establishes the necessary signing certificates and endpoints for secure authentication.

**Add-On Security Settings**

For the digital payments add-on, the primary security settings are handled at the platform level through the trust configuration described in the this section and the role management described in the next section. There is no additional application-specific security required within the add-on for user authentication.

## 3.5 Role Management and Authorizations

*Authentication* confirms a user's identity, and *authorization* determines what they are allowed to do. In SAP BTP, authorization is managed through role collections. The SAP digital payments add-on provides specific role templates that you must assign to your users. The most scalable way to manage this is by mapping SAP BTP role collections to user groups defined in your IdP. In the sections that follow, you will first create role collections that bundle together the specific roles provided by the add-on subscription, establishing the permission structure for your implementation team. You will then become familiar with the two primary add-on roles—`DigitalPaymentsAdministrator` for technical and security-sensitive configuration, and `DigitalPaymentsKeyUser` for business-level configuration—and determine which roles each user needs. Finally, you will assign these role collections to individual users or map them to user groups.

### 3.5.1 Role Collections, Users, and Group Management

The authorization model consists of four parts:

- **Role template**
  The add-on subscription provides technical role definitions, such as for an administrator or key user.
- **Role**
  You create an instance of a role based on a template.
- **Role collection**
  This is a custom bundle that can hold multiple roles. You assign this collection to users.
- **User group**
  This is a label or group defined in your IdP. You map this group to an SAP BTP role collection so that any user added to the group automatically gets the necessary permissions.

The digital payments add-on supports multiple options to manage authorizations for users. You can assign authorizations to users depending on the type of trust configuration enabled in the SAP BTP cockpit. Also, you must decide whether you prefer to maintain the authorizations of individual users in the IdP or in SAP BTP. If you're using the default trust configuration with the default IdP, then you will directly assign users to role collections.

Let's start by creating a new role collection, as follows:

1. In the SAP BTP cockpit, navigate to your subaccount, then choose **Security • Role Collections**, as shown in Figure 3.38.

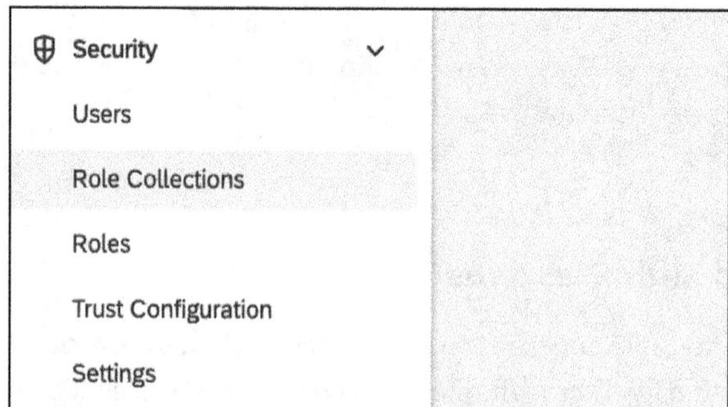


*Figure 3.38: SAP BTP Cockpit: Role Collections Navigation*

2. To create a new role collection, choose the **Create** button in the top right-hand corner, as shown in Figure 3.39.
3. In the window that opens, enter a **Name** and **Description**, as shown in Figure 3.40, and choose **Create**. You will now see a new role collection in the list of role collections.

*Figure 3.39: SAP BTP Cockpit: Create New Role Collection*

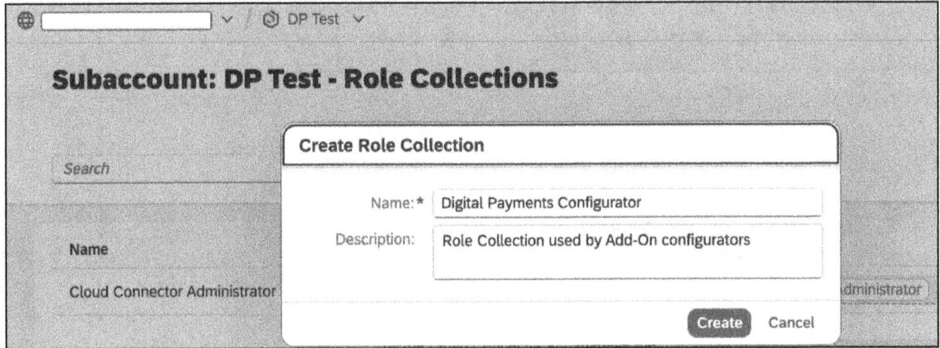


*Figure 3.40: Create Role Collection: Enter Name and Description*

4. You can also copy an existing role collection by clicking the **Copy** link at the end of the row of an existing role collection, as shown in Figure 3.41. This will prompt you to enter a new name and description; you can see the included roles in the newly copied role collection, and you can save your changes.

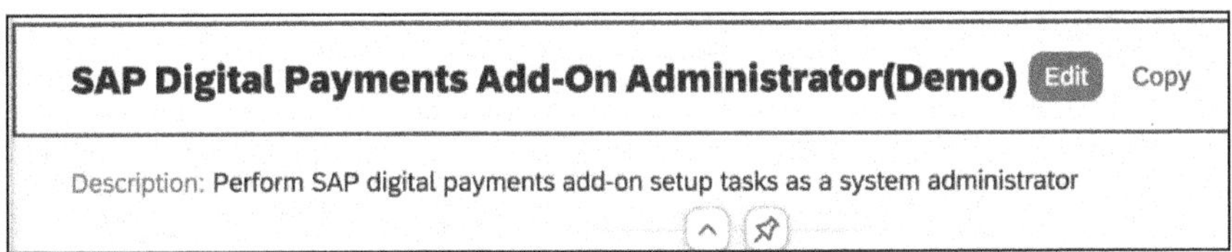


*Figure 3.41: SAP BTP Cockpit: Copy Role Collection*

You can now add or remove roles and assign users to the role collection. If you're using a custom trust configuration—for example, with SAP Cloud Identity Services—you have options to map role collections to user groups. Before configuring SAP BTP, you should create the corresponding user groups in your IdP.

### 3.5.2 Add-On-Specific Roles

The digital payments add-on provides the following two primary role templates that you can use to add roles to the role collection:

- `DigitalPaymentsAdministrator`
  This role grants access to technical and security-sensitive UIs. Users with this role can activate PSP adapters and manage the secure credentials (API keys) required to connect to them.
- `DigitalPaymentsKeyUser`
  This role grants access to business configuration UIs. Users with this role can define routing rules, manage merchant ID aliases, and configure the layout of the hosted payment page.

Let's now look at the steps to add these two roles to the role collection:

1. Select the **Digital Payments Configurator** role collection that you created in the previous section, then choose the **Edit** button, as shown in Figure 3.42.

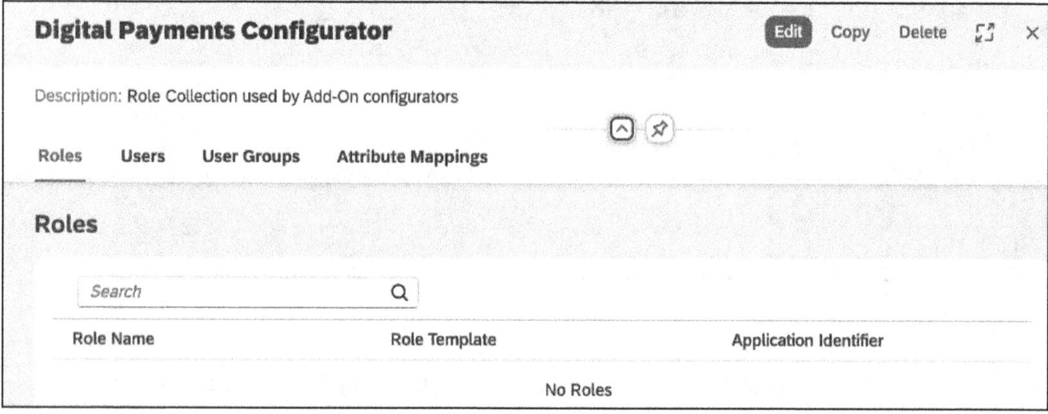


*Figure 3.42: Role Collection: Edit Button*

2. Select the role line item. In the window that opens, select **DigitalPaymentsAdministrator** in the **Role Template** dropdown, as shown in Figure 3.43.

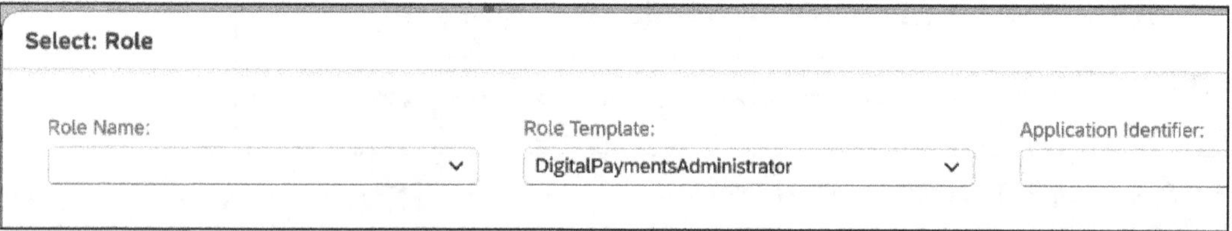


*Figure 3.43: Role Template: Search*

3. Select the checkbox next to the **DigitalPaymentsAdministrator** role name, as shown in Figure 3.44.

Roles

| Role Name | Role Template | Application Identifier |
|---|---|---|
| DigitalPaymentsAdministrator | DigitalPaymentsAdministrator | digitalpayments-test!t21537 |

*Figure 3.44: Role Name: Checkbox Selection*

4. You will see the role name appear in the selected roles as shown in Figure 3.45. Now click the **Add** button to add the role to the role collection.

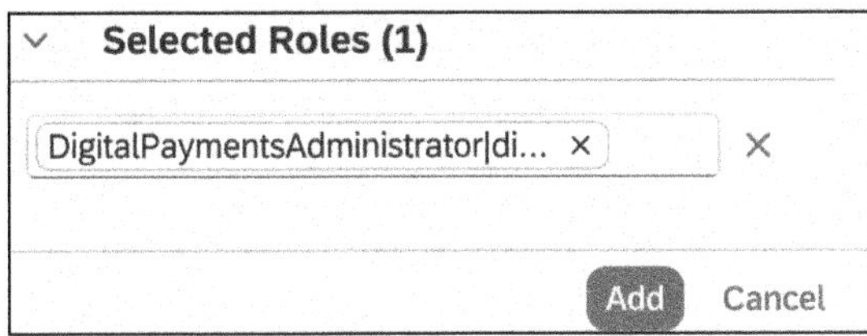


*Figure 3.45: Add Role to Role Collection*

5. Next, select the **DigitalPaymentsKeyUser** role template, as shown in Figure 3.46, and add it to the role collection.

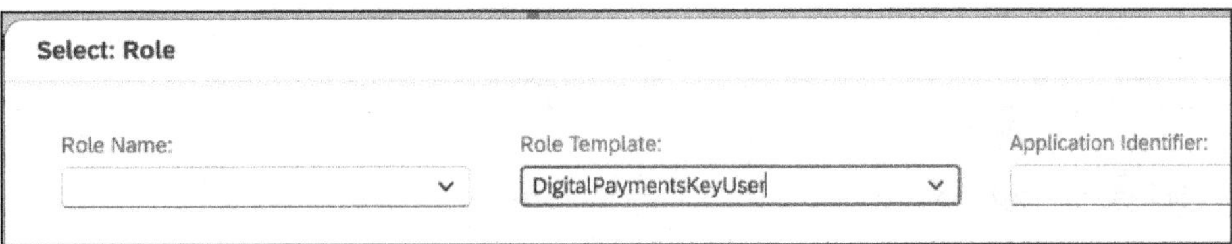


*Figure 3.46: DigitalPaymentsKeyUser Role Template Selection*

6. Now you will see that both roles have been added to the role collection (Figure 3.47). Click **Save** to save the role collection.

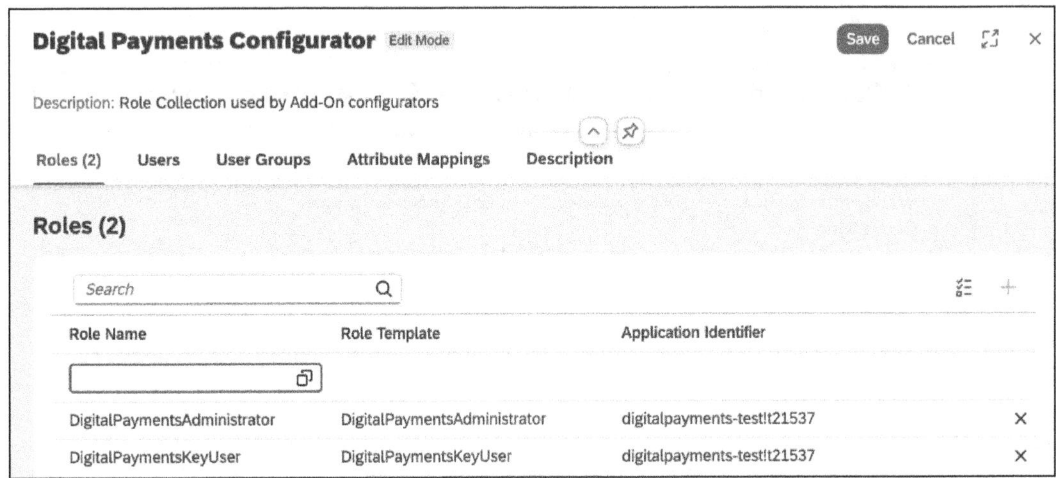


*Figure 3.47: Role Collection: Successful Addition of Both Roles*

Now that the roles have been added to the role collection, it's time to add the necessary users to the role collection.

### 3.5.3 Authorization Setup

Let's look at the steps to create new users and assign role collections to them:

1. Navigate to the **Security • Users** menu option in the subaccount explorer, as shown in Figure 3.48.

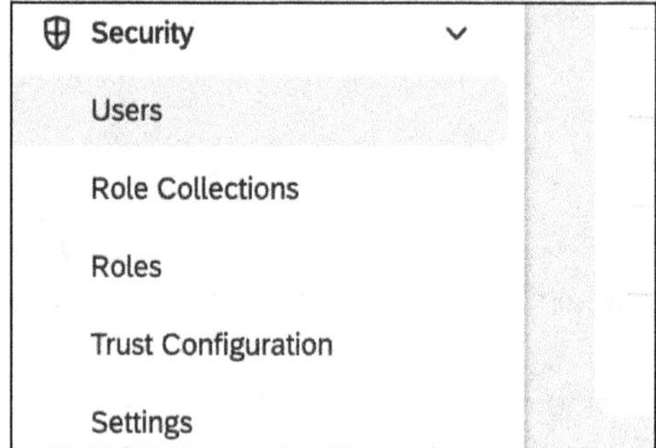


*Figure 3.48: Add-On Subaccount: Navigation to Users*

2. Select the **Create** button shown in the top right-hand corner of Figure 3.49.

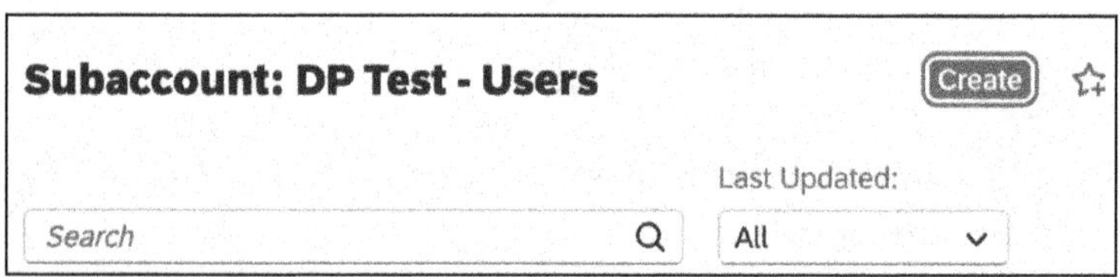


*Figure 3.49: Add-On Subaccount: Create Button*

3. In the window that opens, enter the **User Name** and **E-Mail** address, and select the **Identity Provider** (in this case, **Default identity provider**), then select **Create** to create the user, as shown in Figure 3.50.

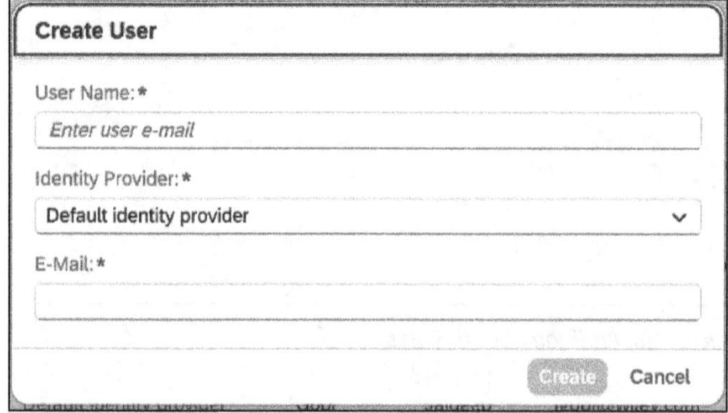


*Figure 3.50: Add-On Subaccount: Create Using Default Identity Provider*

4. Now go to the role collection to which you want this new user added. In the **Users** section, search for a user (see Figure 3.51) and select that user to assign the role collection to them.

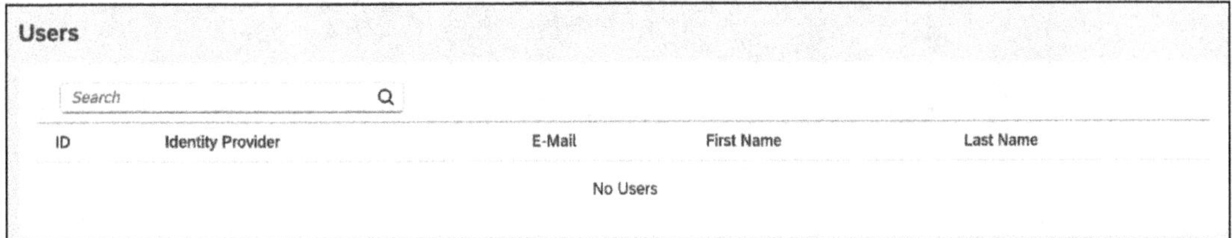


*Figure 3.51: Add-On Role Collection: Search for New Users*

Your users are now fully authorized. When they log in, they will be granted the permissions defined in the role collection. Remember to repeat these steps for the production subaccount as well, as necessary.

## 3.6 Testing the Initial Setup

After completing the user authorization configuration, it is important to validate the environment before attempting to connect external PSPs. In the following sections, we'll show you how to perform the validation and then walk through some common setup issues that you may encounter.

### 3.6.1 Validation Procedures

You can verify user access to the add-on configuration UIs by following the steps shown here. Test access with a user that belongs to the Digital Payments Configurator role collection, for both the DP Test and Production subaccounts.

1. In your subaccount left-hand menu, navigate to **Instances and Subscriptions • Subscriptions.** Select the three dots (ellipsis) next to the **Subscribed** status and choose **Go to Application,** as shown in Figure 3.52.

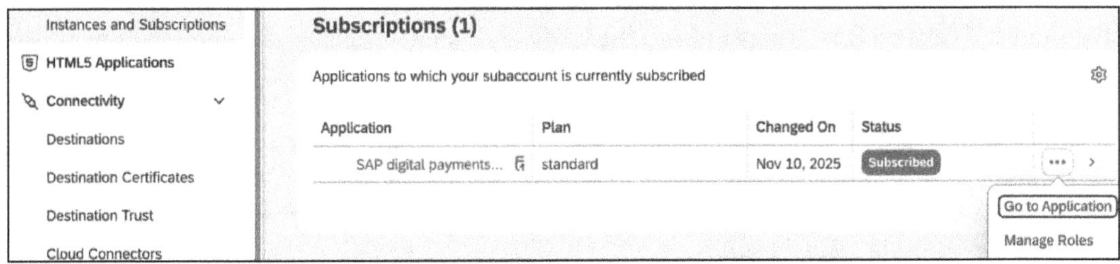


*Figure 3.52: Application Link: Go to Application*

2. Upon successful authentication, you will see a browser screen like the one shown in Figure 3.53. This confirms that the new user is authenticated to log into the add-on application home page, which is the expected behavior.

*Figure 3.53: Application Login Screen for New Add-On Configurator Users*

The URL portion that appears before the "/index.html" is called the base URL. Table 3.1 shows the URL paths that can be added to the base URL to access different configuration screens.

| Add-On Config UI URL Path | Configuration Screen |
|---|---|
| */pspStatus/index.html* | **Payment Service Provider Status**; use this screen to activate different DP Adapters. |
| */pspDetermination/index.html* | **Payment Service Provider Determination**; use this screen to create PSP determination rules based upon different criteria in the digital payments request—for example, company code. |
| */merchantMapping/index.html* | **Merchant Mapping**; use this screen to set merchant aliases for different merchant IDs. |

*Table 3.1: Add-On Config UI Screens and Corresponding URLs*

Use the new user that was created to test that access is provided to these URLs.

### 3.6.2 Common Setup Issues

If validation fails, you can review the following common configuration problems and their solutions:

- **Authentication and authorization issues**
  *Problem:* Users cannot log into the add-on application:
  - Verify that trust configuration is established correctly between the SAP BTP subaccount and the IdP.
  - Confirm that the user exists in the IdP with the correct email address.
  - Check that the user's email in the IdP exactly matches the email used when assigning role collections.

  - For SAP ID users, ensure the user has completed the self-registration process and verified their email.

  *Problem:* Users can log in but see "access denied" or empty configuration screens:

  - Confirm the user is assigned to the correct role collection (Digital Payments Configurator).
  - Verify that the role collection contains both required roles: `DigitalPaymentsAdministrator` and `DigitalPaymentsKeyUser`.
  - Check that the role assignment has propagated (there can be a few minutes delay after assignment).
  - Have the user log out completely and log back in to refresh their authorization token.

- **Service instance and service key issues**

  *Problem:* Service key creation fails or returns an error:

  - Verify that Cloud Foundry is enabled in the subaccount.
  - Confirm that the user creating the service key has both Space Developer and Space Manager roles in the digitalpayments space.
  - Check that the correct service plan is selected (test/demo M2M for test subaccounts; productive M2M for production).
  - Ensure the M2M service instance was created successfully before attempting to create a service key.

  *Problem:* Service key credentials appear incomplete or are missing fields:

  - Regenerate the service key by deleting and recreating it.
  - Verify you are viewing the correct service key in JSON format (not form view).
  - Confirm you are looking at the service key for the M2M instance, not the application subscription.

- **Connectivity and network issues**

  *Problem:* Cannot access the add-on application URL:

  - Verify the subscription status shows **Subscribed** in the SAP BTP cockpit.
  - Confirm you are using the correct subdomain in the URL (the one specified during subaccount creation).
  - Check that your browser is not blocking third-party cookies, which are required for SAP BTP applications.
  - Clear the browser cache and cookies, then try accessing the application again.
  - Try accessing from an incognito/private browser window to rule out cached credentials.

- **Entitlement and subscription issues**

  *Problem:* Cannot find SAP digital payments add-on in the **Service Marketplace**:

  - Verify that entitlements are assigned to the correct specific subaccount, not just the global account.

- Confirm you are looking in the correct subaccount (test vs. production).
- Check that the global account has an available quota for the service plans.
- Ensure you are searching for the exact service name in the **Service Marketplace**.

*Problem:* Subscription creation fails:

- Verify that the Cloud Foundry environment is enabled in the subaccount.
- Confirm that the entitlement quota is available (not already consumed by another subaccount).
- Check that you selected the correct plan (demo/test plans for test subaccounts; standard plans for production).
- Wait a few minutes and retry, as temporary service availability issues can occur.

- **Configuration data issues**

  *Problem:* Configuration changes are not visible or not persisting:

  - Allow up to five minutes for configuration changes to propagate across the distributed system.
  - Verify you are making changes in the correct tenant (test vs. production).
  - Confirm you have the `DigitalPaymentsKeyUser` role for business configuration changes.
  - Clear the browser cache and reload the configuration UI.
  - Check that you clicked **Save** or **Submit** after making configuration changes.

When encountering any setup issue, consider the following general troubleshooting tips:

- **Check the SAP BTP service status**

  Check out *https://support.sap.com/en/my-support/systems-installations/cac/maintenance-windows.html* to confirm there are no platform-wide service disruptions affecting SAP BTP or the specific region you are using.

- **Review audit logs**

  In the SAP BTP cockpit, check the audit logs to review recent authentication attempts and authorization decisions for detailed error information.

- **Test with default IdP**

  If using a custom IdP, temporarily test with SAP ID to isolate whether the issue is with the custom IdP configuration.

- **Compare test and production**

  If one environment works but the other does not, systematically compare the configurations to identify differences.

- **Consult SAP support**

  If issues persist after following these troubleshooting steps, open a support ticket with SAP under component FIN-FSCM-DP-DP. Provide the following:

  - Screenshots of error messages
  - Service key details (with credentials redacted)

- RFC destination configuration (with credentials redacted)
- Audit log entries related to the issue
- Exact steps to reproduce the problem

By working through these common issues and their solutions, most setup problems can be resolved without requiring SAP support intervention. The key is to validate each layer of the configuration independently before moving to the next integration point.

## 3.7 Summary

This chapter provided comprehensive, step-by-step instructions for establishing the technical foundation of the SAP digital payments add-on within your SAP BTP environment. The setup activities covered are essential prerequisites for subsequent configuration and integration work. We validated the commercial and technical prerequisites at the global account level, confirming administrative access, verifying Cloud Foundry runtime availability, selecting the appropriate SAP BTP region for data residency, and, most importantly, ensuring that the necessary service plan entitlements were allocated to your global account.

This chapter also described in detail provisioning the dedicated subaccounts based on the recommended landscape strategy: a nonproductive (test) environment for implementation and quality assurance activities and a productive environment for live transaction processing. The separation between these environments is enforced by the distinct service plans (demo vs. production) that cannot coexist within a single subaccount. Each subaccount was configured with a unique subdomain identifier that becomes part of the tenant's URL structure for both administrative access and API endpoints.

A major portion of this chapter focused on establishing secure M2M communication channels. We enabled the Cloud Foundry runtime within each subaccount, created the digital payments Cloud Foundry space, and provisioned M2M service instances for both test and production environments. The service keys generated from these instances contain the critical OAuth 2.0 credentials—client ID, client secret, token endpoint URL, and service API endpoint—that enable consumer applications to authenticate and communicate securely with the add-on.

This chapter also described setting up comprehensive user authentication and authorization by configuring trust relationships between the SAP BTP subaccount and an IdP. We demonstrated using SAP ID as the default provider and discussed the approach of using Identity Authentication versus integrating directly with corporate identity providers through SAML 2.0 federation. This flexibility allows organizations to leverage existing user directories and enable SSO across their SAP landscape.

This chapter also described validation procedures that confirm each layer of the configuration is working correctly. We tested user access to the add-on's UI applications by logging in with properly authorized users and accessing the configuration UIs for PSP status, PSP determination, and merchant mapping. These validation steps catch the most common configuration errors, such as missing role assignments or incorrect trust configurations, at a point where they are still easy to correct.

With the SAP BTP tenant fully provisioned, the communication channels secured through OAuth 2.0, and user authorizations properly configured, your environment is now ready for the next phase of implementation. In the next chapter, we will discuss connecting to external PSPs by activating PSP adapters and configuring their API credentials.

Chapter 4

# Integration with Payment Service Providers

*With the SAP BTP environment configured, the SAP digital payments add-on tenants provisioned, and key components activated, the next critical step is connecting the add-on to the payment service providers that will process actual customer payments. This chapter bridges the gap between the technical infrastructure and the external payment ecosystem, providing you the knowledge and hands-on procedures to integrate Stripe and other certified PSPs so that your organization can begin accepting and processing payments at scale.*

This chapter provides comprehensive guidance for integrating payment service providers (PSPs) with the SAP digital payments add-on, using Stripe as the primary example. It covers the PSP integration architecture, detailed configuration steps using Stripe's certified adapter, and setup of payment methods including cards and digital wallet options. You will learn webhook configuration for real-time payment events, handling authorization, capture, and refund scenarios using actual Stripe implementation examples. The chapter includes production-tested configurations, troubleshoots common Stripe integration issues, and describes best practices for optimizing payment processing performance.

## 4.1 Understanding PSP Integration Architecture

With the SAP BTP environment configured (Chapter 2) and the digital payments add-on provisioned (Chapter 3), the next important step is connecting the add-on to PSPs. While Chapter 2 introduced the high-level architecture of PSP adapters, this section explains the technical patterns, communication flows, and integration approaches that enable the add-on to communicate with external payment processors.

Understanding these architectural patterns is important before starting the hands-on configuration steps; it helps implementation teams make informed decisions about PSP selection, troubleshoot integration issues effectively, and design robust payment flows that align with business requirements.

### 4.1.1 Payment Gateway Fundamentals

PSPs handle the technical complexity of payment processing between merchants and the card payment ecosystem. This section focuses on the gateway mechanisms and API architecture that enable PSP integration with the digital payments add-on. To accomplish this, we first examine the authentication and security layer that protects API communication with PSPs. We then examine the request router and validator layer that ensures data integrity before transmission. Next, we explore the API endpoint layer and how different PSPs expose distinct endpoints for authorization, capture, refunds, and tokenization. Finally, we review the communication protocol requirements and regional API endpoints that ensure your payments infrastructure meets enterprise security and compliance standards.

### API Gateway Architecture

Modern PSPs expose their payment processing capabilities through REST API gateways that provide standardized endpoints for payment operations. The digital payments add-on integrates with these gateways through PSP-specific adapters. Let's review the various technical layers that are part of the PSP infrastructure:

- **Authentication and security layer**
  PSPs implement different authentication approaches for API security. The digital payments add-on adapter framework abstracts these authentication differences. Each adapter manages its PSP's authentication requirements using credentials securely stored in SAP Credential Store. Table 4.1 details the primary authentication and security mechanisms as well as any alternate authentication mechanisms for your understanding. The digital payments add-on adapter framework abstracts these details from you, but it is important to understand these mechanisms. That way, you can answer any questions that arise from your internal security and compliance team when you plan to go live with the add-on and any of the PSPs listed here.

| Payment Service Provider | Primary Auth Method | Security Mechanism | Alternate Mechanisms | Details |
|---|---|---|---|---|
| Stripe | API key authentication | HTTP authorization header | ▪ Strong Customer Authentication (3D Secure, CVV checks)<br>▪ Dashboard multi-factor authentication (MFA)<br>▪ Internal mutual TLS (mTLS) | ▪ Uses secret API keys passed as bearer tokens<br>▪ Offers robust customer-facing authentication methods |

*Table 4.1: Payment Service Providers: API Authentication and Security Mechanisms*

| Payment Service Provider | Primary Auth Method | Security Mechanism | Alternate Mechanisms | Details |
|---|---|---|---|---|
| Novalnet | API key authentication | HTTP authorization header | ■ Hosted payment pages<br>■ Strong Customer Authentication (two-factor authentication [2FA]) | ■ Uses specific activation and access keys<br>■ Offers hosted solutions for secure card data handling |
| Cyber-source | Certificate-based authentication | mTLS | ■ HTTP signature (API key alternative)<br>■ Payer authentication (3D Secure) | ■ Client certificates used for server-to-server auth<br>■ Multiple options for merchant API access and customer verification |
| Worldpay/ Paymetric | Token-based authentication | OAuth 2.0/JSON Web Tokens (JWT) | ■ API keys<br>■ Session keys<br>■ Transaction session keys | ■ Uses OAuth-style JWT bearer tokens |

*Table 4.1: Payment Service Providers: API Authentication and Security Mechanisms (Cont.)*

- **Request router and validator layer**
  This layer performs technical validation before forwarding requests to payment processing engines. Key functions include the following:
  - Merchant ID resolution: Maps the merchant account alias to the PSP's technical merchant identifier.
  - Schema validation: Verifies request payloads conform to the PSP's API schema.
  - Idempotency control: Detects duplicate requests to prevent accidental double-charging.
- **API endpoint layer**
  PSPs provide distinct endpoints for different payment operations. Let's compare endpoint patterns across two certified PSPs (see Table 4.2) to see the subtle differences in their API endpoint layers. The add-on adapter framework shields implementation teams from having to deal with these complexities; they can instead integrate using one unified API framework provided by the add-on, which accelerates the implementation schedule. Consumer applications only integrate with the add-on API endpoints; the add-on takes care of managing the integration differences with each of these PSP API endpoints.

| Operation | Stripe V2 | CyberSource |
|---|---|---|
| Create authorization | POST /v1/payment_intents | POST /pts/v2/payments |
| Capture payment | POST /v1/payment_intents/{id}/capture<br>Where {id} is the transaction ID from the authorization request | POST /pts/v2/payments/{id}/captures<br>Where {id} is the transaction ID from the authorization request |
| Refund ppayment | POST /v1/refunds | POST /pts/v2/refunds |
| Tokenize card | POST /v1/payment_methods | POST /tms/v2/tokens |

*Table 4.2: PSP API Endpoint Comparison*

**Regional API Endpoints**

Global PSPs operate multiple regional API gateways for performance and data residency compliance. Stripe maintains a global endpoint (*api.stripe.com*) that automatically routes to regional data centers based on merchant account configuration. Cybersource provides distinct regional endpoints. Table 4.3 shows the regional endpoints available for widely used PSPs. It is important to understand these details so that you can choose the right PSP based on your company's global and regional policies and security requirements tied to payment processing.

| PSP | Endpoint Strategy | Data Center (EU) | Data Center (US) |
|---|---|---|---|
| Stripe | Global endpoint with intelligent routing | Yes (via global endpoint) | Yes (via global endpoint) |
| Novalnet | Single primary data center location | Yes (primary data centers in Germany) | No (data is in EU) |
| Cybersource | Distinct regional endpoints required | Yes (via specific EU endpoint) | Yes (via specific NA endpoint) |
| Worldpay/Paymetric | Multiple regional endpoints/platforms | Yes (multiple EU endpoints exist) | Yes (multiple US endpoints exist) |

*Table 4.3: PSP Data Center Strategy*

**Canonical Request and Response Formats**

All certified PSPs use JSON for request payloads and responses and have their own API specification. The add-on API specification defines the canonical request structure to be used. PSP adapters translate digital payments add-on canonical request structures into individual PSP-specific formats. Implementation teams do not have to spend time trying to define a common canonical format for integrating with multiple PSPs; instead, they can leverage the add-on-provided canonical format to accelerate the implementation.

Let's look at an example digital payments add-on canonical request structure for a payment authorization of USD 15,000, shown in Listing 4.1.

```
json
{
  "AuthorizationByDigitalPaytSrvc": "DPA-AUTH-12345",
  "AmountInAuthorizationCurrency": "15000",
  "AuthorizationCurrency": "USD",
  "PaytCardByDigitalPaymentSrvc": "DPA-CARD-67890",
  "MerchantAccount": "merchant-alias-001"
}
```

*Listing 4.1: Add-On: Example Canonical Request Structure*

The PSP adapter translates this canonical JSON into the PSP-specific JSON format. For Stripe, this becomes what you see in Listing 4.2.

```
json
{
  "amount": 15000,
  "currency": "usd",
  "payment_method": "pm_1NAbCD2eZvKYlo2C",
  "capture_method": "manual",
  "metadata": {
    "dpa_auth_token": "DPA-AUTH-12345"
  }
}
```

*Listing 4.2: Stripe PSP: JSON Example Request Payload*

For another PSP, its JSON format would be different; however, implementation teams do not have to manage this complexity. Instead, they just define the routing conditions in the add-on configuration UI to route traffic to individual PSPs based on their company requirements. The add-on takes care of the mapping between canonical and PSP-specific formats.

### Communication Protocol Requirements

It is recommended that all PSP communications use the following:

- Transport protocol: HTTPS (port 443)
- TLS version: 1.2 or higher (mandatory)
- Certificate validation: Required for production environments

For SAP S/4HANA and SAP ERP systems, these requirements are implemented through RFC destinations (Type G, HTTP connection to external server), configured in Transaction SM59. The DigiCert TLS RSA4096 Root G5 certificate must be installed in Trust Manager (Transaction STRUST). For other applications, you can refer to their administration and user guides for the configuration steps to enable the recommended communication protocols so that your payment processing will be secure and meet your enterprise standards.

### 4.1.2 API-Based Integration Patterns

Payment processing through the add-on follows industry-standard gateway integration patterns. Understanding these patterns is important for designing effective order-to-cash and invoice-to-cash processes. In this section, we examine the two-step *authorization and capture* pattern, which separates the fund reservation phase from the settlement phase, a typical requirement for scenarios in which the final invoice amount may differ from the initial order amount. We then explore the direct capture payment flow, which combines authorization and settlement into a single operation for use cases requiring immediate payment without preauthorization. Finally, we explore the three distinct Stripe API integration patterns that accommodate different payment collection scenarios, from standard add-on card registration to external collection with token preparation to centralized refund operations.

#### Two-Step Authorization and Capture

The most common pattern for business-to-business (B2B) and business-to-consumer (B2C) scenarios involves separating authorization from settlement. Figure 4.1 shows the two-step flow.

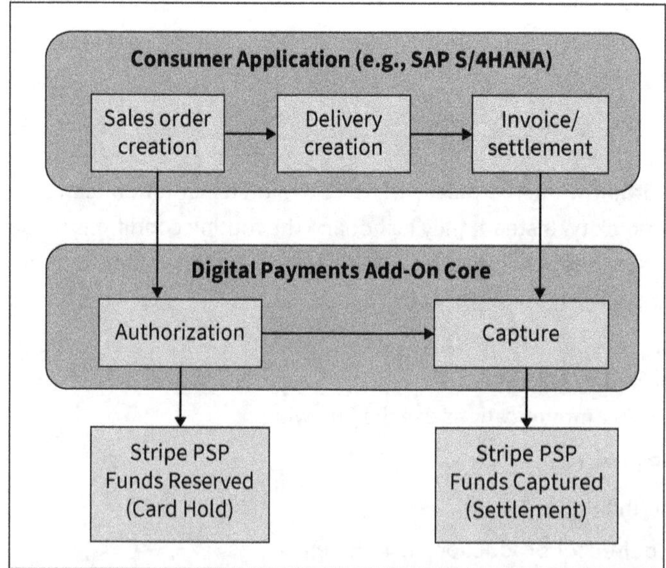


*Figure 4.1: Authorization and Capture: SAP S/4HANA Two-Step Scenario*

The following are the steps involved to authorize a $1,000 payment when placing a sales order:

1. A customer places an order totaling $1,000.
2. The consumer application calls the digital payments add-on */v1/authorizations* API.
3. The digital payments add-on routes the authorization request to the appropriate PSP adapter (e.g., Stripe).
4. The adapter translates the consumer application authorization request into a PSP-specific authorization request.
5. The PSP reserves funds on the customer's card (as a hold; not yet charged).
6. The PSP returns an authorization code (e.g., ch_3ABC...).
7. The digital payments add-on creates an internal authorization token and returns it to the consumer application.
8. The order proceeds to delivery processing.

In the capture phase (billing document settlement), if there are physical goods involved, capture happens after the goods are shipped. The following are the steps during the capture phase:

1. Goods are shipped, and an invoice is generated in the consumer application.
2. Finance runs the payment settlement job.
3. The consumer application calls the digital payments add-on */v1/charges* API with an authorization token.
4. The digital payments add-on retrieves the PSP authorization reference.
5. The adapter sends a capture request to the PSP.
6. The PSP transfers funds from the customer to the merchant account.
7. The settlement is posted in financial accounting.

This pattern is essential for scenarios in which the final invoice amount may differ from the initial order amount due to delivery variances, price adjustments, or partial shipments.

### Direct Capture Payment Flow

For scenarios requiring immediate payment without preauthorization, the add-on supports direct capture. Common use cases are as follows:

- Subscription billing, in which cards are charged on renewal dates
- Invoice payment links sent to customers
- E-commerce checkout with immediate fulfillment

Figure 4.2 shows the simplified direct capture flow.

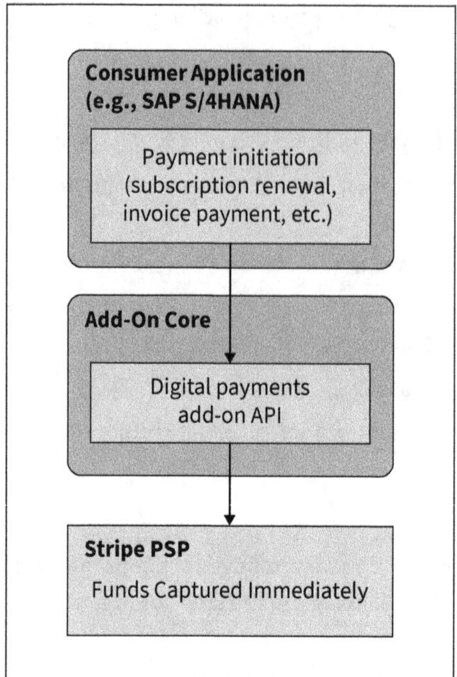


*Figure 4.2: Direct Capture Flow*

The key distinction is that a single API call performs both authorization and settlement simultaneously at the PSP.

### Stripe API Integration Patterns

The Stripe V2 adapter implements three distinct integration patterns depending on where the initial payment interaction occurs:

- **Pattern A: Standard digital payments add-on card registration**
  In this pattern, payment card registration is initiated from the consumer application (e.g., SAP S/4HANA's business partner maintenance). The Stripe V2 adapter receives the registration request, then returns the correct registration URL for the consumer application to securely capture the payment card details. Subsequently, the consumer application can leverage the poll API endpoint to poll the card details and get the digital payments add-on's internal token. This token is then stored in the application for future use, if customer consent to do so is obtained during card entry. Subsequent transactions from the same customer can reuse the digital payments add-on token, thereby simplifying payment integration requirements. Figure 4.3 depicts a simple logical flow to illustrate this pattern.

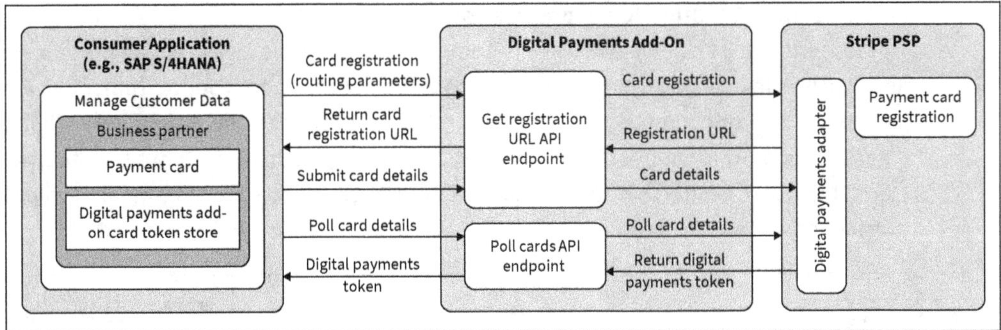


*Figure 4.3: Standard Digital Payments Add-On Card Registration Pattern*

- **Pattern B: External collection with token preparation**
  This pattern supports scenarios in which an external e-commerce storefront collects payment information by directly integrating with Stripe but needs to hand off the transaction to an SAP system (e.g., SAP S/4HANA) for order processing and fulfillment. The `getforpaymentcardwithauthorization` endpoint validates that the external payment intent exists and has status `requires_capture`, then wraps it in a digital payments add-on authorization token that SAP S/4HANA can use.

  After the e-commerce application sends the order information (along with the digital payments add-on authorization ID) to the SAP system, the order document follows the typical document flow steps as necessary to be fulfilled (including, for physical goods, shipment to the customer). After the fulfilment is complete, the SAP application triggers the end of business day settlement operation to capture the funds on the customer payment card. It is possible for the customer to request a refund if, say, the goods were damaged or lost in transit or for another reason; in that situation, the SAP application can trigger the refund endpoint as part of the credit memo settlement process. Figure 4.4 shows the step-by-step flow.

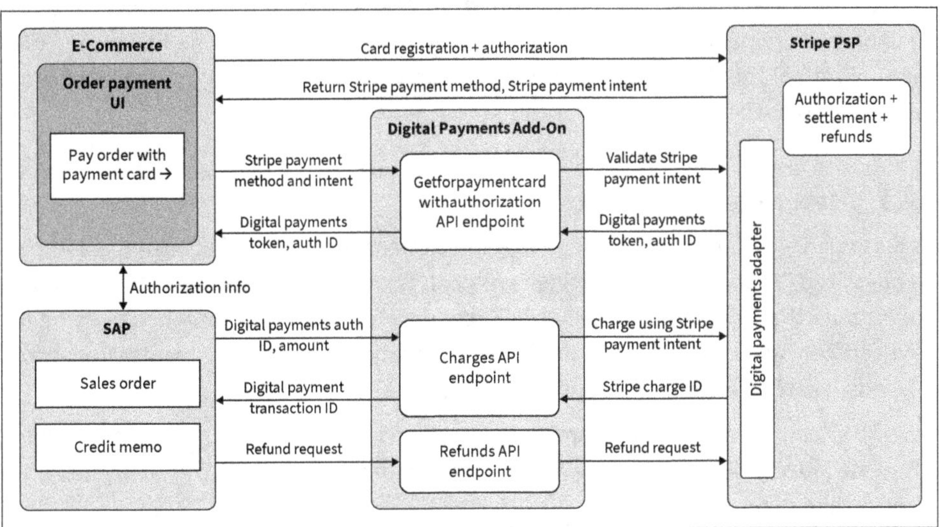


*Figure 4.4: External Collection with Token Preparation*

- **Pattern C: Direct capture token preparation**
  This pattern is used when payments are completed outside the digital payments add-on flow (such as in a mobile app or a self-service web application for your customer invoice payments) but your company wants to centralize refund processing through the digital payments add-on for consistent financial reconciliation. Figure 4.5 shows the step-by-step sequence of events that can facilitate centralizing refund processing for all payment transactions in your enterprise.

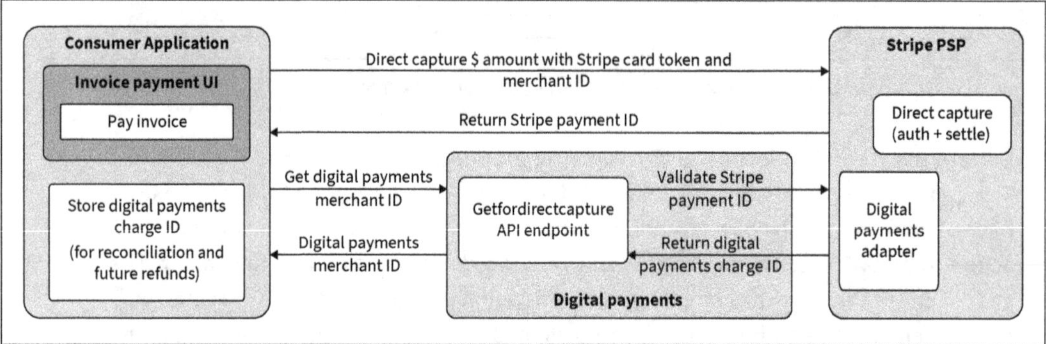


*Figure 4.5: Centralized Refund Operations Through Digital Payments Add-On*

The consumer applications, after they complete their payment flows outside of digital payments add-on, can integrate to the `getfordirectcapture` add-on API endpoint and get the digital payments add-on charge ID from the PSP. The digital payments add-on charge ID can then be used to centrally manage refund processing. As shown earlier in Figure 4.4, the consumer applications can send the digital payments add-on charge ID to the `refunds` API endpoint to trigger a refund (either partial or full, based on your company's needs). This pattern avoids the need for your implementation team to integrate all your company's applications with the digital payments add-on for all payment processing. Those applications can continue to integrate directly with the PSP and manage payment processing; just a minor enhancement is needed to get and store the digital payments add-on charge ID in the application. This digital payments add-on charge ID then can be used to manage refund processing as needed.

### 4.1.3 PSP Adapter Framework

As discussed in Chapter 2, PSP adapters serve as translation layers between the add-on's standardized API and each PSP's unique protocol. The adapter framework follows a plug-in architecture that allows different PSPs to integrate with the add-on without modifying the core service. Figure 4.6 illustrates the logical separation between the adapter framework and individual PSP implementations.

Each adapter implements a standard contract defined by the framework while maintaining complete independence in how it communicates with its target PSP. This design provides some architectural advantages:

- **Version independence**
  Adapters can be updated to support new PSP API versions without requiring changes to the digital payments add-on core service or other adapters. For example, Stripe's transition from the V1 to the V2 API was handled entirely within the Stripe adapter.
- **Error isolation**
  Failures in one adapter's communication with its PSP do not affect other adapters. If Stripe experiences an API outage, for example, transactions routed to Cybersource or Worldpay continue processing normally.
- **Independent certification**
  Each PSP can develop and certify its own adapter following SAP's adapter specification, enabling a growing ecosystem of payment providers without SAP needing to build every integration directly.

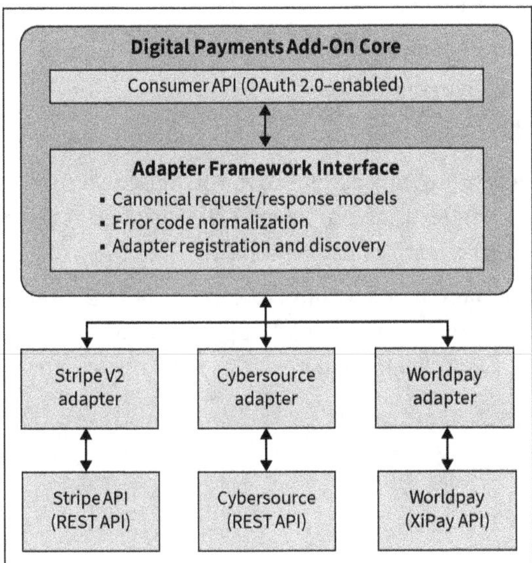


*Figure 4.6: Adapter Framework: Logical Architecture*

Adapters perform these essential functions:

- **Protocol and format translation**
  - Transform the digital payments add-on's canonical request structure into PSP-specific formats.
  - Convert PSP responses back into the add-on's standardized representation.
  - Handle PSP-specific API versioning (e.g., the Stripe PaymentIntents API vs. the legacy Charges API).
- **Secure communication**
  - Manage outbound HTTPS connections to PSP API endpoints.
  - Retrieve PSP API credentials from SAP Credential Store.
  - Handle TLS 1.2+ encryption.

- **PSP-specific processing**
  - Support unique PSP features (e.g., Level 2/3 data for corporate cards, 3D Secure authentication).
  - Manage PSP-specific metadata and custom fields.
  - Handle regional API variations and payment method availability.

Administrators can control which PSP adapters are available for payment routing through the PSP status management UI. Activating a PSP makes the adapter available to the routing engine. When an adapter is activated, the following is true:

- The routing engine includes this PSP when evaluating routing rules
- The adapter becomes eligible to process payment transactions
- Merchant account credentials must be configured

When an adapter is deactivated, the following is true:

- The routing engine excludes this PSP from consideration
- Existing merchant configurations remain but are not used
- No new transactions can be routed to this PSP

Activating an adapter does not automatically route transactions to it. You must also configure PSP determination rules that specify when to use a particular PSP, based on routing parameters such as company code, currency, country, or payment type.

Global organizations often require multiple merchant accounts to support different business units, brands, or geographic regions. The adapter framework supports multimerchant scenarios through the merchant mapping functionality. Figure 4.7 shows how merchant aliases flow through the system.

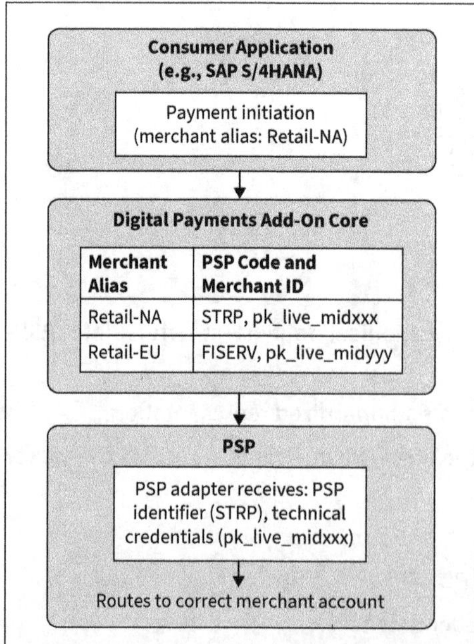


*Figure 4.7: Add-On: Merchant Alias Mapping Mechanism*

This merchant alias approach provides two key benefits:

- Consumer applications use logical names (e.g., virtual bank account IDs) rather than PSP-specific technical identifiers.
- Switching PSPs requires updating only the merchant mapping configuration, not the consumer application master data.

One of the critical adapter framework responsibilities is normalizing PSP-specific responses into the digital payments add-on's canonical format. Adapters translate PSP status codes into standardized digital payments add-on responses that consumer applications can handle consistently. Stripe PSP, as an example, uses codes like *card declined*, *insufficient funds*, and *expired card*, while other PSPs may use different terminology for the same conditions. The adapter framework maps these into digital payments add-on standard codes so that consumer applications can manage payment processing logic-related functions identically, regardless of which PSP processes the transaction.

Stripe is one of the most widely deployed PSP adapters for the digital payments add-on. PSP adapters follow versioning independent from the digital payments add-on core service. The Stripe adapter versioning demonstrates this:

- Stripe V1: Legacy adapter using the Stripe Charges API (deprecated)
- Stripe V2: Current production adapter using the Stripe PaymentIntents API (recommended)

Both versions can coexist within the same digital payments add-on tenant, differentiated by PSP code configuration. This enables phased migration strategies in which existing integrations continue operating on V1 while new implementations use V2.

Unlike the V1 adapter, which was tightly coupled to the digital payments add-on core, Stripe V2 operates as a fully independent microservice with its own dedicated infrastructure. Table 4.4 illustrates the Stripe V2 adapter's internal architecture.

| **Stripe V2: Four Key Internal Components** | |
|---|---|
| Adapter API gateway:<br>■ Request validation<br>■ Rate limiting<br>■ Request/response logging | Stripe configuration service:<br>■ Merchant onboarding UI<br>■ API key management<br>■ Webhook endpoint registration |
| Payment processing engine:<br>■ Payment intent management<br>■ Payment method handling<br>■ 3D Secure orchestration<br>■ L2/L3 data processing | Webhook event handler:<br>■ Event validation (signature check)<br>■ Event processing queue<br>■ Retry logic |

*Table 4.4: Stripe V2 Adapter: Internal Architecture*

The key architectural improvements in V2 are as follows:

- **Independent user interface services**
  Stripe V2 provides dedicated UI applications hosted at region-specific URLs (e.g., *us-prod-uiservices-dpa.sap.stripeconnectors.com*). These UIs handle merchant onboarding, configuration management, and advice scheduling without requiring access to the main digital payments add-on configuration interfaces.
- **Enhanced PaymentIntent API**
  V2 leverages Stripe's modern PaymentIntent API instead of the legacy Charge API used in V1. PaymentIntents provides better support for strong Customer Authentication (SCA) requirements under the Revised Payment Services Directive (PSD2), handles complex payment flows more gracefully, and offers improved idempotency controls.
- **Payment method support**
  V1 primarily supported credit/debit cards, but V2 natively integrates with Stripe's full payment method catalog, including the following:
  - Digital wallets (Apple Pay, Google Pay, Link) and buy now, pay later options
  - Bank transfers (ACH, SEPA, BACS, BECS, ACSS) and direct debit schemes
  - Regional payment methods (iDEAL, Bancontact, Giropay, EPS, P24, Alipay, WeChat Pay)
- **Separate webhook infrastructure**
  V2 implements a dedicated webhook event-processing system with signature verification, automatic retry logic, and event deduplication. This ensures that payment state changes (successful captures, failed authorizations, refund completions) are reliably propagated back to the digital payments add-on core service.

## 4.2 Setting Up PSP Accounts and API Credentials

This section provides practical, step-by-step guidance for creating PSP accounts and obtaining the API credentials required for adapter configuration. We'll use Stripe as the primary example (with production-tested procedures). You'll learn how to set up both test and production PSP environments, generate the necessary API keys, and prepare webhook endpoints for event handling.

### 4.2.1 PSP Account Creation

Before the digital payments add-on can process transactions through a PSP, you must establish a commercial relationship with that provider and create the necessary merchant accounts. This section covers the account creation process for Stripe and highlights the differences for other certified PSPs.

Before creating PSP accounts, ensure the prerequisites listed in Table 4.5 are met.

| Category | Prerequisite |
|---|---|
| Business registration documents | PSPs require the following business verification documents:<br>▪ Business registration certificate or articles of incorporation<br>▪ Tax identification number (EIN in US, VAT number in EU)<br>▪ Business bank account information for settlement<br>▪ Authorized signatory identification documents |
| Commercial agreement | Review and accept the PSP's service agreement, including the following:<br>▪ Transaction processing fees<br>▪ Monthly minimum fees (if applicable)<br>▪ Chargeback fees<br>▪ Currency conversion fees for multicurrency processing |
| Technical contact information | Identify the team members who will receive the following:<br>▪ API credential notifications<br>▪ Webhook event notifications<br>▪ Technical support communications<br>▪ Security alerts |

*Table 4.5: PSP Account Setup: Prerequisites*

### Stripe Account Creation

Stripe offers a streamlined self-service onboarding process, as described in Table 4.6.

| Step | Details |
|---|---|
| Initial registration | ▪ Navigate to *https://stripe.com* and select **Get Started** or **Sign up with Google.**<br>▪ Enter your business email address and create a password.<br>▪ Verify your email address by clicking the confirmation link sent to your inbox.<br>▪ Complete your profile by providing the following:<br>  – Business legal name and address<br>  – Industry/business type<br>  – Website URL (if applicable)<br>  – Expected processing volume |

*Table 4.6: Stripe Account Creation Steps*

| Step | Details |
|---|---|
| Business verification | Stripe's automated verification typically requests the following business details:<br>■ Tax ID (EIN for US entities, VAT for European entities)<br>■ Business structure (LLC, corporation, partnership, sole proprietorship)<br>■ Date of incorporation<br>■ Legal name and date of birth of authorized representative<br>■ Government-issued ID verification (passport, driver's license)<br>■ Social Security Number or equivalent (for US entities) |
| Bank account details | ■ Bank account number and routing number for settlement deposits<br>■ Account ownership verification (typically via microdeposit confirmation) |

*Table 4.6: Stripe Account Creation Steps (Cont.)*

Stripe may request additional documentation for certain industries or higher-risk business models.

Next, you'll need to install the SAP Digital Payments Add-On V2 app. This step registers your Stripe account with the digital payments add-on ecosystem. You can install the SAP-specific app from Stripe's App Marketplace, as follows:

1. Navigate to the Stripe marketplace (*https://marketplace.stripe.com/*), as shown in Figure 4.8, and search for "SAP Digital Payments Add-On" in the search bar.

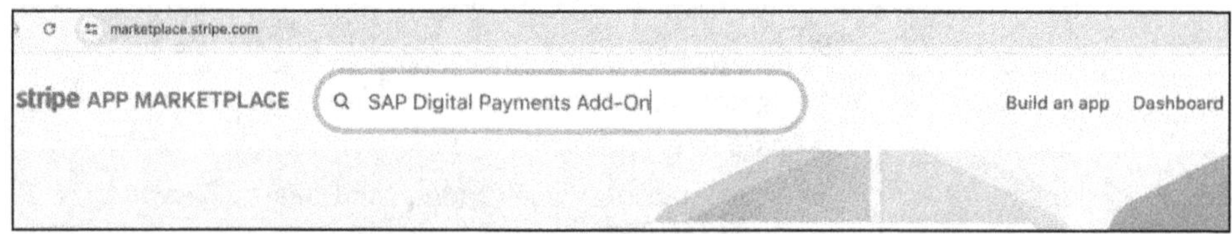


*Figure 4.8: Stripe App Marketplace*

2. Select the **SAP Digital Payments Add-On V2** app from the search results (see Figure 4.9).



*Figure 4.9: Add-On App*

3. Select **Install app** (see Figure 4.10). When installing, you can choose the installation mode:
   - **Install in test mode**: For development and quality assurance activities (connects to your test subaccount)
   - **Install in live mode**: For production transaction processing (connects to your production subaccount)

   You must install the app separately in both test and live modes.

*Figure 4.10: Add-On: Install App*

4. In most implementations, you will install test mode first, validate the complete integration, then install live mode before production go-live. Review the app permissions (read and write access to payment methods, read and write access to payment intents, read access to charges and refunds, and webhook event subscription permissions), then click **Install app in test mode** (see Figure 4.11) to complete the installation.

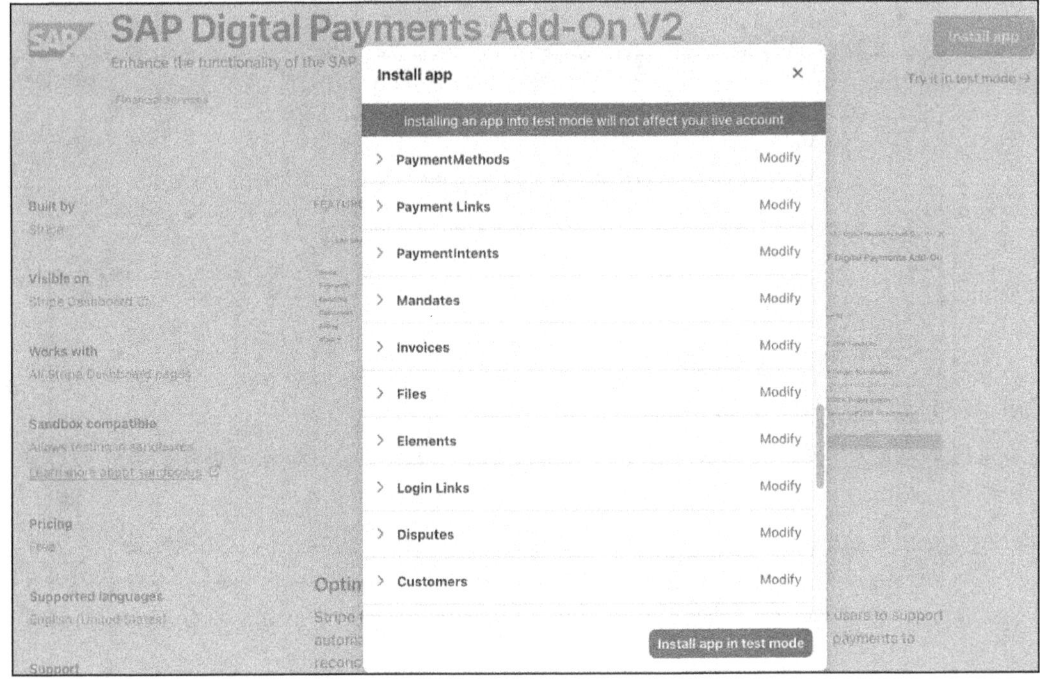


*Figure 4.11: Add-On App Installation: Review Permissions*

5. After installation, you will see a message prompting you to **Continue to app settings**, as shown in Figure 4.12.

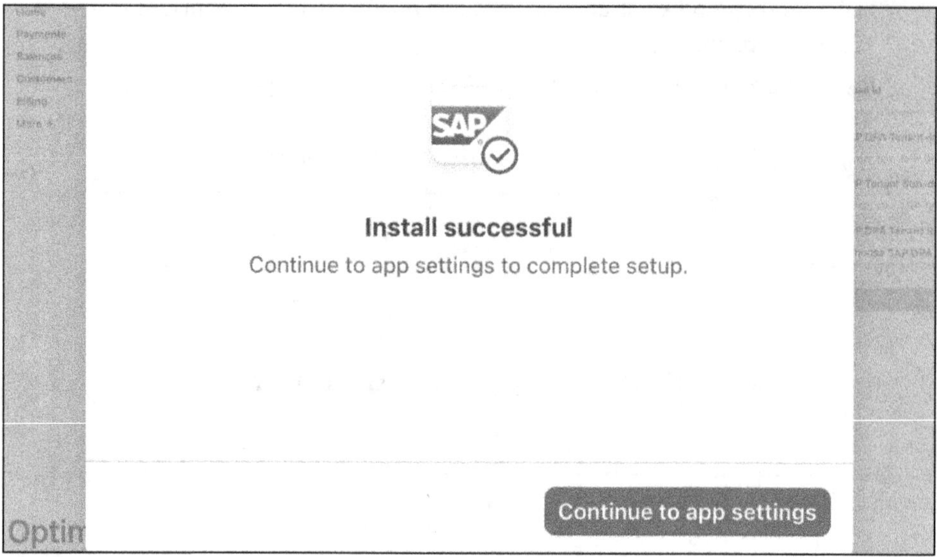


*Figure 4.12: Postinstall Step: App Settings*

6. The app displays your API credentials (publishable key and secret key), as shown in Figure 4.13. Store the keys in a safe place where you won't lose them. The best practice for security is to copy both keys to a secure password manager or encrypted vault. Do not store them in plaintext files or email them to team members.

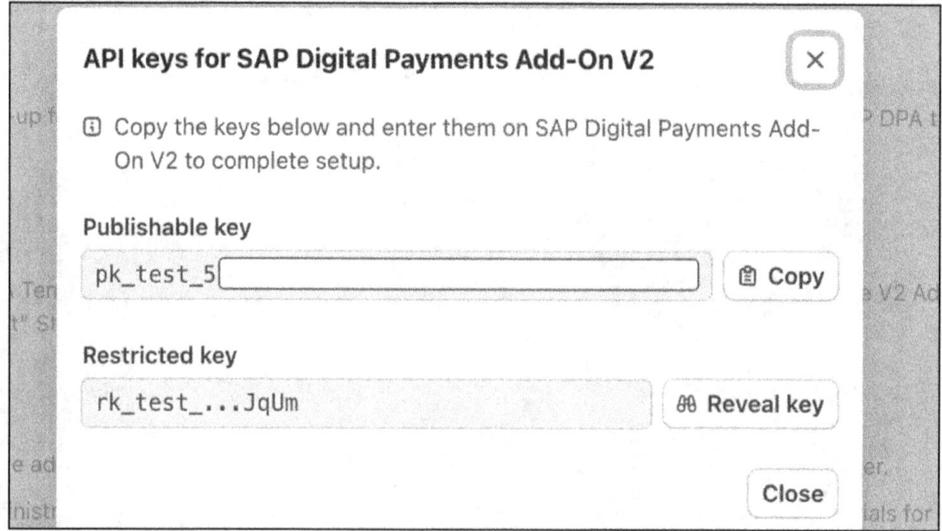


*Figure 4.13: Add-On Installation: API Keys*

7. Select **Close**. Figure 4.14 shows the Stripe dashboard view after successful app installation, where **SAP Digital Payments Add-On V2** appears in the installed apps list.

Settings ›
Team and security
Team | Single sign-on (SSO) | SCIM provisioning | Two-step authentication | Security history | Installed apps
Stripe Apps
SAP Digital Payments Add-On V2 ···

*Figure 4.14: Stripe Dashboard: Add-On App Successful Installation*

To retrieve or regenerate API keys after the initial installation, navigate to the **Apps** section in the Stripe dashboard, select **SAP Digital Payments Add-On V2**, and click **View API keys** (see Figure 4.15) to display the current credentials.

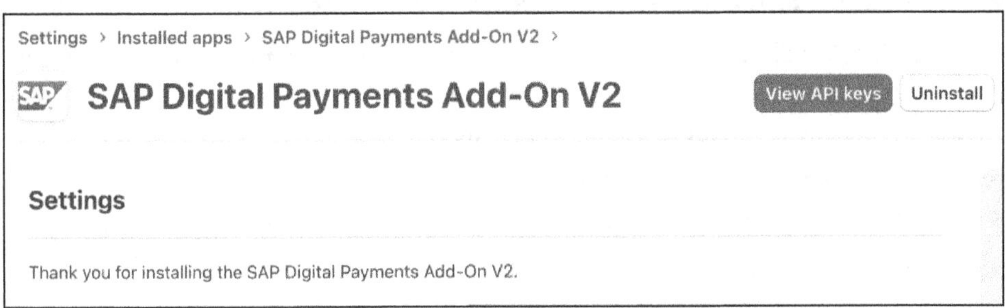


*Figure 4.15: Postinstall: View API Keys*

Large companies often require multiple Stripe accounts for different business units or geographic regions. The following are the most common multimerchant scenarios:

- Brand separation: Separate merchant accounts for different customer-facing brands to maintain distinct settlement reporting.
- Geographic separation: Different accounts for US versus European operations to comply with data residency requirements.
- Business unit independence: Separate profit and loss tracking for divisions or subsidiaries.

When you see a need to create multiple Stripe merchant accounts, follow this checklist:

- Use a different email address for each account (Stripe requires a unique email per account).
- Complete the business verification process for each entity.
- Install the SAP Digital Payments Add-On V2 app in each account separately.
- Configure separate merchant mappings in the digital payments add-on.
- Configure PSP determination parameters in the digital payments add-on configuration UI.

### Account Creation for Other Certified PSPs

Unlike Stripe's self-service model, Cybersource requires assisted onboarding. The typical steps to work through are as follows:

- Contact the Cybersource sales team or your acquiring bank partner.
- Complete a merchant application with business verification documents.
- Sign the merchant service agreement.
- Receive your merchant ID assignment (typically within 10–15 business days).
- Gain access to your Cybersource Business Center portal credentials.
- For digital payments add-on integration, do the following:
  - Request the tokenization feature activation on your account.
  - Collect your SAP BTP tenant ID.
  - Email the merchant ID and tenant ID mapping to *GDLSAPDPA@visa.com*, clearly indicating if this is for the test or production environment.

Worldpay integration (formerly Paymetric) requires multiple steps, as follows:

- Establish a Worldpay merchant account by visiting *https://www.sap.com/products/crm/partners/worldpay-uk-limited-worldpay-b2b-payments-formerly-paymetric.html* and click **Request a Quote** to initiate a quote for a contract.
- Contract with WorldPay for XiSecure tokenization services.
- Onboard with the token exchange service if multi-PSP token portability is required.
- Request 3DS Flex onboarding if using 3D Secure authentication.
- Receive your credential package, including the following:
  - Merchant account ID
  - Web Services API (WSA) user name and password
  - TMS profile name and merchant identifier
  - Shared key for webhook signature validation

The following is a quick checklist for setting up a PayPal Braintree account; this is for reference only as the onboarding steps vary. Braintree follows PayPal's standard merchant onboarding:

- Create a PayPal Business account at paypal.com.
- Navigate to the **Braintree** section within the PayPal Business dashboard.
- Complete the Braintree-specific verification (additional underwriting beyond PayPal).
- Receive your Braintree merchant ID and API credentials.
- Generate sandbox credentials for testing.

The Novalnet account setup checklist is EU-region focused, and the steps are as follows:

- Complete a merchant application with European business verification.

- Select the desired payment methods from the available options.
- Configure settlement currency preferences.
- Receive API credentials and project ID assignments.

### 4.2.2 API Keys and Webhooks Overview

PSP API credentials consist of multiple components, each serving a specific security and operational purpose. Again, we'll split our focus between Stripe and other PSPs.

#### Stripe API Credential Types

Stripe uses a dual-key architecture for API authentication. Table 4.7 lists key features of the publishable key that you can use as a best practices reference.

| Characteristic | Description |
|---|---|
| Visibility | Can be embedded in client-side code (JavaScript, mobile apps) |
| Permissions | Limited to creating tokenized payment methods and setup intents |
| Security level | Safe to expose in frontend applications |
| Usage in digital payments add-on | Used as the `MerchantAccount` parameter in digital payments add-on API calls |
| Rotation | Can be regenerated without affecting existing tokens |

*Table 4.7: Publishable Key Characteristics*

Table 4.8 lists key features of the secret key that you can use as a best practices reference.

| Characteristic | Description |
|---|---|
| Visibility | Must be kept strictly confidential, never exposed in client-side code |
| Permissions | Full account access including charges, refunds, and customer data access |
| Security level | Equivalent to root credentials, must be stored encrypted |
| Usage in digital payments add-on | Used by Stripe adapter to authenticate backend API calls |
| Rotation | Requires updating the adapter configuration |

*Table 4.8: Secret Key Characteristics*

### Credential Types for Other PSPs

The credential types for other certified PSPs are as follows:

- **Cybersource**
  Cybersource uses an authentication model based on merchant profiles. Cybersource credentials are not created through self-service;l they are generated by the account manager or requested through the Cybersource Business Center portal with appropriate administrative permissions. You can obtain more details by visiting the Cybersource developer website at *https://developer.cybersource.com/technology-partners/sap-dpa.html.*
- **Worldpay/Paymetric**
  The Worldpay integration requires multiple credential sets due to its multicomponent architecture. You can obtain more details by visiting the Worldpay website at *https://docs.paymetric.info/sap-digital-payments-add-on-pr.htm.*
- **Novalnet**
  Novalnet's payment module for the SAP digital payments add-on provides a unified dashboard that acts as a central platform to accept digital payments, process payment transactions in a secure manner, and manage payments. You can obtain more details by visiting the Novalnet website at *https://www.novalnet.com/integration/sap-digital-payment-module/.*

## 4.2.3 Test and Production Environments Setup

All certified PSPs provide separate test and production environments to enable safe implementation and testing without risking live transactions or customer data. This section covers separation requirements for environments and best practices.

### Stripe Environments

Stripe implements environment separation through "modes" rather than separate systems (see Table 4.9).

| Test Mode Characteristics | Live Mode Characteristics |
|---|---|
| ▪ API endpoint: Same base URL (api.stripe.com) for both modes.<br>▪ Transaction processing: No real card networks involved; uses Stripe's test card numbers.<br>▪ Settlement: No actual funds transfer; simulated settlement for testing.<br>▪ Data isolation: Separation from live mode data; test transactions never appear in live reporting. | ▪ API endpoint: Same base URL (api.stripe.com).<br>▪ Transaction processing: Real card network communication with actual authorizations and declines.<br>▪ Settlement: Actual funds transferred to business bank account (typically T+2 business days).<br>▪ Production data: All transactions appear in live reporting and reconciliation. |

*Table 4.9: Stripe Test Mode Versus Live Mode*

The Stripe dashboard includes a mode toggle in the top-left corner (see Figure 4.16). This toggle controls which environment you're viewing and configuring. When installing the SAP Digital Payments Add-On V2 app, you must explicitly choose which mode to install it in; the credentials generated are specific to the selected mode.

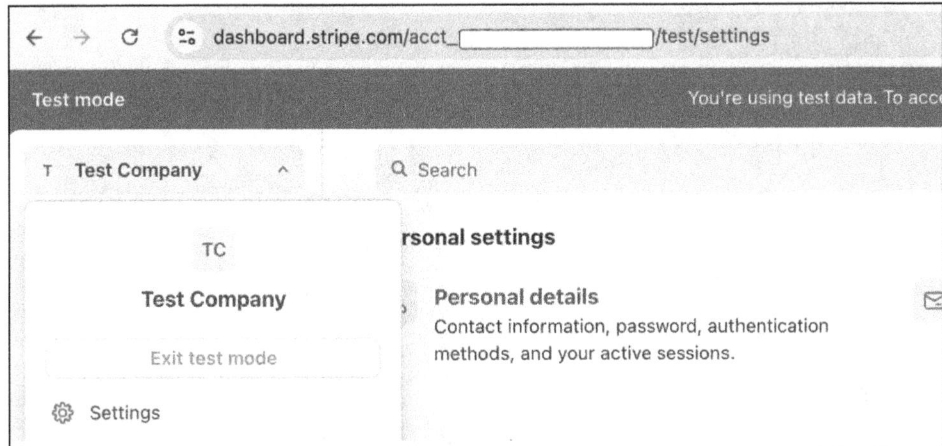


*Figure 4.16: Test and Live Mode Toggle*

Your SAP BTP landscape includes separate subaccounts for test and production. These must be mapped to corresponding PSP environments. Table 4.10 shows the correct mapping.

| SAP BTP Subaccount | Subscription Plan | Stripe Environment | Stripe Credentials | Consumer Application |
|---|---|---|---|---|
| DP Test | SAP digital payments add-on demo | Test mode | pk_test_..., sk_test_... | Development and quality assurance SAP S/4HANA systems |
| DP Production | SAP digital payments add-on | Live mode | pk_live_..., sk_live_... | Production SAP S/4HANA system |

*Table 4.10: PSP Environment Mapping to SAP BTP Subaccounts*

Do not configure live mode PSP credentials in the DP Test subaccount or test mode credentials in DP Production. Mixing environments can result in test transactions being charged to real cards or production transactions failing in test mode.

After installing the app, you must connect your SAP BTP tenant to the Stripe adapter, as follows:

1. From the Stripe dashboard, navigate to **Apps • SAP Digital Payments Add-On V2**, then click the app to open the onboarding interface. A window opens as shown in Figure 4.17.

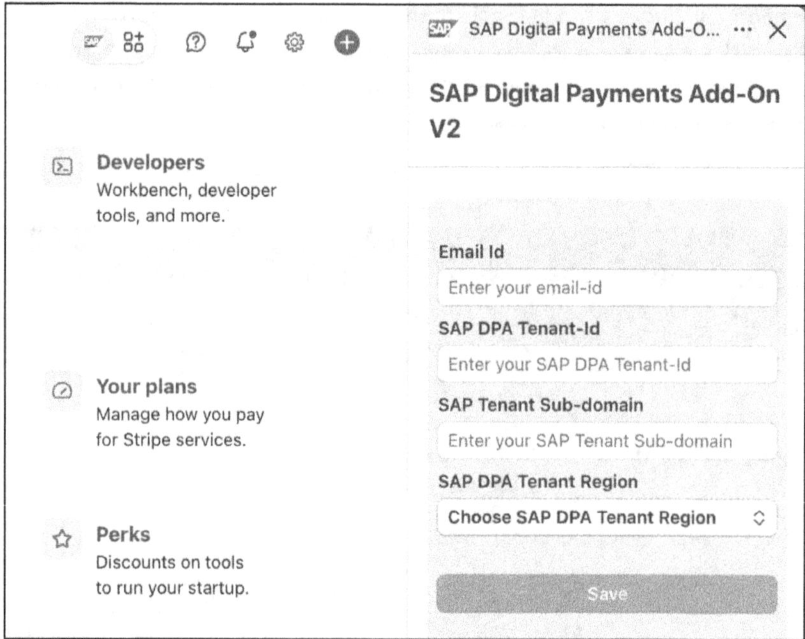


*Figure 4.17: Stripe Dashboard: Add-On Onboarding UI*

2. Enter the necessary information; see Table 4.11 for a guide.

| Field | Description |
|---|---|
| **Email** | Your administrator email for account verification and recovery |
| **SAP DPA Tenant-Id** | Found in your SAP BTP subaccount URL (the subdomain portion) |
| **SAP Tenant Sub-domain** | The subdomain you created in Chapter 3 |
| **SAP DPA Tenant Region** | Select from dropdown:<br>■ **US10 Test** (for US East Virginia test environment)<br>■ **EU10 Test** (for Europe Frankfurt test environment) |

*Table 4.11: Add-On Onboarding in Stripe*

3. Click **Save** to create the connection.

You will receive an email notification from *no-reply@verificationemail.com* with your user name and temporary password. These credentials are distinct from your Stripe login and are used specifically to access the Stripe configuration UI (for onboarding merchants) and the Stripe advice UI (for scheduling payment advice retrieval).

Maintaining strict separation between the test and production environments prevents critical issues. Table 4.12 lists the advantages of environment isolation.

| Category | Advantages |
|---|---|
| Data contamination prevention | ▪ Test transaction tokens never mix with production payment records.<br>▪ Customer card data registered in production never appears in test systems.<br>▪ Financial reconciliation remains clean, with no test data in production reports. |
| Testing safety | ▪ Developers can safely test refund scenarios without affecting real customer payments.<br>▪ Quality assurance teams can simulate payment failures without impacting production authorization rates.<br>▪ Integration testing can use high transaction volumes without triggering PSP fraud alerts. |
| Compliance separation | ▪ The PCI DSS audit scope clearly distinguishes test versus production systems.<br>▪ Production credential access logs are separate from test environment access logs.<br>▪ Security incident response procedures can isolate test environment compromises. |

*Table 4.12: Environment Isolation: Advantages*

#### Other PSP Environment Models

Cybersource uses separate test and production instances; that is test and production are completely isolated systems, as follows:

- Test environment: Separate URL (*apitest.cybersource.com*) with distinct test merchant IDs.
- Production environment: Production URL (*api.cybersource.com*) with live merchant IDs.

Worldpay, on the other hand, has separate merchant IDs and credentials for each environment:

- Certification environment: For integration development and testing.
- Production environment: For live transaction processing.

## 4.3 Configuring PSP Adapter in the Add-On

With PSP accounts created and API credentials obtained, the next step is configuring the PSP adapter within your digital payments add-on tenant. This section provides details for activating the Stripe V2 adapter, onboarding merchant accounts, and establishing the technical connection between your SAP BTP tenant and Stripe's payment infrastructure. Stripe serves as the primary example; configuration differences will be highlighted for other certified PSPs.

### 4.3.1 Adapter Installation

*PSP adapter installation* in the digital payments add-on context refers to *activation* rather than deploying new software components. Adapters are predeployed within your SAP BTP tenant when you subscribe to the add-on service. Activation makes the adapter available to the routing engine for transaction processing.

Steps for accessing the PSP status management UI are as follows:

1. Log into the SAP BTP cockpit (*https://cockpit.btp.cloud.sap*), navigate to your DP Test subaccount, and select **Services • Instances and Subscriptions** from the left navigation pane. Under **Subscriptions**, locate the **SAP digital payments add-on Demo** subscription, then click **...** and select the **Go to Application** link, as shown in Figure 4.18.

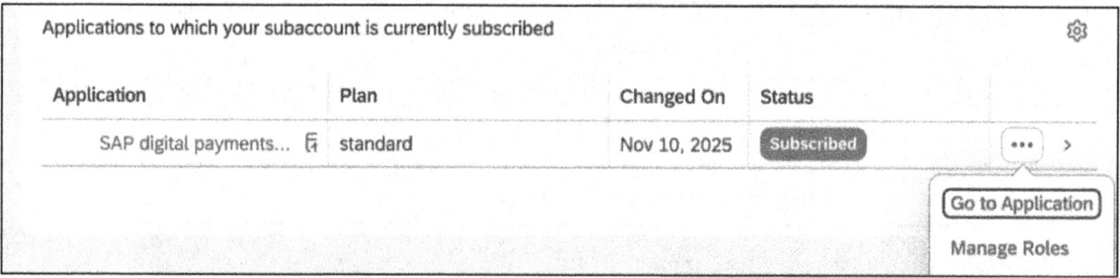


*Figure 4.18: Add-On Subaccount: Go to Application Navigation*

2. You'll be directed to the base application URL. The format depends on your SAP BTP region:
   - US10 Test:
     *https://{subdomain}.test-digitalpayments-sap.cfapps.us10.hana.ondemand.com*
   - EU10 Test:
     *https://{subdomain}.demo-digitalpayments-sap.cfapps.eu10.hana.ondemand.com*

   Here, *{subdomain}* is the subdomain you specified in Chapter 3 (e.g., *yourcompany-dp-test*). Figure 4.19 shows an example base application URL if your SAP BTP region is US10.

*Figure 4.19: Add-On Subaccount: Base Application URL*

3. Append */pspStatus/index.html* to the base URL to access the PSP status management UI, as shown in Figure 4.20.

*Figure 4.20: Add-On Subaccount: PSP Status Configuration UI*

When accessing the PSP status UI, you may be prompted to authenticate through your identity provider, and your user must be assigned to a role collection containing the `DigitalPaymentsAdministrator` role. The UI displays all available adapters for your tenant. Figure 4.21 shows the typical view with multiple PSP adapters listed.

-test.test-digitalpayments-sap.cfapps.us10.hana.ondemand.com/pspStatus/index.html

**Payment Service Provider Status**

| Payment Service Provider |
|---|
| CardEasy |
| PayFabric from EVO Payments |
| Paymetric |
| Paypal |
| PayPal Braintree |
| Stripe V1 |
| Stripe V2 |
| Worldpay B2B, Paymetric |

*Figure 4.21: PSP Status: Configuration UI*

Locate the row for Stripe V2 (initially shows as **Inactive**), click the **Activate** button, and confirm the activation in the dialog that appears, in which the status should be updated to **Active** (Figure 4.22).

| | |
|---|---|
| Stripe V1 | Inactive |
| Stripe V2 | Active |
| Worldpay B2B, Paymetric | Inactive |

*Figure 4.22: Add-On Subaccount: Stripe Adapter Activation*

The activated status confirms that the routing engine will now consider Stripe when evaluating PSP determination rules, the Stripe adapter is available for merchant configuration, and transactions can be routed to Stripe once merchant accounts are onboarded. Activating the adapter does not automatically route transactions to Stripe. You must still configure PSP determination rules and merchant account mappings with your API credentials.

The production adapter activation follows an identical process with few environment-specific differences. Navigate to your DP Production subaccount, access the SAP digital payments add-on subscription, use the production base URL, append */pspStatus/index.html*, and activate Stripe V2 using the same procedure (see Figure 4.23).

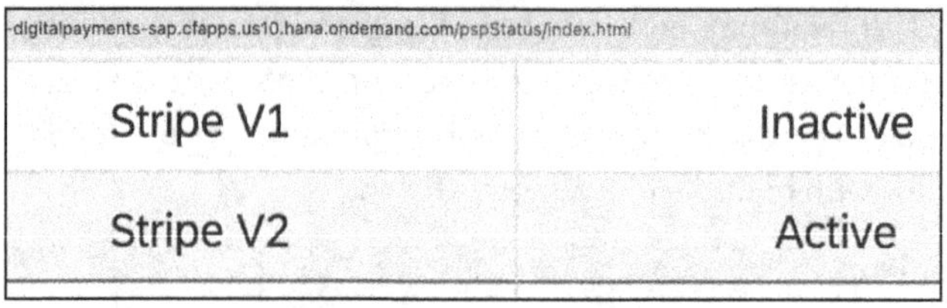

-digitalpayments-sap.cfapps.us10.hana.ondemand.com/pspStatus/index.html

| | |
|---|---|
| Stripe V1 | Inactive |
| Stripe V2 | Active |

*Figure 4.23: Add-On Production Subaccount: Stripe V2 Adapter Activation*

**Activating Multiple PSP Adapters**

If your organization uses multiple PSPs, activate each adapter separately. You must repeat the activation process for each required PSP, and each adapter requires its own merchant account onboarding. Make sure that you configure routing rules to direct transactions to the appropriate PSP based on the transaction characteristics.

### 4.3.2 Configuration Parameters

After activating the Stripe adapter, you must onboard your merchant account(s) by configuring the API credentials you obtained (covered in Section 4.2.2). This establishes the actual connection between the digital payments add-on and your Stripe account, enabling payment processing. Stripe provides a dedicated web application for merchant configuration separate from the digital payments add-on configuration UIs. This is different from other PSPs, which use the standard digital payments add-on merchant mapping interface. We begin by accessing the Stripe configuration UI using region-specific URLs and authenticating with credentials generated during the tenant connection process. We then configure the merchant account by entering publishable and secret API keys, selecting the default currency for settlement, and optionally enabling Level 2/3 data transmission for optimized interchange rates. Next, we create merchant aliases that map Stripe technical merchant IDs to user-friendly identifiers used by consumer applications, enabling multimerchant scenarios across different business units and currencies. Finally, we configure PSP determination rules that route transactions to the appropriate merchant based on company code, currency, payment method, and other routing parameters, ensuring payments are processed by the correct merchant account.

### Accessing the Stripe Configuration UI

The Stripe configuration UI application requires separate authentication credentials generated during the tenant connection process. Based on your SAP BTP region and environment, construct the appropriate URL. Table 4.13 shows example URLs for US and Europe that you can use as a reference.

| SAP BTP Region and Environment | Stripe Configuration UI URL |
|---|---|
| US East (VA) test | *https://us-test-uiservices-dpa.sap.stripeconnectors.com/stripeConfiguration/index.html?subdomain={subdomain}* |
| US East (VA) production | *https://us-prod-uiservices-dpa.sap.stripeconnectors.com/stripeConfiguration/index.html?subdomain={subdomain}* |
| Europe (Frankfurt) test | *https://eu-demo-uiservices-dpa.sap.stripeconnectors.com/stripeConfiguration/index.html?subdomain={subdomain}* |
| Europe (Frankfurt) production | *https://eu-prod-uiservices-dpa.sap.stripeconnectors.com/stripeConfiguration/index.html?subdomain={subdomain}* |

*Table 4.13: SAP BTP Region and Environment: Stripe Configuration UI URLs*

Replace *{subdomain}* with your SAP BTP subaccount subdomain (e.g., *yourcompany-dp-test*). For example, for the US10 test region, *https://us-test-uiservices-dpa.sap.stripeconnectors.com/stripeConfiguration/index.html?subdomain=yourcompany-dp-test.*

You will have access to the Stripe configuration UI for merchant onboarding. The merchant onboarding process connects your Stripe merchant account to the digital payments add-on tenant. You can configure multiple merchant IDs for a given digital payments add-on tenant. Figure 4.24 shows the Stripe configuration home page with existing example merchant configurations.

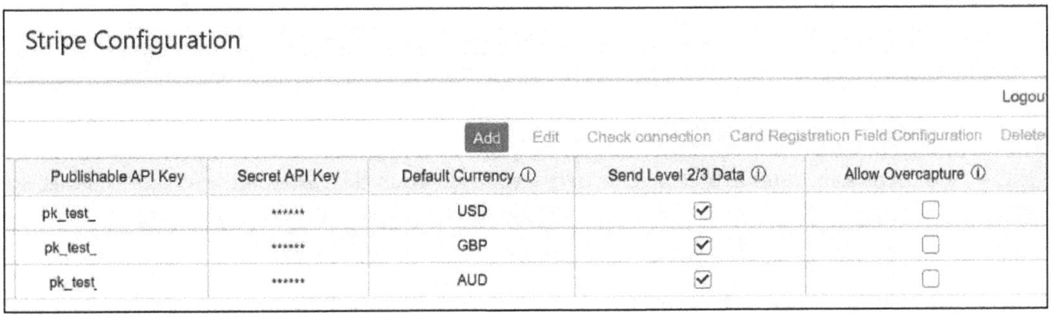

*Figure 4.24: Stripe Merchant Configuration*

You can add new merchants by clicking the **Add** button in the top-right corner. A configuration form appears, requesting the following information:

- **Publishable API Key**
  Enter the publishable key (obtained in Section 4.2.2).
- **Secret API Key**
  Enter the secret key (obtained in Section 4.2.2).
- **Default Currency**
  Select from the dropdown list of ISO currency codes; this represents the primary currency for this merchant account. This should match the settlement currency configured in your Stripe dashboard account settings.
- **Send Level 2/3 Data**
  Check this checkbox if you want to send Level 2/3 data to Stripe during payment capture. Level 2/3 data includes line-item details, tax amounts, and customer purchase order numbers and is required for optimized interchange rates on corporate and purchasing cards.

After you add a new merchant account, you can select the newly onboarded merchant from the list and click the **Check Connection** button at the top of the page. The system performs a live API call to Stripe using your credentials and a successful connection shows a message: **Connection to Stripe merchant account successful.**

#### Troubleshooting Merchant Onboarding Issues

The following are some possible causes if you observe errors such as invalid API credentials, and troubleshooting steps to follow:

- Credentials copied incorrectly: Recopy from Stripe dashboard, ensure no leading/trailing spaces.
- Wrong mode credentials: Verify test keys (pk_test_/sk_test_) are used in test environment, live keys (pk_live_/sk_live_) in production.
- Keys from wrong Stripe account: Confirm you're using credentials from the account where the SAP Digital Payments Add-On V2 app is installed.
- Keys revoked or rolled: Generate new keys from **Stripe Dashboard • Apps • SAP DPA V2 • App Settings.**

Another potential issue is that the merchant saves successfully but the check connection fails. Possible causes and solutions are as follows:

- Network connectivity: Verify outbound HTTPS (port 443) from SAP BTP to *api.stripe.com*.
- Stripe account restricted: Check the Stripe dashboard for account verification issues.
- API key permissions insufficient: Ensure the SAP Digital Payments Add-On V2 app installation granted all required permissions.
- Regional endpoint mismatch: Verify that your Stripe account region matches your SAP BTP region.

If you can't access the Stripe configuration UI, some possible causes and solutions are as follows:

- Incorrect subdomain in URL: Verify the subdomain matches your SAP BTP subaccount exactly.
- Wrong region prefix: Ensure *us-test*, *us-prod*, *eu-demo*, or *eu-prod* matches your environment.
- Authentication credentials not received: Check your spam folder for email from *no-reply@verificationemail.com*.
- Tenant not onboarded: Verify that you completed the tenant connection as described in Section 4.2.3.

#### Configuring Merchant Aliases

After onboarding merchants in the Stripe configuration UI, you must create merchant aliases in the digital payments add-on merchant mapping interface to make these merchants available to consumer applications. When using multiple merchant IDs, you must define merchant ID aliases for both payment card payments and external payments. This is important for generating unique digital payment transaction advice for each merchant ID.

Follow the steps mentioned in Section 4.3.1, in which we walked through accessing the PSP status to get the base application URL of your add-on. Navigate to *{baseURL}/merchantMapping/index.html* to access the merchant mapping configuration UI. An example URL for US10 looks like this: *https://yourcompany-dp-test.test-digitalpayments-sap.cfapps.us10.hana.ondemand.com/merchantMapping/index.html*.

To create the merchant alias, follow these steps:

1. Before creating merchant aliases, it is a good practice to create a simple table like the one in Table 4.14. Here you can gather all the merchant accounts, with their corresponding technical merchant IDs and the user-friendly merchant aliases that will be configured and used in the consumer applications. We have gathered information to configure merchant aliases for a company that has three business units, one in North America (BU1), one in Europe (BU2), and one in Australia (BU3). We have also assumed that for BU2 we would like to support two different currencies: GBP and EUR.

| PSP | Technical Merchant ID | Currency | Merchant Alias |
|---|---|---|---|
| Stripe V2 | pk_test_1xxx... | USD | DPBU1NA01 |
| Stripe V2 | Pk_test_2xxx... | GBP | DPBU2EUR01 |
| Stripe V2 | Pk_test_3xxx... | EUR | DPBU2EUR02 |
| Stripe V2 | Pk_test_4xxx... | AUD | DPBU3APAC01 |

*Table 4.14: Merchant Aliases: Example Table to Gather Required Details*

2. In the merchant mapping UI, click **New Alias** and enter the following configuration:
   - **Payment Service Provider**: Stripe V2
   - **Merchant**: pk_test_1xxx... (publishable key from Stripe configuration UI)
   - **Merchant Alias**: DPBU1NA01 (user-friendly identifier used by consumer applications)
3. Click **Save**. Repeat the steps for all the four merchant alias entries. Figure 4.25 shows the merchant mapping UI with multiple aliases configured.

| Payment Service Provider | Merchant | Merchant Alias |
|---|---|---|
| Stripe V2 | pk_test_... | DPBU1NA01 |
| Stripe V2 | pk_test_... | DPBU2EUR01 |
| Stripe V2 | pk_test_... | DPBU2EUR02 |
| Stripe V2 | pk_test_... | DPBU3APAC01 |

*Figure 4.25: Merchant Mapping: Example Company*

These aliases are then used in PSP determination rules, virtual bank account configuration, payment advice reconciliation processes, and so on.

### Configuring PSP Determination Rules

After merchant aliases are created, you must configure routing rules that direct transactions to the appropriate PSP and merchant. PSP determination is typically needed in these scenarios:

- Register a payment card: Identify the PSP to be used to tokenize the payment card.
- Authorize a payment card: Identify the PSP to be used to perform an authorization.
- Refund a payment: Determine the PSP's merchant account in certain refund cases in which the merchant account is not yet known.

Follow the steps mentioned in Section 4.3.1 to access the PSP status and get the base application URL of your add-on. Navigate to *{baseURL}/ pspDetermination /index.html* to access the PSP determination UI. An example URL for US10 looks like this: *https://yourcompany-dp-test.test-digitalpayments-sap.cfapps.us10.hana.ondemand.com/pspDetermination /index.html.*

Before configuring the routing rules, it is important to understand the mandatory and optional parameters that are supported and the sequencing and prioritization of the routing conditions (see Figure 4.26):

- **Sequence Number**
  This field determines the priority and sequence in which the routing condition is determined at runtime. In Figure 4.26, you can see that there are four routing conditions with sequence numbers 1 through 4. At runtime, the routing rules will be evaluated from 1

through 4. If you want to change the sequence of the routing conditions, you can use the **Move Up** or **Move Down** option.

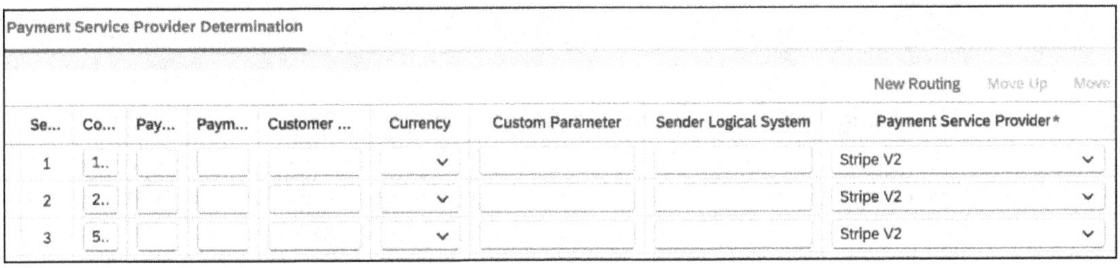

Payment Service Provider Determination

New Routing Move Up Move

| Se... | Co... | Pay... | Paym... | Customer ... | Currency | Custom Parameter | Sender Logical System | Payment Service Provider* |
|---|---|---|---|---|---|---|---|---|
| 1 | 1.. | | | | | | | Stripe V2 |
| 2 | 2.. | | | | | | | Stripe V2 |
| 3 | 5.. | | | | | | | Stripe V2 |

*Figure 4.26: Routing Conditions: Mandatory and Optional Parameters*

- **Company Code**
  *Company code* typically refers to your company's legal entity, which is legally registered to operate your business in a certain region. When an example organization has three business units, its corresponding company codes are configured in the consumer application as 1000 (BU1), 2000 (BU2), and 3000 (BU3). The company code is an optional parameter, but in most implementations you will see that this field is used to enforce a clear separation between legal entities and their corresponding payment transactions and bank accounts.
- **Payment Method**
  This is an optional parameter; you can choose **Credit Card (CC)** or **External Payment (EP)**. We will discuss payment methods and their significance further in Section 4.3.3.
- **Customer country/Region**
  This is an optional parameter in which you can specify two-character ISO country codes (e.g., US, DE, GB).
- **Currency**
  This is an optional parameter in which you can specify three-character ISO currency codes (e.g., USD, EUR, GBP, and AUD). Figure 4.27 shows an example configuration in which a company with three business units wants to support four currencies for payment transactions using Stripe as the PSP.

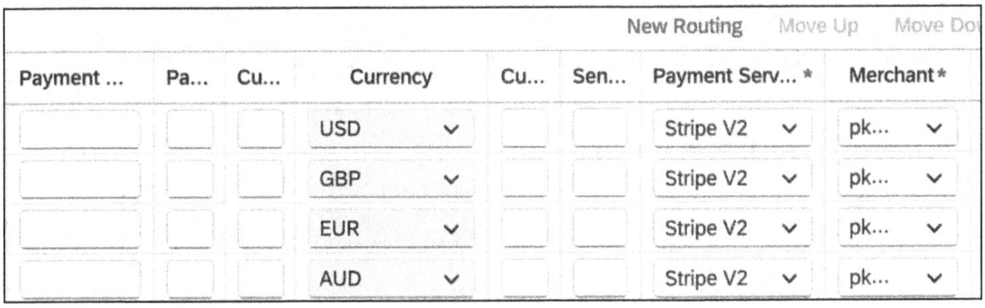

New Routing Move Up Move Do

| Payment ... | Pa... | Cu... | Currency | Cu... | Sen... | Payment Serv... * | Merchant* |
|---|---|---|---|---|---|---|---|
| | | | USD | | | Stripe V2 | pk... |
| | | | GBP | | | Stripe V2 | pk... |
| | | | EUR | | | Stripe V2 | pk... |
| | | | AUD | | | Stripe V2 | pk... |

*Figure 4.27: PSP Determination: Currency Codes Usage*

- **Custom Parameter**
  This is an optional, user-defined routing parameter that can be used to specify routing conditions in special cases in which your implementation demands that a user-defined parameter be used to route payment transactions, rather than the standard parameters.
- **Sender Logical System**
  **Sender Logical System** serves as a technical identifier for the consumer application system and may function as a routing parameter. The value format is contingent upon the specific consumer application. Details for deriving the sender logical systems are as follows:
  - **SAP S/4HANA**
    The sender logical system is constructed by concatenating the backend system's installation number with the logical system ID. To locate the installation number, navigate to **System • Status** and reference the **Installation Number** field. For the logical system ID, use Transaction SE16, enter table `T000`, select **Table Contents**, then select **Execute**. Double-click the relevant client; the logical system ID can be found in the **LOGSYS** field.
  - **SAP S/4HANA Cloud Public Edition**
    In this case, the sender logical system corresponds to the logical system ID. Access your system with administrator credentials, search for "Communication Arrangements", and select the arrangement associated with the SAP digital payments add-on integration (scenario ID SAP_COM_0216). The logical system ID is listed in the **Own SAP Cloud System** field within the **Common Data** section.
- **Payment Service Provider**
  This and the corresponding technical merchant ID are mandatory parameters in the routing conditions. You will see that only PSPs whose adapters have been already activated are visible in the payment service provider determination user interface. Make sure you choose the correct technical merchant ID along with the PSP so that the transactions are routed to the correct merchant account for processing.

The table entries are evaluated using these guidelines:

- Rows are checked from top to bottom, one at a time.
- The first row that matches is used to select the PSP or merchant account.
- Every parameter listed in a row must match exactly for that row to be chosen.
- Empty cells act as wildcards and will match anything. However, a cell containing a space character does not count as empty.
- If none of the rows match, an error message will indicate that no PSP was found.

Only fill in a field if you want it to influence PSP selection; otherwise, leave it blank. It's best to include a final catchall row at the bottom with your default PSP and all other fields left empty. This ensures that transactions go through even if nothing else matches.

### 4.3.3 Payment Method Setup

With merchant accounts onboarded, the next configuration step defines which payment methods are available for transaction processing. The Stripe V2 adapter supports over 40 payment methods through Stripe's Payment Elements framework, but not all methods are automatically enabled.

The following payment methods are available by default after merchant onboarding (no additional configuration required):

- Credit cards: Visa, Mastercard, American Express, Discover, JCB, Diners Club, UnionPay
- Debit cards: All major debit card networks
- Digital wallets: Apple Pay, Google Pay, Link (Stripe's native wallet)

These card-based methods work immediately after merchant onboarding because they use the standard card-processing flow.

The SAP digital payments add-on enables support for *external payments*, which refers to transactions handled through external channels such as bank debits or redirects instead of payment cards. You can process external payments using either a one-step process (*direct capture*), in which funds are charged straight to the payer's account without prior authorization, or a two-step process (*manual capture* or separate authorization and capture), in which the payment is first authorized before being charged.

The Stripe V2 connector supports the following business processes:

- **External payment direct capture**
  When a user places an order and opts to pay through an external payment channel, approval for the transaction must be secured from the PSP. The consumer application submits a direct capture request to the SAP digital payments add-on. Depending on the selected payment type, the appropriate PSP is identified, and the initiation request is transmitted in the required format. The user then authorizes the direct capture using the PSP's interface, which is called by the consumer application. Upon receiving authorization, the execution request for direct capture is issued, and the resulting outcome is returned to the consumer application through the SAP digital payments add-on.
- **External payment capture with authorization**
  This process has the following elements:
  - **External payment authorization**
    If a user chooses to pay for an order via an external payment channel, transaction approval from the PSP is necessary. The consumer application dispatches an authorization request to the SAP digital payments add-on. Following this, the corresponding PSP is selected according to payment type, and the initiation request is sent in the prescribed format. Approval is granted by the user through the PSP's interface, accessed by the consumer application. The execution request for authorization is sent

upon approval, and the result is relayed back to the consumer application via the SAP digital payments add-on.

  - **External payment authorization cancellation**
    Should a user cancel an order that has already received payment approval, the consumer application issues an authorization cancellation request to the SAP digital payments add-on. Based on payment type, the relevant PSP is identified, and the request is sent in the correct format. The PSP processes the cancellation and communicates the result to the consumer application through the SAP digital payments add-on. If the cancellation cannot be executed, the consumer application is notified accordingly.
  - **External payment settlement**
    The consumer application periodically initiates settlement runs to finalize authorized external payments. The authorization data is sent to the SAP digital payments add-on, which identifies the relevant PSP and forwards the settlement request in the appropriate format. The PSP completes the settlement process and returns the outcome to the consumer application via the SAP digital payments add-on.

- **External payment refund**
  In instances in which a settled external payment must be refunded—such as product returns—the consumer application requests a refund, citing reference information for the original payment. The suitable PSP is determined, and the refund data is transmitted in the required format. The PSP processes the refund and returns the result to the consumer application through the SAP digital payments add-on.
- **Digital payment advice**
  To obtain the details of processed digital payments, the consumer application regularly executes reports requesting advice information from the SAP digital payments add-on. The add-on collects advice details from all applicable PSPs, consolidates the information, and delivers it to the consumer application.
- **External payment authorization preparation**
  If the consumer application sends an external payment authorization request directly to the PSP, the PSP processes and returns the result straight to the consumer application. To facilitate subsequent processing within other systems, the consumer application communicates the external payment authorization to the SAP digital payments add-on via a preparation request. The authorization is recorded by the add-on, and an authorization reference is supplied to the consumer application.
- **External payment direct capture preparation**
  For external payments captured directly at the PSP, the PSP processes the capture and returns the result directly to the consumer application. To support further system integration, the consumer application submits the capture information to the SAP digital payments add-on through a direct capture preparation request. The add-on stores the direct capture, providing a corresponding payment reference to the consumer application.

## 4.4 Payment Methods and Currency Configuration

Once the adapter is connected and the merchant account is onboarded, the next important step is defining how customers can pay. The Stripe V2 adapter for the SAP digital payments add-on supports a wide range of payment methods beyond standard credit cards, including digital wallets; buy now, pay later (BNPL) options; and other payment schemes. This section guides you through the supported payment methods, the architectural differences between direct capture and authorization, and the specific configuration required to handle multicurrency settlements and Level 2/3 data processing. We first review the complete catalog of Stripe payment methods, from card-based options to bank redirects to direct debits, and explain which methods support two-step authorization/capture versus direct capture only; this distinction determines your order-to-cash process design. We then configure multicurrency support by creating distinct merchant aliases for each currency you need to process, enabling your global organization to settle funds into different bank accounts without conversion fees. Next, we walk through how to enable Level 2/3 data transmission to reduce interchange costs on corporate and purchasing cards. Finally, we activate specific payment methods in both the Stripe dashboard and the payment page configuration UI to make them available to customers.

### 4.4.1 Stripe Payment Methods Overview

The Stripe V2 adapter leverages Stripe's PaymentIntents API architecture. This allows it to support payment methods that require complex authentication flows (like 3D Secure) or asynchronous confirmation (like bank redirects). However, not all payment methods behave the same way within the order-to-cash process. It is critical to distinguish between methods that support authorization and capture (two steps) and those that require direct capture (one step), as follows:

- **Card-based payments (two steps or one step)**
  These are the most flexible methods. They support preauthorization (holding funds before shipping), which is the standard requirement for SAP S/4HANA sales order processing. This includes the following payment methods:
  - Credit/debit cards: Visa, Mastercard, American Express, Discover, Diners Club, JCB, and UnionPay
  - Digital wallets: Apple Pay and Google Pay (treated as *wallet cards* in the digital payments add-on)
- **BNPL**
  These payment methods allow customers to finance purchases. The integration behavior varies by provider:
  - Klarna/Afterpay/Clearpay: Support authorization and capture. You can authorize the funds at order creation and capture them upon delivery.
  - Affirm: Typically supports direct capture only in this integration.

- **Bank redirects and real-time payments (direct capture only)**
  These methods require the customer to log into their bank environment to approve the payment. Because the funds are pushed immediately, they do not support preauthorization. If you use these methods, your SAP process must handle immediate settlement documentation. These methods include the following:
  - European local methods: iDEAL (Netherlands), Bancontact (Belgium), Giropay (Germany), EPS (Austria), P24 (Poland)
  - Asian wallets: Alipay and WeChat Pay
- **Direct debits (asynchronous)**
  - Schemes: ACH (USA), SEPA (Europe), BACS (UK), BECS (Australia), and ACSS (Canada).
  - Behavior: Unlike cards, these methods have a settlement delay. The status may remain Pending for several days before turning to Success or Failed.

Use Table 4.15 to plan which digital payments add-on scenario to use for each method.

| Payment Method | Support |
|---|---|
| Credit/debit cards (Code: CC) | Supports direct capture and separate auth/capture. |
| Apple Pay / Google Pay (Code: WC) | Supports direct capture and separate auth/capture. |
| PayPal (Code: PP) | Supports direct capture and separate auth/capture. |
| Klarna (Code: KL) | Supports direct capture and separate auth/capture. |
| Afterpay/Clearpay (Code: A2) | Supports direct capture and separate auth/capture. |
| iDEAL (Code: ID) | Supports direct capture only. |
| Bancontact (Code: BA) | Supports direct capture only. |
| Giropay (Code: G1) | Supports direct capture only. |
| Alipay (Code: AP) | Supports direct capture only. |
| ACH/eCheck (Code: AC) | Supports direct capture only. |

*Table 4.15: Payment Methods and Support for Two-Step and One-Step Payments Process*

### 4.4.2 Currency Support and Multimerchant Strategy

Global organizations often require multiple currencies (e.g., USD, EUR, and GBP) to settle funds into different bank accounts without incurring conversion fees. The SAP digital payments add-on handles this through merchant aliases.

When you onboard a merchant in the Stripe configuration UI, you will typically select a default currency (e.g., USD). This selection tells the adapter which currency to use for reporting and default settlement behavior for that specific configuration entry.

If your consumer application (e.g., SAP S/4HANA) needs to process payments in multiple currencies, you cannot simply use one generic merchant alias. You must create a distinct configuration for each currency, even if they all point to the same underlying Stripe account.

Stripe treats certain currencies (like JPY, KRW, or CL) as zero-decimal currencies. While the digital payments add-on core handles the normalization of amounts (e.g., converting 1,000 JPY to the correct minor unit integer for Stripe), you should ensure your SAP S/4HANA currency decimal configurations (Transaction OY04) match the ISO standards to prevent settlement errors.

### 4.4.3 Level 2 and Level 3 Data Configuration

For B2B transactions, passing additional data fields to the card issuer can significantly reduce interchange fees. This is known as Level 2 (L2) and Level 3 (L3) data processing:

- Level 2 data includes the tax amount and customer reference code.
- Level 3 data includes line-item details such as product codes, unit quantities, unit of measure, and discount amounts.

The Stripe V2 adapter has a built-in capability to map SAP billing data to Stripe's L2/L3 fields:

- Open the Stripe configuration UI (refer to Section 4.3.2 for detailed, step-by-step instructions to navigate to the Stripe configuration UI).
- Locate your merchant configuration.
- Check the box labeled **Send Level 2/3 Data**.
- Click **Save**.

Note that this feature only works if the upstream consumer application (e.g., SAP S/4HANA) sends the line-item data in the payload to the digital payments add-on. If SAP S/4HANA sends a summary total only, the adapter cannot fabricate L3 data. Also make sure that your Stripe account is configured to accept L3 data.

### 4.4.4 Mail Order/Telephone Order and Offline Mandates Support

Many companies operate call centers in which agents take card details over the phone (a mail order/telephone order [MOTO] process) or process offline mandates for recurring billing. The Stripe V2 adapter includes specific support for these scenarios through virtual terminals.

MOTO transactions are exempt from SCA because the cardholder is not present to perform biometric or one-time passcode (OTP) verification. To enable this in the digital payments add-on, perform the following tasks:

- Enable the feature with Stripe: You can contact Stripe Support to enable MOTO processing on your Stripe account. By default, Stripe may decline high volumes of MOTO transactions as fraud risks.

- Configure the digital payments add-on: When the call center agent enters the card in the SAP system, the digital payments add-on adapter flags the transaction as MOTO.
- Virtual terminal: The adapter utilizes a specialized card registration virtual terminal that allows the agent to manually key in the primary account number (PAN) and card verification code (CVC) without triggering 3D Secure.

For recurring payments (like subscriptions), you may collect a physical signature from a customer authorizing you to debit their account. This is an offline mandate, and the following should be noted:

- The adapter supports flagging these transactions, so Stripe knows that a physical agreement exists.
- This is important for SEPA Direct Debit to avoid chargebacks.

### 4.4.5 Enable Specific Methods in the Add-On

Although the Stripe adapter technically supports the methods listed in Section 4.4.1, they will not appear on your checkout page until you explicitly configure them in the SAP digital payments add-on. This is done using the Payment Page Configuration app. Begin by navigating to the Payment Page Configuration interface and selecting or creating a configuration for your region. You will then activate individual payment methods, assigning each method to the Stripe V2 adapter, selecting whether to use authorization or direct capture transaction types, and assigning merchant aliases and weighting values to control the visual display order. Finally, you will enable those same payment methods in the Stripe dashboard, ensuring that both the add-on configuration and the PSP account are aligned before customers can use these payment options.

Let's look at these processes in more detail.

#### Enabling Payment Methods

Follow these steps to navigate to the user interface where you can enable the payment types and transaction types and assign them to the correct PSP:

1. Log into the SAP BTP cockpit (*https://cockpit.btp.cloud.sap*), navigate to your DP Test subaccount, and select **Services • Instances and Subscriptions** from the left navigation pane. Under **Subscriptions,** locate the SAP digital payments add-on demo subscription, click **...**, and select the **Go to Application** link as shown in Figure 4.28.

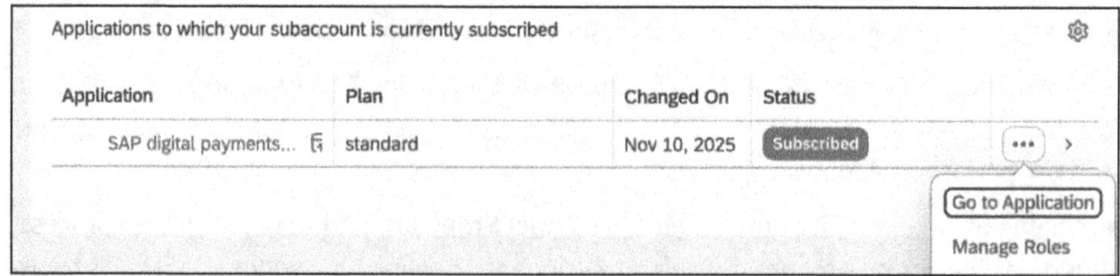


*Figure 4.28: DP Test Subaccount: Application Link Navigation*

2. You'll be directed to the base application URL. The format depends on your SAP BTP region:
   - US10 Test:
     *https://{subdomain}.test-digitalpayments-sap.cfapps.us10.hana.ondemand .com*
   - EU10 Test:
     *https://{subdomain}.demo-digitalpayments-sap.cfapps.eu10.hana.ondeman .com*

   Here, *{subdomain}* is the subdomain you specified in Chapter 3 (e.g., *yourcompany-dp-test*). Figure 4.29 shows an example base application URL if your SAP BTP region is US10.

*Figure 4.29: Add-On Application: Home Page*

3. Append */paymentPageConfiguration/index.html* to the base URL to access the PSP status management UI, as shown in Figure 4.30.

*Figure 4.30: Add-On: Payment Page Configuration URL*

4. Select an existing configuration or click **Create** to start a new one. Figure 4.31 shows an example existing configuration for the EMEA region.

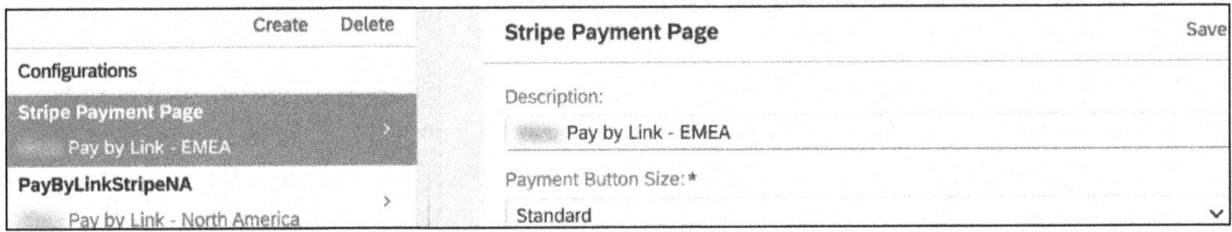


*Figure 4.31: Payment by Link*

5. You will see a list of generic **Digital Payment Types** (e.g., payment card, PayPal, Klarna, Apple Pay). Check the **Active** box for the desired method; for this example, this will be Klarna. In the **Payment Service Provider** column, select **Stripe V2** (see Figure 4.32).

**Configuration Details**

| Digital Payment Type | Active | Payment Service Provider | Transaction Type | Merchant | Weighting |
|---|---|---|---|---|---|
| ACH/eCheck | ☐ | Select | Select | Select | |
| Afterpay | ☐ | Select | Select | Select | |
| Alipay | ☐ | Select | Select | Select | |
| Bancontact | ☐ | Select | Select | Select | |
| eps Online Transfer | ☐ | Select | Select | Select | |
| giropay | ☐ | Select | Select | Select | |
| Ideal | ☐ | Select | Select | Select | |
| Klarna | ☑ | Stripe V2 | Authorization | pk_ | |

*Figure 4.32: Payment Type Activation: Klarna Example*

6. Select either **Authorization** (to reserve funds) or **Direct Capture** (for immediate charge) as the **Transaction Type**, as shown in Figure 4.33.

| | | | |
|---|---|---|---|
| Bancontact | ☐ | Select | Select |
| eps Online Transfer | ☐ | Select | Select |
| giropay | ☐ | Select | Select |
| Ideal | ☐ | Select | Select |
| Klarna | ☑ | Stripe V2 | Authorization |
| Payment Card | ☑ | Stripe V2 | Informative entry |
| PayPal | ☐ | Select | Authorization |
| Przelewy24 | ☐ | Select | Direct Capture |
| Wallet Cards | ☐ | Select | Select |

*Figure 4.33: Payment Page Configuration: Transaction Types*

7. Assign the **Merchant**. Select the merchant alias to be used and, for the weighting, assign a number (0–99) in the **Weighting** column (see Figure 4.34). This determines the visual sort order of the buttons on the checkout page (higher numbers appear first).
8. Now you are ready to process the Klarna payment type.

| Configuration Details | | | | | |
|---|---|---|---|---|---|
| **Digital Payment Type** | **Active** | **Payment Service Provider** | **Transaction Type** | **Merchant** | **Weighting** ⓘ |
| ACH/eCheck | ☐ | Select | Select | Select | |
| Afterpay | ☐ | Select | Select | Select | |
| Alipay | ☐ | Select | Select | Select | |
| Bancontact | ☐ | Select | Select | Select | |
| eps Online Transfer | ☐ | Select | Select | Select | |
| giropay | ☐ | Select | Select | Select | |
| Ideal | ☐ | Select | Select | Select | |
| Klarna | ☑ | Stripe V2 | Authorization | pk_ | 98 |
| Payment Card | ☑ | Stripe V2 | Direct Capt... | pk_ | 99 |

*Figure 4.34: Payment Page Configuration: Merchant Alias and Weighting Assignment*

**Enabling Payment Methods in Stripe Dashboard**

Although the Stripe adapter supports multiple payment methods, they must be explicitly enabled in your Stripe account before becoming available for transaction processing, as follows:

1. Log into the Stripe dashboard (*dashboard.stripe.com*) and ensure you're in the correct mode using the mode toggle: either test mode (test subaccount integration) or live mode (production subaccount integration). Navigate to **Settings • Payments** (see Figure 4.35).

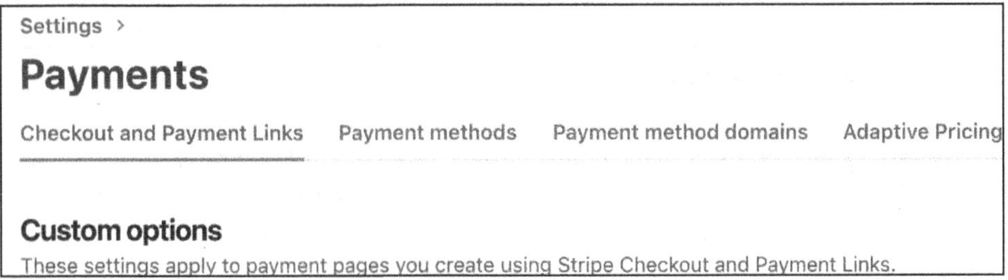


*Figure 4.35: Stripe Payment Methods: Navigation UI*

2. Review the currently enabled methods. The **Payment methods** page displays four sections (see Figure 4.36):
   - **All**: These are payment methods you can enable based on your account country and business type.
   - **Enabled:** These are the payment methods currently active for your account.
   - **Disabled**: These are the payment methods currently inactive for your account.
   - **Requires action**: Some payment methods may require follow-up actions. If so, they will be listed here.

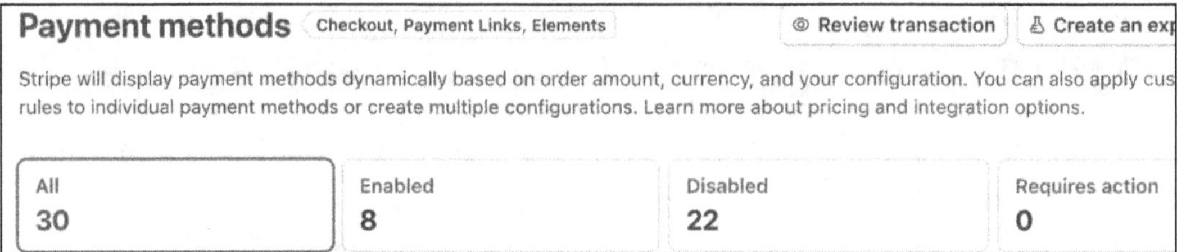


*Figure 4.36: Stripe Payment Methods UI*

3. Most new Stripe accounts have some payment methods enabled by default (e.g., card payments, Apple Pay, Link). Figure 4.37 shows an example Stripe account.

*Figure 4.37: Default Payment Methods: Stripe Payments*

4. To enable additional wallet payments, scroll to the **Disabled** section (Figure 4.38) and locate them (e.g., Alipay, Amazon Pay). Select Alipay from the list of **Payment methods**.

*Figure 4.38: Stripe Payment Methods: Inactive Example*

5. To enable Alipay as a new payment method. Click the **Turn on** button (see Figure 4.39). Once turned on, the payment method immediately moves to the **Enabled** section and becomes available for processing.

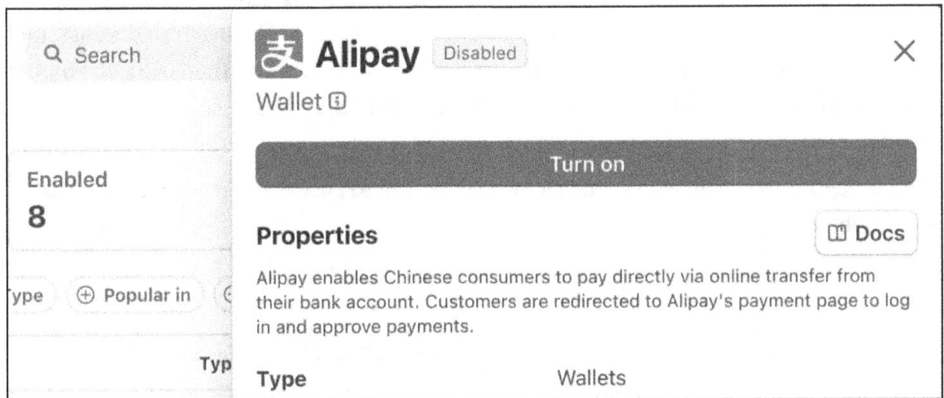


*Figure 4.39: Alipay Payment Method: Enablement*

6. Let's look at another example, for SEPA Direct Debit (for European Economic Area [EEA] merchants). Like you did for Alipay, locate **SEPA Direct Debit** in the **Disabled** payments method section and click **Turn on**. You will see that it is now enabled (see Figure 4.40).

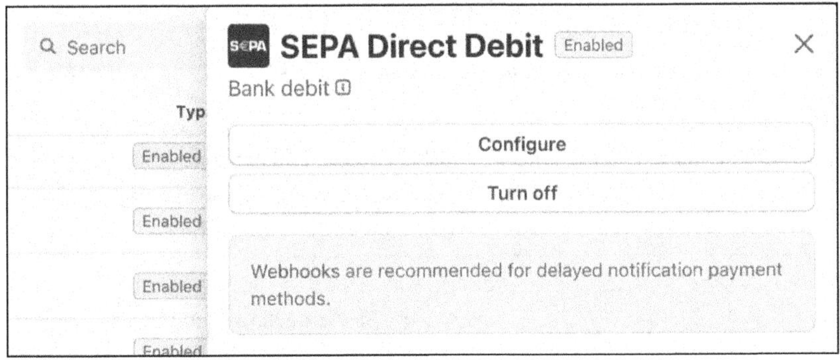


*Figure 4.40: SEPA Direct Debit Enrollment*

7. If required in your jurisdiction, configure your creditor **Identifier** (see Figure 4.41) by selecting the **Configure** button.

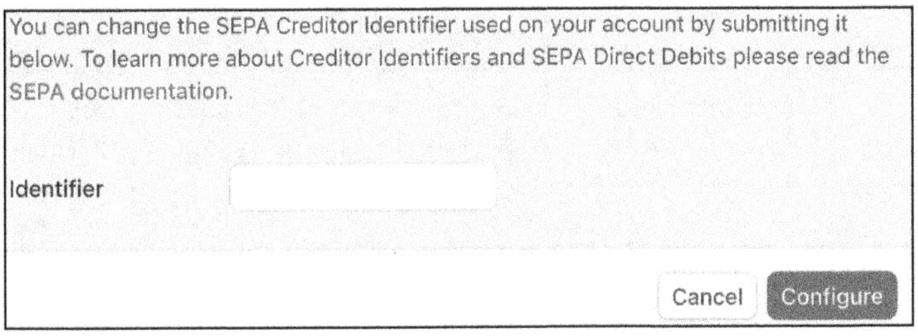

You can change the SEPA Creditor Identifier used on your account by submitting it below. To learn more about Creditor Identifiers and SEPA Direct Debits please read the SEPA documentation.

Identifier

Cancel Configure

*Figure 4.41: SEPA Direct Debit: Configure Creditor Identifier UI*

## 4.5 Webhook Setup and Event Management

Payment processing communication is typically asynchronous. While the SAP digital payments add-on initiates transactions, the final status is often determined later by the PSP. This is particularly true for redirect methods (like 3D Secure or iDEAL) in which the customer must perform actions on a banking page. To handle these updates, the Stripe V2 adapter utilizes webhooks. A *webhook* is a notification sent from Stripe to the digital payments add-on whenever an event occurs (e.g., `payment_intent.succeeded`).

The Stripe V2 adapter utilizes a thin event architecture to ensure performance and reliability, as follows:

- Notification: Stripe sends a lightweight notification to the adapter containing just the event ID and object ID.
- Verification: The adapter validates the signature to ensure the request is genuine.
- Retrieval: The adapter calls back to the Stripe API to fetch the full, latest status of the object. This prevents race conditions that might cause a webhook to contain stale data.
- Normalization: The status is mapped to SAP standard codes (e.g., P for pending or S for success) and pushed to the digital payments add-on core.

For the integration to function, the adapter must subscribe to specific Stripe events. The SAP Digital Payments Add-On V2 app automatically registers these during installation, but you should verify they are active. Table 4.16 lists the critical events.

| Event Name | Process | Description |
|---|---|---|
| `payment_intent.succeeded` | Authorization/ capture | Confirms funds are authorized or captured. Updates the digital payments add-on token status to **Success**. |
| `payment_intent.payment_failed` | Exception handling | Signals a decline. The adapter maps the Stripe decline code to an SAP error code. |
| `payment_intent.requires_action` | 3D Secure | Signals that the customer must perform SCA. |
| `charge.refunded` | Refund | Confirms a refund has processed. Essential for updating financial accounting. |
| `payment_method.attached` | Card registration | Confirms a new card/wallet has been saved to a customer profile. |
| `mandate.updated` | SEPA/direct debit | Tracks changes to the digital mandate required for bank debits. |

*Table 4.16: Stripe PSP: Webhook Events List*

When the SAP Digital Payments Add-On V2 app is installed, it programmatically creates a webhook endpoint in your Stripe account that points to the digital payments add-on infrastructure. You can verify that the webhook is active and working by following this verification procedure:

1. Log into the Stripe dashboard (ensure you are in the correct mode, either test or live) and navigate to **Settings • Developers • Webhooks**.
2. Look for an endpoint URL that matches the digital payments add-on connector domain:
   - US Region: *https://us-prod-gateway.sap.stripeconnectors.com*
   - EU Region: *https://eu-prod-gateway.sap.stripeconnectors.com*
3. Validate the status; ensure that it is **Enabled**.
4. Check the event subscriptions. Click the endpoint and ensure the events listed in Table 4.16 are present.

In payment processing, network timeouts can cause uncertainty about whether the charge request failed or the response was just lost. If the consumer application retries a charge request that succeeded the first time, the customer could be double-charged. Stripe supports idempotency keys to solve this. An *idempotency key* is a unique value generated by the client (in this case, the digital payments add-on adapter) and sent with the API request.

The digital payments add-on handles idempotency as follows:

- Key generation: When a consumer application initiates a payment, the digital payments add-on adapter generates a universally unique identifier (UUID) for that specific transaction attempt.
- Stripe check: When Stripe receives the request, it checks if it has already seen a request with that specific idempotency dey in the last 24 hours.
- Replay: If Stripe recognizes the key, it does not create a new charge. Instead, it simply replays the original response (e.g., **Success** or **Decline**).

This mechanism ensures that even if the digital payments add-on encounters a network error and automatically retries the API call, the customer is never charged twice for the same order.

The digital payments add-on can periodically fetch a settlement file from Stripe, which consumer applications can use to clear open items in accounts receivable. The Stripe V2 Adapter includes a dedicated user interface, the Stripe advice UI, to manage this process:

1. Log into the SAP BTP cockpit (*https://cockpit.btp.cloud.sap*), navigate to your DP Test subaccount, and select **Services • Instances and Subscriptions** from the left navigation pane. Under **Subscriptions**, locate the SAP digital payments add-on demo subscription, click **...**, and select the **Go to Application** link as shown in Figure 4.42.

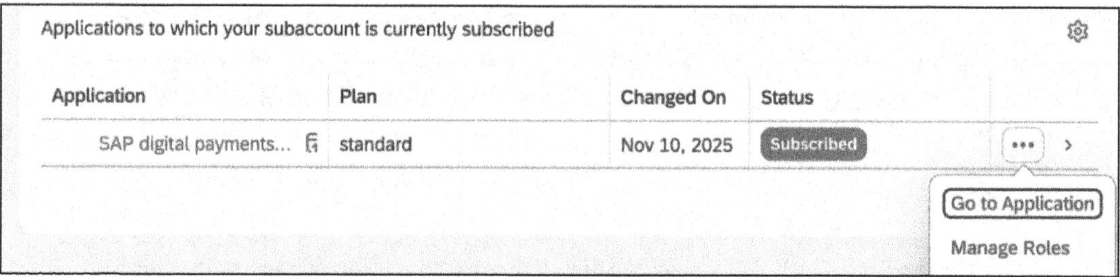


*Figure 4.42: Add-On Subaccount: Go to Application Link*

2. You'll be directed to the base application URL. The format depends on your SAP BTP region:
   - US10 Test:
     *https://{subdomain}.test-digitalpayments-sap.cfapps.us10.hana.ondemand. com*
   - EU10 Test:
     *https://{subdomain}.demo-digitalpayments-sap.cfapps.eu10.hana.ondemand. com*

   Here, *{subdomain}* is the subdomain you specified in Chapter 3 (e.g., *yourcompany-dp-test*). Figure 4.43 shows an example base application URL if your SAP BTP region is US10.

*Figure 4.43: Add-On Application: Home Page*

3. Append */stripeAdvice/index.html* to the base URL to access the Stripe advice UI, as shown in Figure 4.44.

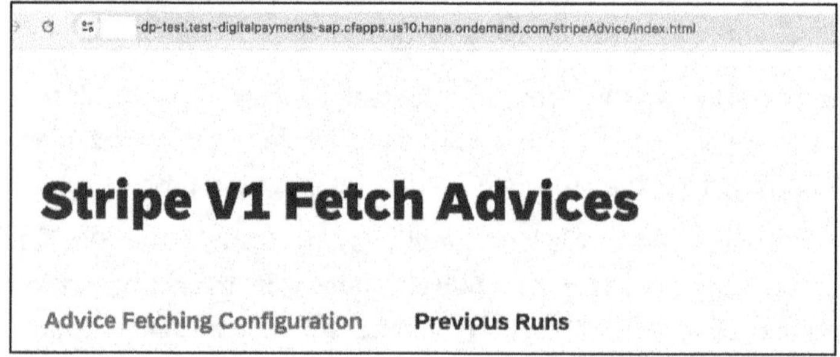


*Figure 4.44: Stripe Advice UI*

The scheduler allows you to automate the retrieval of settlement data. Instead of manually downloading spreadsheets from Stripe, the adapter pulls the data and pushes it into the digital payments add-on core, which then forwards it to consumer applications (e.g., SAP S/4HANA). When setting up a schedule, you must define the following parameters:

- **Recurrence Pattern**: Defines the frequency of the job, as follows:
  - **Daily**: Runs every day. Best for high-volume, B2C businesses.
  - **Weekly**: Runs on specific days (e.g., every Monday).
  - **Monthly**: Runs on specific dates (e.g., the 1st and 15th of the month).
- **Time Slot** (UTC): Specifies the exact time the job should run. It is critical to set this time for after Stripe has closed its daily settlement batch (usually midnight UTC) to ensure you capture the full day's data.
- **Starting Date**: The date from which the scheduler should become active.

To set a daily schedule, follow these steps:

1. Navigate to the **Advice Fetching Configuration** section (see Figure 4.45).

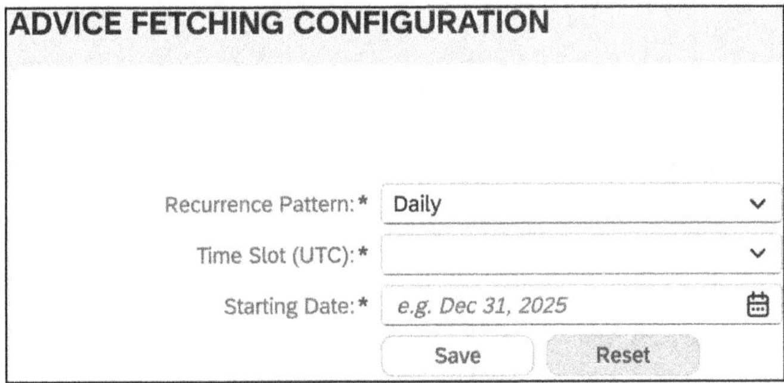


*Figure 4.45: Advice Fetching Configuration UI*

2. Set **Recurrence Pattern** to **Daily** and select a **Time Slot** (e.g., **02:00 - 03:59** UTC). This ensures the previous day's settlements are ready (see Figure 4.46).

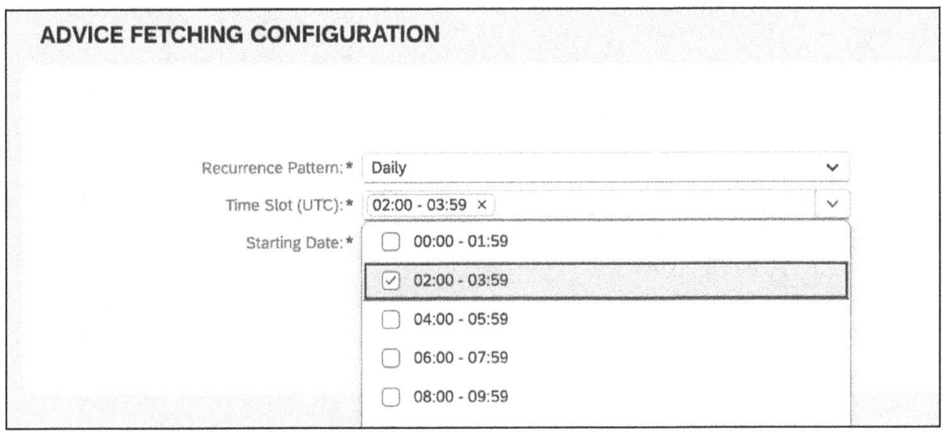


*Figure 4.46: Advice Fetching Configuration: Daily Fetch Pattern Example*

3. Set the **Starting Date** to today (e.g., December 4), as shown in Figure 4.47.

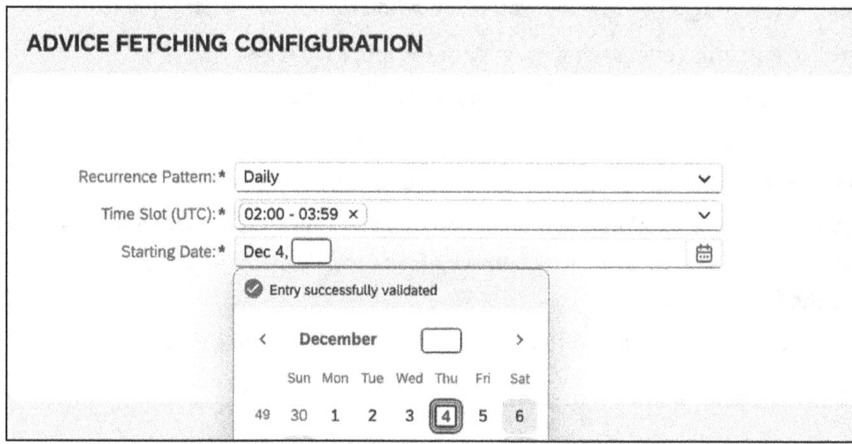


*Figure 4.47: Fetch Payment Advice: Choosing Current Date Example*

4. Click **Save** (see Figure 4.48).

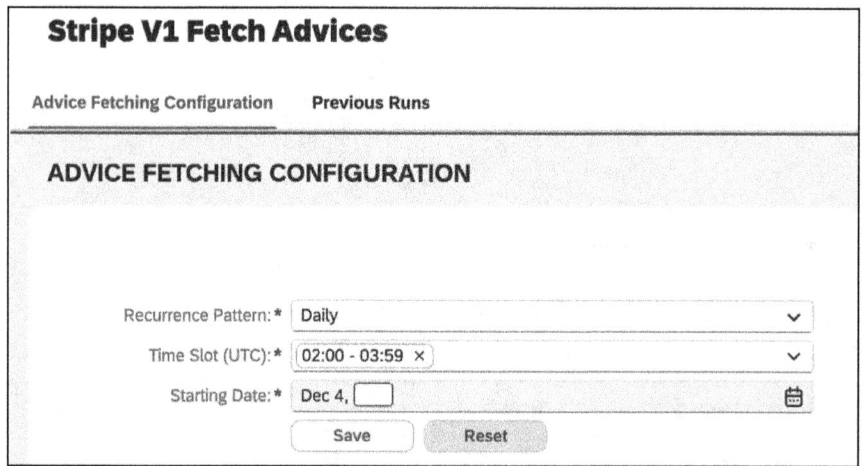


*Figure 4.48: Advice Fetching Configuration: Daily Schedule Example*

For testing purposes or period-end closing, you may need to fetch advice data immediately without waiting for the scheduler. Click the **Run Immediately** button in the top-right corner of the advice UI (see Figure 4.49).

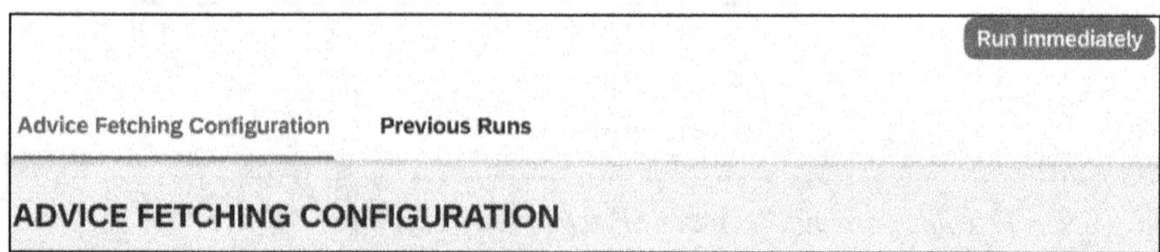


*Figure 4.49: Stripe Advice: Ad Hoc Run*

The system will trigger a background job to fetch the data. You can view the status in the **Previous Runs** table on the main dashboard. A green status icon indicates the advice was successfully fetched and handed off to the digital payments add-on core (see Figure 4.50).

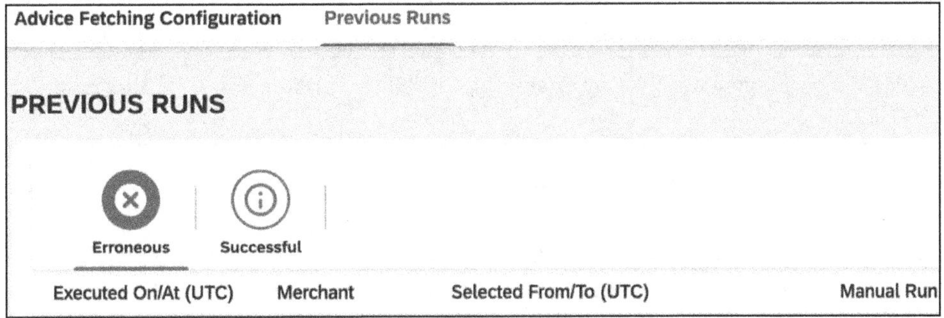


*Figure 4.50: Stripe Advice Schedule Viewer UI*

## 4.6 Testing PSP Integration

Stripe provides a comprehensive set of test card numbers that simulate various payment scenarios without processing real transactions or charging actual cards. These test cards work exclusively in test mode and enable validation of both successful payments and various decline scenarios.

Implementation teams typically have three broad categories of test scenarios: one for success or happy path payment card tests, one for error or decline scenarios; and one for certain regions that enforce additional authentication steps, like 3D Secure.

Table 4.17 lists standard test cards that can be used for testing success scenarios.

| Card Number | Brand | CVC | Expiry | 3D Secure Required | Test Scenario |
|---|---|---|---|---|---|
| 4242 4242 4242 4242 | Visa | Any three digits | Any future date | No | Successful payment |
| 5555 5555 5555 4444 | Mastercard | Any three digits | Any future date | No | Successful payment |
| 3782 822463 10005 | American Express | Any four digits | Any future date | No | Successful payment |
| 6011 1111 1111 1117 | Discover | Any three digits | Any future date | No | Successful payment |
| 3056 9300 0902 0004 | Diners Club | Any three digits | Any future date | No | Successful payment |

*Table 4.17: Stripe PSP: Standard Test Cards*

| Card Number | Brand | CVC | Expiry | 3D Secure Required | Test Scenario |
|---|---|---|---|---|---|
| 3566 0020 2036 0505 | JCB | Any three digits | Any future date | No | Successful payment |
| 6200 0000 0000 0005 | UnionPay | Any three digits | Any future date | No | Successful payment |

*Table 4.17: Stripe PSP: Standard Test Cards (Cont.)*

Table 4.18 lists test cards that can be used for negative test cases, so that implementation teams can test how the application behaves when card processing errors out.

| Test Card Number | Decline Reason | API Error Code | Test Scenario |
|---|---|---|---|
| 4000 0000 0000 0002 | Generic decline | `card_declined` | Generic decline (no specific reason) |
| 4000 0000 0000 9995 | Insufficient funds | `insufficient_funds` | Customer account has insufficient balance |
| 4000 0000 0000 9987 | Lost card | `lost_card` | Card reported lost by cardholder |
| 4000 0000 0000 9979 | Stolen card | `stolen_card` | Card reported stolen by cardholder |
| 4000 0000 0000 0069 | Expired card | `expired_card` | Card expiration date has passed |
| 4000 0000 0000 0127 | Incorrect CVC | `incorrect_cvc` | CVC verification failed |
| 4000 0000 0000 0119 | Processing error | `processing_error` | Generic PSP processing error |
| 4000 0000 0000 3220 | 3D Secure required | `authentication_required` | 3DS authentication must be completed |

*Table 4.18: Stripe PSP: Error and Decline Test Cards*

With the enforcement of SCA in Europe and other regions, testing the 3D Secure flow is important. Table 4.19 lists test cards that simulate 3D Secure-enabled card processing steps.

| Test Card Number(s) | 3D Secure Behavior | Test Scenario |
|---|---|---|
| 4000 0027 6000 3184 | 3D Secure required; authentication succeeds | Successful SCA completion |
| 4000 0082 6000 3178 | 3D Secure required; authentication fails | Customer abandons or fails authentication |
| 4000 0025 0000 3155 | 3D Secure supported but not required | Frictionless flow (no challenge) |
| 4242 4242 4242 4242 | 3D Secure not required | Standard authorization (non-EEA merchant) |

*Table 4.19: 3D Secure Test Cards*

Table 4.20 lists common issues implementation teams face, potential causes, and resolution mechanisms.

| Issue | Cause | Resolution |
|---|---|---|
| No PSP found error | The PSP determination rules in the digital payments add-on are not matching the data sent from the consumer application. | Check the Payment Service Provider Determination app. Ensure there is a catchall row at the bottom with empty conditions pointing to your default Stripe alias. |
| Transaction is pending indefinitely | Webhooks are failing or not firing. | Check the Stripe dashboard under **Developers • Webhooks**. If webhooks are failing (non-200 response), verify that the digital payments add-on tenant is online. If no webhooks are generating, ensure the event types are subscribed. |
| Level 3 data not appearing in Stripe | The upstream SAP system (e.g., SAP S/4HANA) is not sending line-item details, or the **Send Level 2/3 Data** box is unchecked in the adapter configuration. | Verify the checkbox in the Stripe configuration UI. If enabled, verify that the XML/JSON payload from SAP S/4HANA contains item-level details, not just header totals. |

*Table 4.20: Troubleshooting Guide*

| Issue | Cause | Resolution |
|---|---|---|
| Advice job fails to run | The **Run Immediately** job option was triggered for a time window in which no transactions occurred, or the API token for the advice service has expired. | Ensure transactions exist in Stripe for the requested time window. If the error persists, refresh your API credentials in the Stripe configuration UI. |

*Table 4.20: Troubleshooting Guide (Cont.)*

## 4.7 Summary

This chapter provided comprehensive guidance for integrating PSPs with the SAP digital payments add-on, using Stripe as the primary implementation example. You learned about the PSP integration architecture (adapter framework, API patterns, certified PSP ecosystem); created and configured PSP accounts with API credentials in both test and production environments; activated and configured the Stripe V2 adapter with merchant onboarding and routing rules; enabled new payment methods across cards, wallets, and external payment types with multicurrency support; configured webhook endpoints for asynchronous event handling and payment advice scheduling; and validated the complete integration through comprehensive test scenarios using Stripe's test cards.

Your PSP integration environment is now operational, with adapters activated in both test and production SAP BTP subaccounts, merchant accounts onboarded with verified connectivity to Stripe, merchant aliases mapped for routing configuration, PSP determination rules directing transactions to appropriate merchants based on company code and currency, payment methods enabled in the Stripe dashboard and configured through the payment page setup UI, webhook endpoints registered with confirmed delivery, and the payment advice scheduler configured for automated reconciliation.

The adapter framework architecture ensures PSP independence; each adapter operates as an isolated microservice with its own API protocols and webhook endpoints. This isolation enables version updates without affecting the digital payments add-on core service, allows adding new PSPs through configuration rather than code changes, and ensures that failures in one PSP don't impact others.

The subsequent chapters build on this PSP foundation: Chapter 5 covers the SAP ERP implementation requiring custom ABAP development for payment authorization during order entry, capture during billing, and refund handling through credit memos with sample code and batch job configuration. Chapter 6 the addresses the SAP S/4HANA implementation, leveraging standard SAP Fiori applications, out-of-the-box order-to-cash integration, and built-in analytics for payment monitoring.

# Chapter 5
# Implementation in SAP ERP

*While SAP S/4HANA offers streamlined, out-of-the-box integration with the SAP digital payments add-on, a large portion of the global SAP install base continues to run on SAP ERP, so a thorough understanding of the SAP ERP–specific implementation details is essential and relevant. This chapter walks you through the full SAP ERP implementation path, from technical prerequisites and payment card configuration through custom ABAP development, order-to-cash integration, batch job management, and reporting. It provides practical guidance to build a production-ready digital payments solution on the classic SAP ERP platform before we turn to SAP S/4HANA's more modern approach.*

This chapter addresses the unique requirements of implementing the SAP digital payments add-on in SAP ERP environments, where more custom development is required compared to SAP S/4HANA. Unlike SAP S/4HANA's streamlined, out-of-the-box integration, SAP ERP implementations need enhancements and custom development to integrate payment processing into existing business processes. You will learn how to implement payment authorization during order entry, capture during billing, and handle refunds through credit memos. The chapter includes sample ABAP code, configuration of batch jobs, and recommendations for building custom reports.

The implementation approach for SAP ERP differs from that for SAP S/4HANA due to architectural constraints and the absence of native integration points. SAP S/4HANA provides prebuilt CDS views, standard SAP Fiori applications, and streamlined APIs for payment processing, but SAP ERP requires custom development at multiple integration points. This additional development effort is a trade-off you accept when running SAP ERP rather than migrating to SAP S/4HANA. However, the SAP digital payments add-on architecture minimizes the burden by providing standard APIs and adapter frameworks that reduce the custom code footprint compared to traditional payment gateway integrations.

## 5.1 Prerequisites for SAP ERP

Before beginning technical implementation work, you must verify that your SAP ERP system meets all prerequisites for the digital payments add-on integration. These prerequisites span multiple dimensions, including technical system requirements, support package levels, SAP Notes implementation, authorization setup, and network connectivity. Missing prerequisites discovered during implementation create project delays and rework, making thorough validation at the outset essential.

The prerequisite validation should occur during the project preparation phase, before any development or configuration work begins. Create a prerequisites checklist document that tracks the validation status for each requirement, identifies gaps requiring remediation, and assigns responsibility for closing those gaps. This systematic approach prevents surprises during the integration phase and ensures the technical foundation supports successful implementation.

### 5.1.1 SAP ERP System Requirements

The add-on integrates with SAP ERP systems running on the SAP ERP 6.0 platform. Although the add-on itself operates as a cloud service on SAP BTP, your SAP ERP system must meet specific technical requirements to establish secure communication and handle payment processing workflows. In this section, we examine the minimum release level and support package requirements to establish the foundation. The SAP ERP system must be on a supported Enhancement Package (EHP) level and running current support packages so that the digital payments add-on SAP Notes will have the underlying framework they depend on. Unicode compliance comes next, as the add-on's APIs communicate exclusively in UTF-8, making a non-Unicode system a blocker for any integration work. Network connectivity then defines the outbound HTTPS paths that must be open through the corporate firewall to reach SAP BTP and the payment service providers (PSPs). Also, the database and operating system compatibility needs to be confirmed up front. Finally, the system sizing must account for the incremental load that the real-time authorization calls, background settlement jobs, and payment advice processing will place on the system.

#### Minimum Release Requirements

SAP ERP systems must run on SAP ERP 6.0 with a minimum EHP level to support the digital payments add-on integration framework. The integration relies on web service capabilities, enhanced data structures, and security features introduced in later EHPs. Table 5.1 specifies the minimum and recommended release levels.

| Component | Minimum Version | Recommended Version | Verification Transaction | Notes |
|---|---|---|---|---|
| SAP ERP | SAP ERP 6.0 EHP 5 | SAP ERP 6.0 EHP 7 or higher | **System • Status** | EHP 7 required per SAP Note 3003533 |
| SAP_APPL | 605 | 617 or 618 | Transaction SPAM | Application component version |
| SAP_BASIS | 730 | 740 or higher | Transaction SPAM | Controls core ABAP runtime and RFC framework |
| Kernel | 720 | 745 or higher | **System • Status** | Unicode kernel required |

*Table 5.1: SAP ERP Minimum Release Requirements*

The EHP level does impact the implementation complexity. The minimum version needed is EHP 5, but certain SAP Notes require EHP 7 or higher for the complete configuration framework. If your SAP ERP has EHP 5 or EHP 6, you could plan an EHP upgrade before implementing the digital payments add-on to avoid compatibility issues.

To verify your current release levels, first select **System • Status** in your SAP GUI. In the **SAP System Data** section, select **Details** to view the list of software components. The **Installed Software Component Information** section displays your SAP_APPL and SAP_BASIS component versions. Compare these values against those shown in Table 5.1 to confirm compliance with minimum requirements. Figure 5.1 shows an example SAP ERP system that is compliant with the add-on prerequisites.

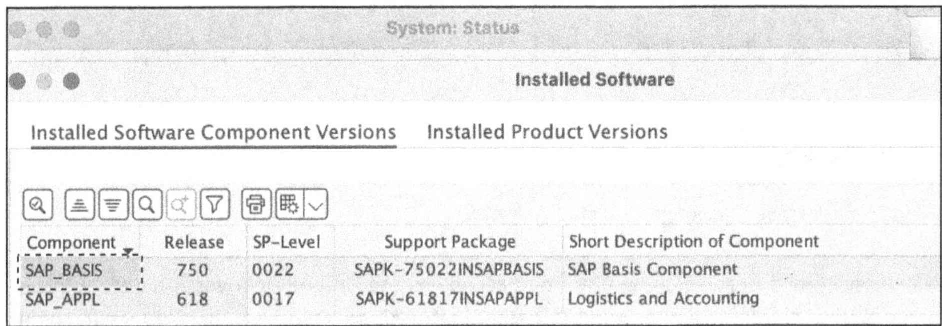


*Figure 5.1: Installed Software Component Versions List*

### Support Package Requirements

Beyond release level requirements, your system must maintain current support package levels to ensure that critical fixes and enhancements are present. SAP releases support packages regularly to address defects, security vulnerabilities, and functional gaps. The digital payments add-on integration relies on specific corrections delivered through support packages.

You can plan for the implementation of new support packages by following standard SAP change management procedures. Your Basis team can download necessary support packages from SAP Service Marketplace and import them into your systems using Transaction SPAM. It is recommended to apply the support packages first in development environments with thorough testing, then transport them to the quality assurance environment to perform necessary end-to-end testing with internal systems and external partners (banks, vendors, etc.) in your landscape. Maintain detailed documentation of all changes for audit purposes.

### Unicode Requirement

Your SAP ERP system must be Unicode-enabled to support the digital payments add-on. The add-on APIs use UTF-8 encoding for all character data, including customer names, addresses, and payment metadata. Non-Unicode systems cannot properly handle international character sets present in global payment data, resulting in data corruption and integration failures.

To verify the Unicode status, there are multiple options. One option is to execute report RSCPINST in Transaction SE38. If the system is Unicode, it will be explicitly stated. Another option is to execute function module `TR_GET_IS_UNICODE_SYSTEM` in Transaction SE37. If the result for `EV_IS_UNICODE_SYSTEM` is `X` (see Figure 5.2), then the system is Unicode-enabled. If it is blank, it is non-Unicode. Alternatively, you can check with your Basis team to get an answer as they have access to check your system profile parameters.

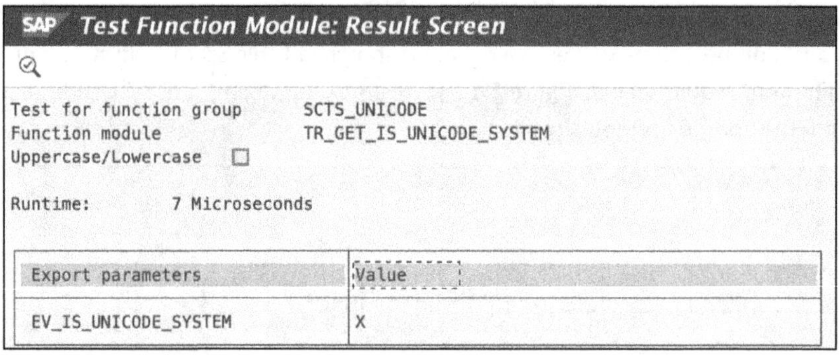

SAP Test Function Module: Result Screen

```
Test for function group      SCTS_UNICODE
Function module              TR_GET_IS_UNICODE_SYSTEM
Uppercase/Lowercase

Runtime:          7 Microseconds
```

| Export parameters | Value |
|---|---|
| EV_IS_UNICODE_SYSTEM | X |

*Figure 5.2: Unicode Check Function Module Result*

If your system is non-Unicode, you must complete the Unicode conversion before implementing the digital payments add-on. A Unicode conversion is a major technical undertaking that requires comprehensive planning, extensive testing, and system downtime. Organizations still operating non-Unicode SAP ERP systems can evaluate two paths: complete the Unicode conversion, followed by the digital payments add-on implementation; or prioritize an SAP S/4HANA migration, which provides both the Unicode and modern payment integration capabilities simultaneously.

**Network Connectivity Requirements**

SAP ERP systems must establish outbound HTTPS connectivity to both SAP BTP endpoints and PSP systems. These connections traverse your corporate firewall and typically require security exception approvals from network and information security teams. Table 5.2 lists the required connectivity endpoints for digital payments add-on integration.

| Destination | Protocol/ Port | Purpose | Firewall Exception Required For |
|---|---|---|---|
| **.cfapps.us10.hana.ondemand.com* | HTTPS/443 | SAP BTP US East region add-on APIs | US10 landscape |
| **.cfapps.eu10.hana.ondemand.com* | HTTPS/443 | SAP BTP EU Frankfurt region APIs | EU10 landscape |

*Table 5.2: Required Network Connectivity Checklist*

| Destination | Protocol/ Port | Purpose | Firewall Exception Required For |
|---|---|---|---|
| **.cfapps.eu11.hana.ondemand.com* | HTTPS/443 | SAP BTP EU Access region APIs | EU11 landscape |
| *api.stripe.com* | HTTPS/443 | Stripe PSP communication | If using Stripe |
| **.authentication.us10.hana.ondemand.com* | HTTPS/443 | SAP BTP OAuth authentication | US10 landscape |
| **.authentication.eu10.hana.ondemand.com* | HTTPS/443 | SAP BTP OAuth authentication | EU10 landscape |

*Table 5.2: Required Network Connectivity Checklist (Cont.)*

You can work with your network and security teams to obtain the necessary firewall approvals. Security teams often require detailed justification for outbound HTTPS exceptions, including the business purpose, data sensitivity classification, regulatory compliance impact, and security control measures. Prepare documentation explaining that the digital payments add-on architecture reduces the PCI DSS compliance scope by preventing cardholder data storage in SAP ERP systems, which typically addresses the primary security concerns.

You can test the network connectivity from your SAP ERP system using Transaction SM59 (RFC Destinations). Create a test RFC destination of type G (HTTP connection to external server), pointing to your add-on tenant URL. Execute the connection test to verify the outbound HTTPS works through your firewall. Note that for digital payments add-on RFC destinations, the standard connection test will fail even with proper configuration due to OAuth authentication requirements; actual validation occurs during the first payment transaction.

If connectivity testing fails, common issues to troubleshoot include the corporate proxy server configuration requiring explicit proxy settings in Transaction SM59, SSL/TLS certificate trust issues requiring certificate import through Transaction STRUST, firewall rules blocking specific destination ports or IP ranges, and network segmentation preventing application servers from reaching internet endpoints.

### Database and Operating System Compatibility

The digital payments add-on does not have specific database platform or operating system requirements beyond those certified for your SAP ERP release level. You can work with your Basis team to analyze additional database-specific optimization considerations for payment workloads, based on expected volumes. Operating system requirements follow standard SAP ERP certifications. You can consult the Product Availability Matrix (PAM; see *https://userapps.support.sap.com/sap/support/pam*) to verify that your specific OS version and patch level are certified for your SAP ERP release.

**System Sizing Considerations**

Payment processing introduces an incremental load on your SAP ERP system through API calls, database operations for payment token storage, and background processing for settlement and reconciliation. Adequate system sizing prevents performance degradation as payment transaction volumes grow. Organizations processing payments through multiple sales channels simultaneously (e-commerce, call centers, customer service portals) should plan to provision incremental capacity during peak processing periods, including holiday seasons and promotional events. Month-end/quarter-end closings may require temporary work process adjustments to handle transaction spikes.

Dialog work processes handle synchronous payment API calls during sales order creation. Background work processes execute batch settlement jobs (Transaction FCC1) and payment advice processing. Memory increases accommodate payment transaction data caching, API response buffering, and enhanced logging for PCI DSS audit requirements.

### 5.1.2 Add-On-Specific SAP Notes Implementation

SAP delivers the digital payments add-on integration for SAP ERP through a series of SAP Notes rather than a standard support package. This delivery approach provides flexibility for organizations to implement payment capabilities on their existing SAP ERP release levels without requiring full upgrade projects. However, it also means that implementation teams must apply multiple SAP Notes in the correct sequence, following detailed manual activities and correction instructions.

The SAP Notes create a comprehensive integration framework, including Data Dictionary objects, ABAP program code, configuration views, and technical connectivity components. Each SAP Note builds upon prerequisites established by previous ones, making the implementation sequence critical. Attempting to implement SAP Notes out of order results in missing dependencies, activation errors, and incomplete integration.

Table 5.3 provides the list of SAP Notes required for the full digital payments add-on integration with SAP ERP, organized by the required implementation sequence. You could choose to implement just a few of these, based on your specific implementation scope.

| Sequence | SAP Note | Short Description | Category | Prerequisites |
|---|---|---|---|---|
| 1 | 2999922 | DDIC Objects | Foundation | None |
| 2 | 3000524 | Program Code | Core logic | SAP Note 2999922 |

*Table 5.3: Add-On-Specific SAP Notes Implementation Sequence*

| Sequence | SAP Note | Short Description | Category | Prerequisites |
|---|---|---|---|---|
| 3 | 3028044 | Message Enhancements | Configuration | SAP Notes 2999922, 3000524 |
| 4 | 3027433 | Central System Messages | Configuration | SAP Note 3028044 |
| 5 | 3028194 | Settlement Module | Business logic | SAP Notes 2999922, 3000524 |
| 6 | 3041924 | Additional Integration | Enhancement | SAP Notes 2999922, 3000524 |
| 7 | 3044268 | German Language Support | Localization | SAP Notes 2999922, 3000524, and 3003533 |
| 8 | 3212765 | Refund Payments | Business logic | SAP Note 3028194 |
| 9 | 3273859 | Include Recreation | Technical fix | SAP Note 3000524 |
| 10 | 3276162 | Index Optimization | Performance | SAP Note 2999922 |
| 11 | 3277101 | Document Search Enhancement | Performance | SAP Note 2999922 |
| 12 | 3290490 | Payment Type Adaptation | Enhancement | Previous SAP Notes |
| 13 | 3341226 | Additional Integration | Enhancement | Previous SAP Notes |
| 14 | 3349208 | Advice Reference Fix | Bug Fix | SAP Note 3028194 |
| 15 | 3003533 | Configuration Guide | Documentation | All previous SAP Notes |

*Table 5.3: Add-On-Specific SAP Notes Implementation Sequence (Cont.)*

You should plan to implement all the SAP Notes (based on your identified scope) as a one-time effort rather than piecemeal over time. This approach provides several advantages: It ensures the dependency resolution occurs in the correct order, enables comprehensive integration testing after all components are in place, simplifies transport management with a single transport request containing all objects, and reduces the overall project timeline compared to iterative implementations.

You can create a dedicated transport request before beginning SAP Note implementation. All manual activities, correction instructions, and configuration changes can be assigned to this transport to maintain traceability and enable clean promotion through your landscape. Use a descriptive transport request name, such as "Digital Payments Add-On Integration," to clearly identify the scope. Now, let's analyze and understand each of these critical SAP Notes, their purpose, and what they do in the SAP ERP system.

#### SAP Note 2999922

SAP Note 2999922 creates the foundational objects for the digital payments add-on's integration with SAP ERP. It creates all Data Dictionary (DDIC) objects required for subsequent SAP Notes to function. Without these objects in place, no other integration SAP Note can be successfully implemented. The SAP Note delivers over 175 repository objects, including data elements, domains, tables, views, search helps, function groups, message classes, and reports.

The SAP Note provides a text file attachment (*note_2999922.txt*) that contains the complete source code for report NOTE_2999922_II. This report, when executed, automatically creates the entire DDIC foundation in your SAP ERP system. This approach ensures consistency across implementations and eliminates manual object-creation errors.

The SAP Note establishes fundamental data elements that define payment-specific field characteristics used throughout the integration, as follows:

- `DP_TRANSID`: Digital payment transaction ID, a unique identifier for each payment transaction
- `DP_PAYMENT_TYPE`: Payment type code (e.g., PC for payment card, EP for external payment)
- `DP_PAYTREQ`: Payment request ID; links payment requests across systems
- `DP_TOKEN`: Payment card token returned by the add-on
- `DP_PAYMENT_SERVICE_PROVIDER`: PSP code (e.g., STRIPE_V2)
- `DP_PAYID`: PSP-specific payment identifier (e.g., Stripe charge ID)
- `DP_PSP_TRANS_ID`: PSP transaction identifier
- `DP_MERCHANT_ALIAS`: User-friendly merchant account alias

The SAP Note creates append structures that extend standard SAP tables with digital payments add-on fields. This nonmodifying enhancement approach preserves SAP standard objects while adding required payment data storage. The append structures are as follows:

- **`DIGITAL_PAYMENTS_BSEG` (append to table `BSEG`—Accounting Document Segment)**
  Table `BSEG` contains line item details for all accounting documents. The append struc-

ture enables each line item to carry payment-specific information, supporting scenarios in which single documents contain both payment card and traditional payment method line items. It has the following components:

- Component DP_TRANSID (data element: DP_TRANSID)
- Component DP_PAYTYPE (data element: DP_PAYMENT_TYPE)
- Component DP_PAYTREQ (data element: DP_PAYTREQ)

- **DIGITAL_PAYMENTS_BSID (append to table BSID—Customer Open Items)**
  Table BSID holds open customer receivables. The append structure links open items to their corresponding payment transactions, enabling payment status tracking and reconciliation. It has the following components:
  - Component DP_TRANSID
  - Component DP_PAYTYPE
  - Component DP_PAYTREQ
- **DIGITAL_PAYMENTS_BSAD (append to table BSAD—Customer Cleared Items)**
  Table BSAD contains cleared customer items. The append structure maintains payment transaction linkage after items are settled, supporting historical reporting and refund processing. It has the following components:
  - Component DP_TRANSID
  - Component DP_PAYTYPE
  - Component DP_PAYTREQ

Searching for specific payment requests in tables BSID and BSAD via field DP_PAYTREQ would normally trigger full table scans, causing performance degradation in systems with large volumes of customer line items. The SAP Note addresses this through extension index creation, as follows:

- **Extension index on table BSID**
  Accelerates payment advice processing when searching for open items:
  - Index fields: DP_PAYTREQ
  - Index type: Nonunique
- **Extension index on table BSAD**
  Speeds reconciliation and historical payment searches:
  - Index fields: DP_PAYTREQ
  - Index type: Nonunique

These indexes are essential for system performance. Payment advice processing executes daily, searching thousands of line items to match PSP settlement data with SAP accounting documents. Without proper indexing, these jobs can consume excessive runtime and delay financial close processes.

The SAP Note creates several configuration tables that store digital payments add-on settings:

- FAR_DP_T100_MAP: Maps PSP error codes to SAP message numbers in message class DIGITAL_PAYMENTS
- T042IPSP: PSP bank account mapping; links virtual banks to merchant accounts
- FAR_DP_REFUND_PC: Refund processing configuration (used by SAP Note 3212765)
- FAR_DP_PT_CCINS: Payment type to payment card type mapping

Standard SAP configuration occurs through maintenance views accessed via Transaction SM30 or through Transaction SPRO (IMG) paths. The note creates several views, as follows:

- FAR_DP_PSP_BANK: Maintain PSP bank account relationships
- FARV_DP_CHRG_TAX: Configure tax codes for payment processing charges
- V_T042IPSP: Account determination for external payments (created by SAP Note 3000524 but depends on the table from this SAP Note)

The SAP Note also establishes message class DIGITAL_PAYMENTS, containing over 160 message numbers. These messages provide standardized error text for payment-processing failures, API communication errors, configuration issues, and business rule violations. Applications display these messages to users and write them to application logs for troubleshooting.

Search helps provide F4 value help functionality in SAP screens. The following search helps integrate with standard SAP dialog processing, enabling users to select valid values rather than memorizing codes:

- F4_BL_PSP_BANK_KEY: Search help for PSP bank selection
- F4_DP_PMT: Payment method search help
- F4_DP_PSP: PSP selection search help

The SAP Note creates two critical administrative reports:

1. **Report FAR_DP_UPDATE_PSP_PAYMENT_METH**
   Synchronizes payment methods from digital payments add-on to SAP ERP as follows:
   - Functionality: Calls add-on API to retrieve current PSP list and payment method catalog; updates local SAP ERP configuration tables to match add-on settings
   - Execution: Run after PSP configuration changes in add-on
   - Frequency: Execute manually after adding new PSPs or activating new payment methods
2. **Report RFDP_ADVICE_V2**
   Processes digital payment advice files as follows:
   - Functionality: Retrieves payment settlement data from add-on; transforms data to CAMT.053 bank statement format; posts settlement entries in SAP ERP Financials (FI)
   - Execution: Background job (daily run recommended)
   - Integration: Works with electronic bank statement processing (Transaction FF_5)

The SAP Note includes XSLT transformation FAR_DP_ADVICE_TO_CAMT053, which converts the digital payments add-on payment advice JSON format to ISO 20022 CAMT.053 XML format. This transformation enables payment advice processing through standard SAP electronic bank statement functionality without custom parsing logic.

Two function groups provide the technical framework for configuration and PSP management:

- FAR_DP_CONF (Digital Payments Configuration): Contains maintenance view function modules, provides configuration dialog processing, and handles validation logic for configuration data.
- APAR_EBPP_T042IP (Electronic Bill Payment Processing): Manages PSP account determination, handles merchant ID resolution, and provides utility functions for payment processing.

Unlike correction instruction SAP Notes that Transaction SNOTE applies automatically, SAP Note 2999922 requires some manual implementation steps, as follows:

1. **Report NOTE_2999922_II**
   Using Transaction SE38 (ABAP Editor), create new report NOTE_2999922_II in package FBCC. Copy the complete source code from attached file *note_2999922.txt* into the report. After creating the report, activate it to check for syntax errors and execute the report with F8 to begin object creation. The report runs in three steps:
   - Creates domains, data elements, and basic structures
   - Creates tables, views, and search helps
   - Generates maintenance dialogs for configuration views

   Monitor the execution log carefully. The report displays a creation status for each object. A green status indicates successful creation. Yellow warnings about existing objects are acceptable if you are rerunning the report. Red errors indicate problems that require investigation.
2. **Create append structures manually**
   The automated report creates most objects, but append structures for tables BSEG, BSID, and BSAD require manual creation to avoid modification conflicts in systems with existing append structures.
3. **Create extension indexes**
   To prevent full table scans during payment advice processing, create extension indexes on tables BSID and BSAD.
4. **Extend BAPI structures (if using BAPI posting)**
   If your implementation uses BAPIs for programmatic document posting (common in interface developments), extend the following structures:
   - Create append structure DIGITAL_PAYMENT_ACCIT for structure ACCIT with components BSEG_DP_TR (DP_TRANSID), DP_PAYTYPE (DP_PAYMENT_TYPE), and DP_PAYTREQ (DP_PAYTREQ).

- Add the following components directly to structures ACCBAPIFD5 and BAPIACAR09 (these are not tables, so append structures aren't used): DP_TRANSID (type DP_TRANSID), DP_PAYTYPE (type DP_PAYMENT_TYPE), DP_PAYTREQ (type DP_PAYTREQ).
- Create append structure DIGITAL_PAYMENT_GL09 for structure BAPIACGL09 with components DP_TRANSID, DP_PAYTYPE, DP_PAYTREQ, PAYS_PROV (COM_WEC_PAYMENT_SRV_PROVIDER), and PAYS_TRAN (FPS_TRANSACTION).

After completing all manual activities, verify successful implementation as follows:

1. Check if the following repository objects are successfully created: function groups FAR_DP_CONF and APAR_EBPP_T042IP, tables FAR_DP_T100_MAP and T042IPSP, and reports FAR_DP_UPDATE_PSP_PAYMENT_METH and RFDP_ADVICE_V2.
2. Validate append structures as follows:
   - Execute Transaction SE11, enter table BSEG, click **Display**, and select the **Append Structures** button. Verify that **DIGITAL_PAYMENTS_BSEG** appears in the list with status **Active** (see Figure 5.3).

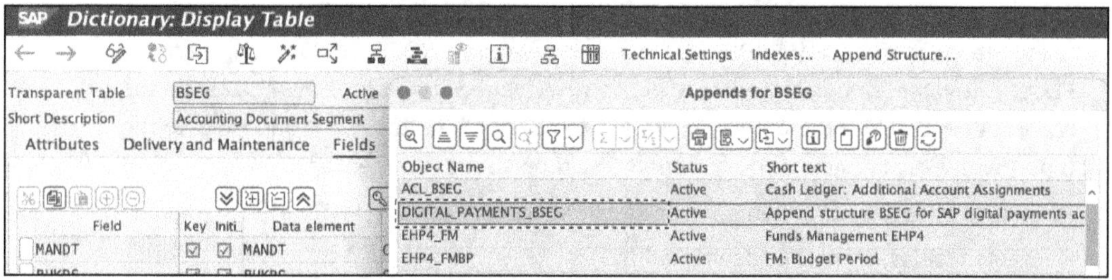

*Figure 5.3: Table BSEG: Append Structure*

   - Open the append structure and confirm that all three fields (**DP_TRANSID**, **DP_PAYTYPE**, and **DP_PAYTREQ**) are present (see Figure 5.4).

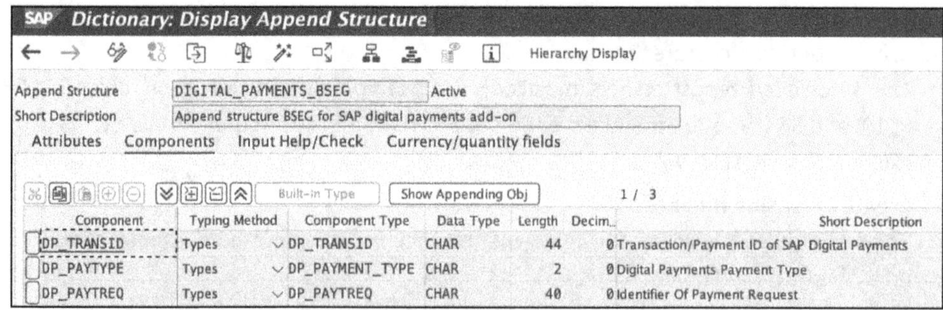

*Figure 5.4: Table BSEG: Append Structure Fields*

   - Repeat for tables BSID and BSAD.
3. Test the configuration views as follows: Run Transaction SM30, enter view name "FARV_DP_CHRG_TAX", and click **Display**. The system should open the maintenance view without

errors (see Figure 5.5). Although no configuration data exists yet, successful view access confirms proper object creation.

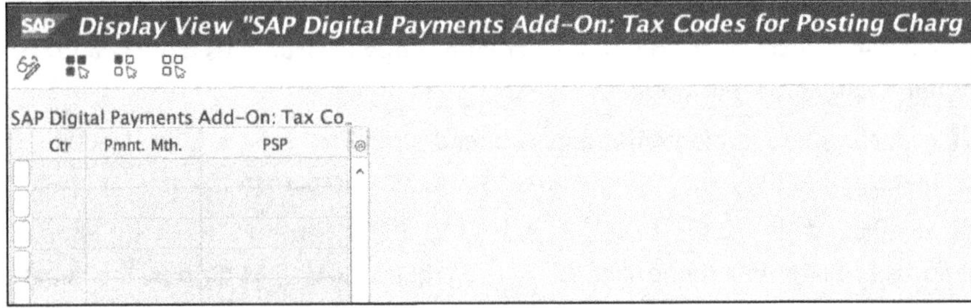


*Figure 5.5: View FARV_DP_CHRG_TAX*

4. Check the extension indexes as follows: Run Transaction SE11, enter table BSID, click **Display**, select the **Indexes** button, and verify that an extension index containing field **DP_PAYTREQ** exists with status **Active** (see Figure 5.6). Repeat for table BSAD.

Status Active Saved Package EBPP_APAR
Index BSID~ZDP exists in database system HDB
Non-unique index
Index on all database systems
For selected database systems
No database index
Unique Index (database Index required)
Table Fields
Index Flds

| Field name | Short Description | DTy. | Length |
|---|---|---|---|
| DP_PAYTREQ | Identifier Of Payment Request | CHAR | 40 |

*Figure 5.6: Table BSID: Extension Index*

Some common implementation issues are as follows:

- **Report NOTE_2999922_II fails during execution**
  The report terminates with a dump or shows numerous red error messages. Causes and resolutions may include the following:
  - Insufficient authorization: The user executing the report needs authorization object S_DEVELOP with activity 01 (create) for object types TABL, DTEL, DOMA, FUGR, and PROG. Request authorization from your security team.
  - Package FBCC doesn't exist: Create package FBCC if it's missing or modify the report to use an alternate customer package (in which case, change all FBCC references to your package name).
  - Existing objects conflict: If objects already exist from a previous attempt, delete them first or modify the report to skip existing objects.

  - Transport request issues: Ensure you have an open, modifiable transport request assigned.

- **Append structure creation fails**
  An **Object already exists** error occurs when creating append structures. Causes and resolutions may include the following:
  - Previous incomplete implementation: If append structures exist but are inactive or incomplete, delete them first using Transaction SE11 and recreate.
  - Name conflict: If the name `DIGITAL_PAYMENTS_BSEG` name is taken by another append, use an alternate name like `DIGITAL_PAYMENTS_BSEG_01` (update the dependent SAP Notes accordingly).
  - Table is locked: Another user may have table `BSEG` open for editing. Wait for lock release or ask the user to release the lock.

- **Extension index creation fails on tables `BSID` and `BSAD`**
  The index cannot be activated or shows errors during creation. Causes and resolutions may include the following:
  - Index number conflict: SAP automatically assigns index numbers. If automatic assignment fails, manually specify the index number (e.g., ZZ1 or ZZ2).
  - Field `DP_PAYTREQ` doesn't exist: Verify the append structure was successfully activated first.
  - Database capacity: Large tables `BSID` and `BSAD` may require substantial time for index building. Schedule this during off-hours.

For documentation and audit purposes, Table 5.4 provides a comprehensive list of key objects created by SAP Note 2999922, organized by object type.

| Object Type | Object Name | Description | Usage |
|---|---|---|---|
| Data Elements | `DP_TRANSID` | Digital Payment Transaction ID | Primary transaction identifier |
| Data Elements | `DP_PAYMENT_TYPE` | Payment Type | Distinguishes CC/EP |
| Data Elements | `DP_PAYTREQ` | Payment Request ID | Request tracking |
| Data Elements | `DP_TOKEN` | Payment Card Token | Tokenized card reference |
| Data Elements | `DP_PAYMENT_SERVICE_PROVIDER` | PSP Code | Identifies PSP (e.g., STRIPE_V2) |
| Data Elements | `DP_PAYID` | PSP Payment ID | PSP-specific identifier |
| Data Elements | `DP_MERCHANT_ALIAS` | Merchant Alias | User-friendly merchant name |

*Table 5.4: Key DDIC Objects Created by SAP Note 2999922*

| Object Type | Object Name | Description | Usage |
|---|---|---|---|
| Tables | FAR_DP_T100_MAP | Error Message Mapping | Maps PSP errors to SAP messages |
| Tables | T042IPSP | PSP Bank Mapping | Links PSPs to virtual banks |
| Tables | FAR_DP_REFUND_PC | Refund Configuration | Refund general ledger account setup |
| Tables | FAR_DP_PT_CCINS | Payment Type Mapping | Maps payment types to card types |
| Views | FAR_DP_PSP_BANK | PSP Bank Maintenance | Configure PSP accounts |
| Views | FARV_DP_CHRG_TAX | Charges Tax Configuration | Tax codes for processing fees |
| Function Groups | FAR_DP_CONF | Configuration Functions | Configuration dialog processing |
| Function Groups | APAR_EBPP_T042IP | PSP Account Determination | Merchant resolution logic |
| Reports | NOTE_2999922_II | Object Creation Report | Self-installing DDIC builder |
| Reports | FAR_DP_UPDATE_PSP_PAYMENT_METH | PSP Sync Report | Updates PSP list from add-on |
| Reports | RFDP_ADVICE_V2 | Payment Advice Processing | Processes settlement files |
| Message Class | DIGITAL_PAYMENTS | Payment Error Messages | Over 160 error/info messages |
| Search Helps | F4_DP_PMT | Payment Method Search | F4 help for payment methods |
| Search Helps | F4_DP_PSP | PSP Search | F4 help for PSP selection |
| Search Helps | F4_BL_PSP_BANK_KEY | PSP Bank Search | F4 help for bank keys |
| XSLT | FAR_DP_ADVICE_TO_CAMT053 | Advice Transformation | Converts JSON to CAMT.053 XML |

*Table 5.4: Key DDIC Objects Created by SAP Note 2999922 (Cont.)*

### SAP Note 3000524

SAP Note 3000524 delivers the core ABAP programming logic required for the digital payments add-on integration with SAP ERP. Whereas SAP Note 2999922 created the data foundation (tables, data elements, structures), this SAP Note implements the business logic that processes payments. This SAP Note provides function modules, classes, interface definitions, and enhancements to standard SAP programs that enable communication with the add-on, transformation of payment data, and integration with the Financial Accounting (FI) and Sales and Distribution (SD) processes. The programming objects created by this note form the technical backbone of every payment transaction processed through the digital payments Add-on.

This SAP Note does the following:

- **Extends core data structures**
  The SAP Note extends critical SAP structures to accommodate digital payments add-on data elements, as follows:
  - Structure CCDATA extensions: CCDATA is the primary structure used throughout SAP for payment card processing. The SAP Note appends seven new fields to this structure (see Figure 5.7). After adding fields to structure CCDATA, you must enter BKPF-WAERS as the currency reference field for field CCDATA-PAYMENT_AMOUNT. This ensures that currency-dependent formatting and validation occurs correctly during payment processing.

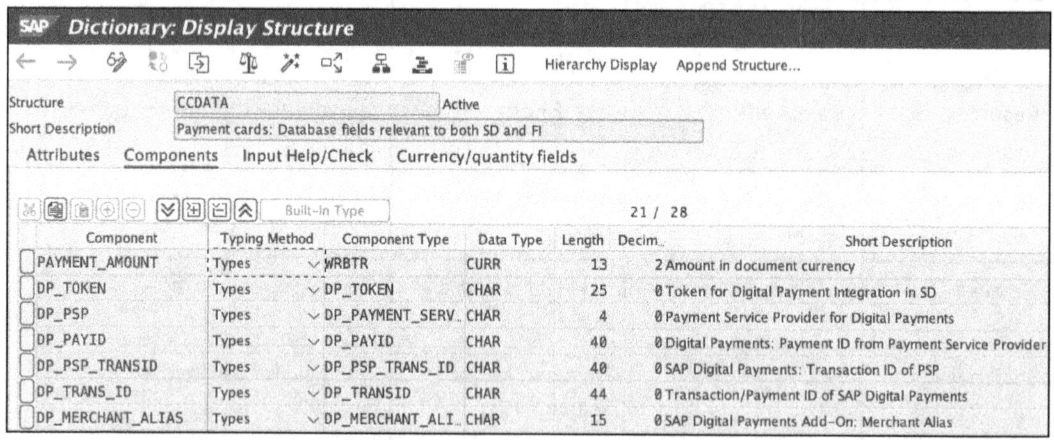


*Figure 5.7: Structure CCDATA Extensions*

  - Table BSEGC extensions: This SAP Note extends this table with two additional fields, DP_ADVREF and DP_RTEXT_LONG (see Figure 5.8). These fields support payment reconciliation by storing detailed PSP response information directly in accounting line items.

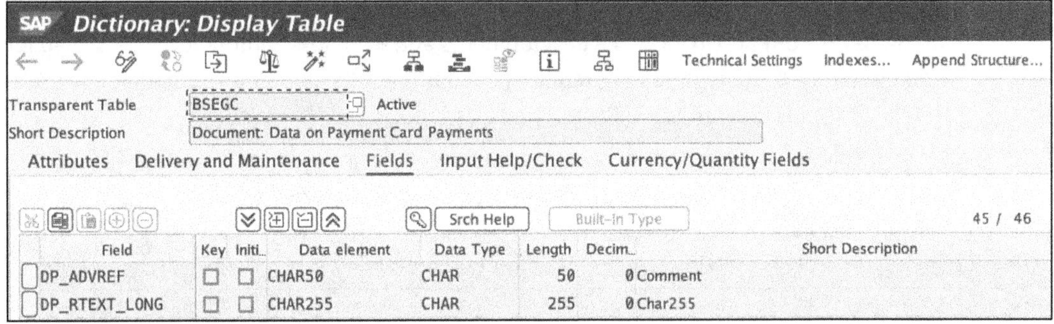


*Figure 5.8: Table BSEGC Extensions*

- Structure CCSET extensions: CCSET defines the payment card settlement data structure. The SAP Note adds two fields:
  - DP_TOKEN (inserted after existing field CCNUM as shown in Figure 5.9; carries tokenized card reference through settlement). The field positioning matters because existing programs read CCSET sequentially. Inserting DP_TOKEN after CCNUM maintains compatibility with custom code that may rely on field order.

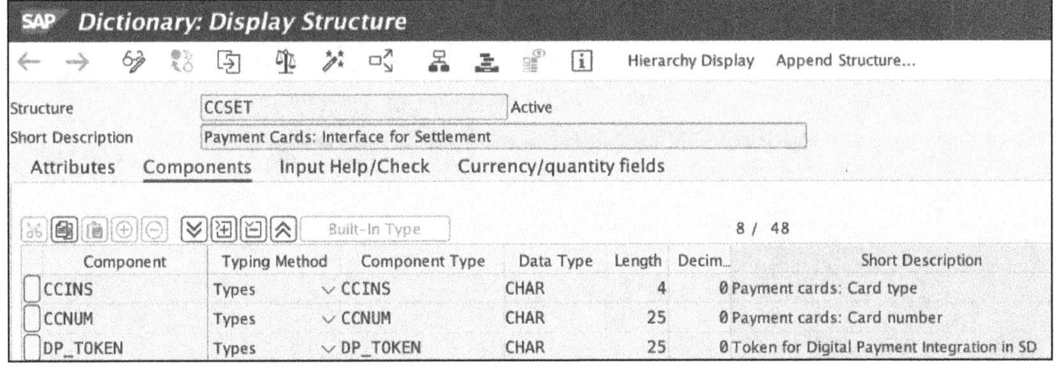


*Figure 5.9: Field DP_TOKEN in Structure CCSET*

  - DP_PSP (identifies PSP used for settlement, as shown in Figure 5.10).

Dictionary: Display Structure

Hierarchy Display | Append Structure...

Structure: CCSET Active

Short Description: Payment Cards: Interface for Settlement

Attributes | Components | Input Help/Check | Currency/quantity fields

Built-In Type 47 / 48

| Component | Typing Method | Component Type | Data Type | Length | Decim... | Short Description |
|---|---|---|---|---|---|---|
| DP_PSP | Types | DP_PAYMENT_SERV... | CHAR | 4 | 0 | Payment Service Provider for Digital Payments |

*Figure 5.10: Field DP_PSP in Structure CCSET*

- **BAPI structure extensions for programmatic posting**
  BAPIs are common in interface developments and workflow automation. BAPIs used for automated document posting require payment data fields in BAPI structures. This SAP Note extends four critical BAPI structures, as follows:
  - Structure `ACCIT` (BAPI Accounting Item): Creates new append structure `DIGITAL_PAYMENT_ACCIT` with three new fields, as shown in Figure 5.11.

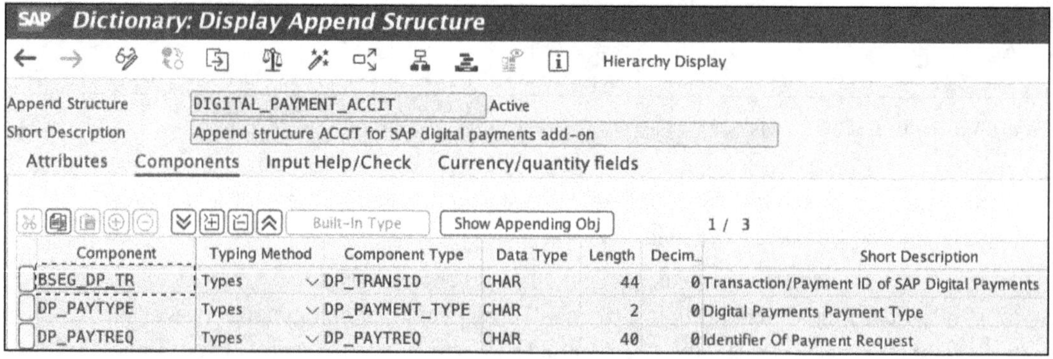


| Component | Typing Method | Component Type | Data Type | Length | Decim... | Short Description |
|---|---|---|---|---|---|---|
| BSEG_DP_TR | Types | DP_TRANSID | CHAR | 44 | 0 | Transaction/Payment ID of SAP Digital Payments |
| DP_PAYTYPE | Types | DP_PAYMENT_TYPE | CHAR | 2 | 0 | Digital Payments Payment Type |
| DP_PAYTREQ | Types | DP_PAYTREQ | CHAR | 40 | 0 | Identifier Of Payment Request |

*Figure 5.11: Append Structure ACCIT Fields*

  - Structure `ACCBAPIFD5`: You cannot use this append structure as it is not a transparent table. Add components `DP_TRANSID`, `DP_PAYTYPE`, and `DP_PAYTREQ` directly (see Figure 5.12).

Dictionary: Display Structure

Hierarchy Display Append Structure...

Structure ACCBAPIFD5 Active

Short Description ACC BAPIs: Surplus Amount of Accounting Document Line Items

Attributes Components Input Help/Check Currency/quantity fields

Built-In Type 233 / 235

| Component | Typing Method | Component Type | Data Type | Length | Decim... | Short Description |
|---|---|---|---|---|---|---|
| DP_TRANSID | Types | DP_TRANSID | CHAR | 44 | 0 | Transaction/Payment ID of SAP Digital Payments |
| DP_PAYTYPE | Types | DP_PAYMENT_TYPE | CHAR | 2 | 0 | Digital Payments Payment Type |
| DP_PAYTREQ | Types | DP_PAYTREQ | CHAR | 40 | 0 | Identifier Of Payment Request |

*Figure 5.12: Structure ACCBAPIFD5: Three New Components Added*

  - Structure `BAPIACAR09`: Add components `DP_TRANSID`, `DP_PAYTYPE`, and `DP_PAYTREQ` directly (see Figure 5.13).
  - Structure `BAPIACGL09`: Create append structure `DIGITAL_PAYMENT_GL09`, containing five data elements, as shown in Figure 5.14.

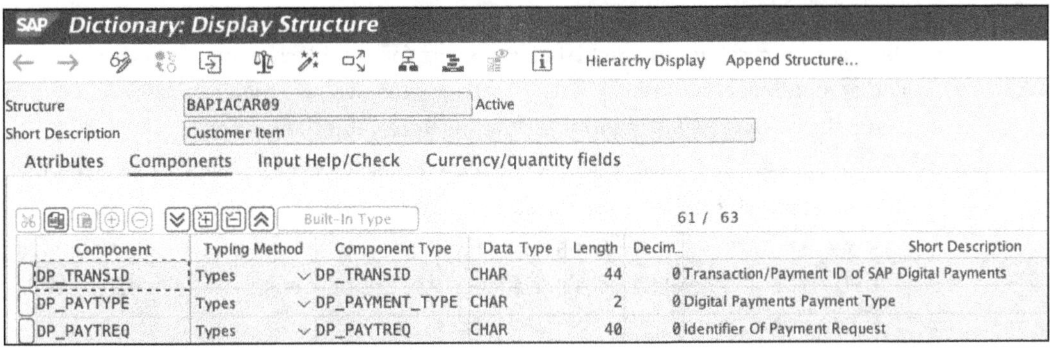

Dictionary: Display Structure

Hierarchy Display Append Structure...

Structure: BAPIACAR09 Active
Short Description: Customer Item

Attributes Components Input Help/Check Currency/quantity fields

Built-In Type 61 / 63

| Component | Typing Method | Component Type | Data Type | Length | Decim... | Short Description |
|---|---|---|---|---|---|---|
| DP_TRANSID | Types | DP_TRANSID | CHAR | 44 | 0 | Transaction/Payment ID of SAP Digital Payments |
| DP_PAYTYPE | Types | DP_PAYMENT_TYPE | CHAR | 2 | 0 | Digital Payments Payment Type |
| DP_PAYTREQ | Types | DP_PAYTREQ | CHAR | 40 | 0 | Identifier Of Payment Request |

*Figure 5.13: Structure BAPIACAR09: Three New Components Added*

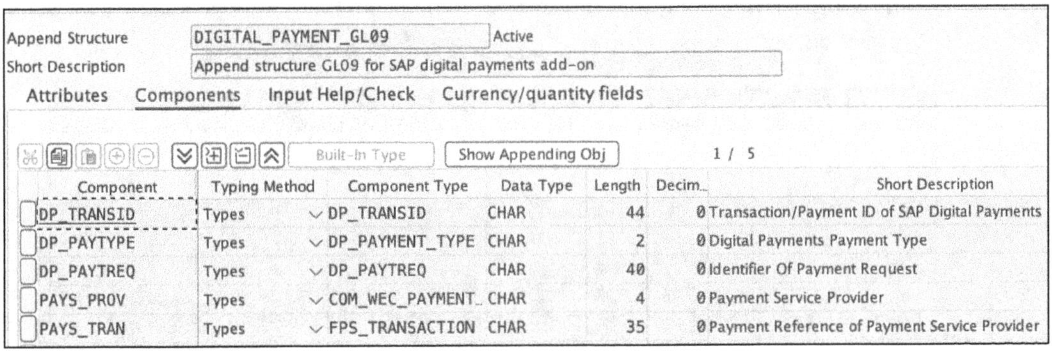

Append Structure: DIGITAL_PAYMENT_GL09 Active
Short Description: Append structure GL09 for SAP digital payments add-on

Attributes Components Input Help/Check Currency/quantity fields

Built-In Type Show Appending Obj 1 / 5

| Component | Typing Method | Component Type | Data Type | Length | Decim... | Short Description |
|---|---|---|---|---|---|---|
| DP_TRANSID | Types | DP_TRANSID | CHAR | 44 | 0 | Transaction/Payment ID of SAP Digital Payments |
| DP_PAYTYPE | Types | DP_PAYMENT_TYPE | CHAR | 2 | 0 | Digital Payments Payment Type |
| DP_PAYTREQ | Types | DP_PAYTREQ | CHAR | 40 | 0 | Identifier Of Payment Request |
| PAYS_PROV | Types | COM_WEC_PAYMENT... | CHAR | 4 | 0 | Payment Service Provider |
| PAYS_TRAN | Types | FPS_TRANSACTION | CHAR | 35 | 0 | Payment Reference of Payment Service Provider |

*Figure 5.14: Structure BAPIACGL09: Five New Components Added*

- **Modifies standard SAP programs for digital payments add-on**
  This SAP Note includes correction instructions that modify three standard SAP programs to integrate digital payment advice processing with electronic bank statement functionality. These modifications are applied automatically through Transaction SNOTE, but understanding the changes helps troubleshoot issues:
  - Main program RFEBKA00 (Electronic Bank Statement Processing): Primary program for processing electronic bank statements from banks. This SAP Note adds a new hidden parameter, `digipay`, to flag when processing digital payment advice files (instead of regular bank statements).
  - Include program RFEBKA03 (Electronic Bank Statement Processing—Global Data Declarations): Contains global variable declarations used by bank statement processing. This SAP Note adds variable `digital_payment` to store the digital payment processing flag.
  - Program RFEKAXML (Electronic Bank Statement Processing—XML Format Handler): Processes bank statements in XML format (including CAMT.053 format). This SAP Note modifies the file upload logic to route digital payment advice to specialized processing and adds new form routine `upload_digital_payment_file`, which calls the payment advice transformation class.

- **Application log configuration**
  Payment processing requires detailed logging for troubleshooting, audit compliance, and operational monitoring. This SAP Note creates two new application log subobjects under the existing FIAR (Financial Accounting Receivables) log object:
  - `FIAR_DP_ADVICE`: Logs all payment advice retrieval and processing activities (API call results, transformation errors, posting success/failure, reconciliation issues)
  - `FIAR_DP_SYNC_PSPS`: Logs PSP synchronization when executing report FAR_DP_UPDATE_PSP_PAYMENT_METH (retrieved PSP list, payment method updates, configuration changes)

  These logs are accessible through Transaction SLG1 (Application Log Display). Filter by object FIAR to view payment-specific entries (see Figure 5.15). During troubleshooting, these logs provide detailed technical information about API communication, data transformation, and processing errors.

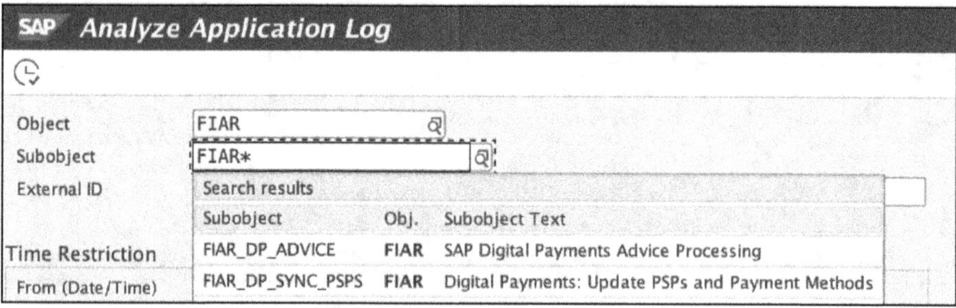


*Figure 5.15: Transaction SLG1: Log Subobjects Newly Added*

- **BAdI implementation for bank statement mapping**
  This note creates a Business Add-In (BAdI) implementation that enables the electronic bank statement processor to recognize and handle digital payment advice data. This BAdI transforms payment advice from the digital payments add-on format into CAMT.053 bank statement format. Run Transaction SE19 to display the technical details (see Figure 5.16).

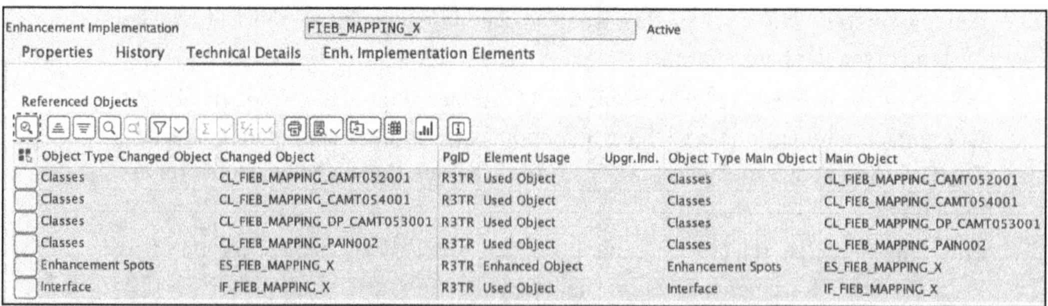


*Figure 5.16: BADI FIEB_MAPPING_X Definition*

- **Format mapping configuration**
  Electronic bank statement processing uses format mappings to determine which transformation to apply for different file formats. This SAP Note requires adding a format-mapping entry for digital payment advice. This configuration tells the bank statement processor that files with format DP_CAMT.053.001.02 should be transformed using the standard CAMT.053 transformation (FIEB_CAMT053_V2_TO_FEB), with the digital payments BAdI implementation handling digital payments-specific extensions.
- **Postimplementation program execution**
  After applying all correction instructions and completing manual activities, execute two standard SAP programs to regenerate derived objects:
  - Program SAPFACCG: Regenerates function group FACI, which contains accounting document posting logic and incorporates new fields (`BSEG_DP_TR`, `DP_PAYTREQ`, `DP_PAYTYPE`) into posting structures
  - Program RGZZGLUX: Regenerates include programs for general ledger account data structures and ensures new table `BSEG` fields are accessible in general ledger reporting and display transactions.
  - Verification after program execution: Run Transaction SE80 (Object Navigator), select **Function Group** from the dropdown, enter "FACI", press `Enter`, expand the function group, open include `LFACIGEN`, and verify that fields `BSEG_DP_TR` (see Figure 5.17), `DP_PAYTREQ`, and `DP_PAYTYPE` (see Figure 5.18) appear within data structure `P_ACC`.

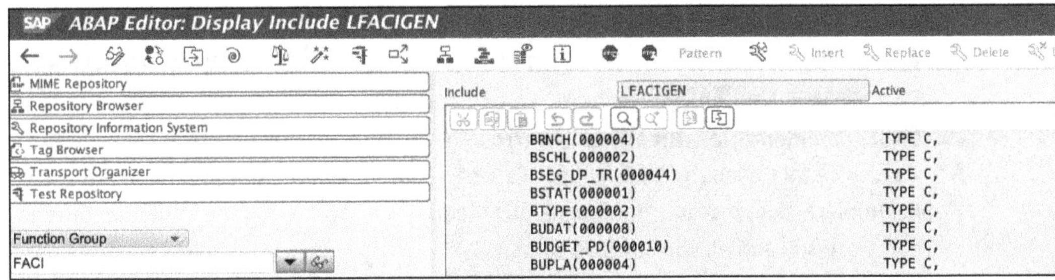


*Figure 5.17: Data Structure P_ACC: Field BSEG_DP_TR Verification*

ABAP Editor: Display Include LFACIGEN

MIME Repository
Repository Browser
Repository Information System
Tag Browser
Transport Organizer
Test Repository
Function Group
FACI

Include LFACIGEN Active

```
DP_PAYTREQ(000040)        TYPE C,
DP_PAYTYPE(000002)        TYPE C,
DP_PSP(000004)            TYPE C,
DP_PSP_TRANSID(000040)    TYPE C,
DP_TOKEN(000025)          TYPE C,
DP_TRANS_ID(000044)       TYPE C,
DTAMS(000001)             TYPE C,
DTAWS(000002)             TYPE C,
```

*Figure 5.18: Data Structure P_ACC: Fields DP_PAYTREQ and DP_PAYTYPE Verification*

- **Configuration views created**
  This SAP Note creates two maintenance views accessible through Transaction SM30, FARV_DP_CHRG_TAX and V_T042IPSP:
  - FARV_DP_CHRG_TAX (Tax Codes for Payment Processing Charges): When PSPs charge merchant fees (typically a percentage of the transaction amount), these fees must be posted to expense accounts with the appropriate tax treatment.
  - V_T042IPSP (Account Determination for External Payments): This maps PSP merchant accounts to virtual bank accounts in SAP by linking the merchant alias used in add-on configuration to a specific house bank and bank account in SAP.
- **Translation (optional)**
  This SAP Note creates texts for messages, selection texts in reports, and field labels—but in the English language only. If you require additional languages, you can manually translate these texts in Transaction SE63.

Some common implementation issues are as follows:

- **Correction instructions fail with Object Not Found errors**
  One cause is that SAP Note 2999922 potentially was not fully implemented. To resolve this problem, return to SAP Note 2999922, execute report NOTE_2999922_II completely, verify all objects were created, then retry SAP Note 3000524.
- **Program SAPFACCG fails during execution**
  The cause is that an append structure in table BSEG contains syntax errors or is not activated. To resolve this problem, execute Transaction SE11, display table BSEG, check the append DIGITAL_PAYMENTS_BSEG activation status, correct any errors, activate, then rerun program SAPFACCG.
- **BAdI implementation already exists error**
  To resolve this, execute Transaction SE19, search for the implementation, delete it if it's incomplete, then recreate it following the manual activity steps. If the implementation is complete and active, you can skip this step.
- **View FARV_DP_CHRG_TAX cannot be displayed**
  To resolve this, return to Transaction SE38, execute report NOTE_2999922_II, ensure all three steps complete successfully, then verify the view access.

#### SAP Note 3028044

SAP Note 3028044 creates database table FAR_DP_T100_MAP, which serves as the central repository for mapping PSP-specific error codes to SAP message numbers. This SAP Note must be implemented before SAP Note 3027433 because the latter populates this table with actual mapping data. Table FAR_DP_T100_MAP enables consistent error messaging across all PSPs. As an example, when Stripe returns error code card_declined, Cybersource returns DECLINE, or PayPal returns DECLINED, the mapping table translates all three to the same SAP message number, ensuring that users see consistent messaging regardless of which PSP processed their transaction.

The note contains ABAP source code for report NOTE_3028044. This report, when executed, creates database table `FAR_DP_T100_MAP` with the appropriate structure, technical settings, and transport layer assignment, as follows:

- **Table `FAR_DP_T100_MAP` structure**
  The created table has the field structure shown in Figure 5.19.

Dictionary: Display Table

Transparent Table: FAR_DP_T100_MAP — Active
Short Description: Mapping from SAP digital payments add-on to T100

Attributes | Delivery and Maintenance | Fields | Input Help/Check | Currency/Quantity Fields

| Field | Key | Initi.. | Data element | Data Type | Length | Decim.. | |
|---|---|---|---|---|---|---|---|
| DP_ERROR_CODE | ☑ | ☑ | CHAR120 | CHAR | 120 | 0 | char120 |
| ARBGB | ☐ | ☐ | ARBGB | CHAR | 20 | 0 | Application Area |
| MSGNR | ☐ | ☐ | MSGNR | CHAR | 3 | 0 | Message number |
| DP_ERR_ATTRIBUTENAME1 | ☐ | ☐ | CHAR200 | CHAR | 200 | 0 | Text field length 200 |
| DP_ERR_ATTRIBUTENAME2 | ☐ | ☐ | CHAR200 | CHAR | 200 | 0 | Text field length 200 |
| DP_ERR_ATTRIBUTENAME3 | ☐ | ☐ | CHAR200 | CHAR | 200 | 0 | Text field length 200 |
| DP_ERR_ATTRIBUTENAME4 | ☐ | ☐ | CHAR200 | CHAR | 200 | 0 | Text field length 200 |

*Figure 5.19: Table FAR_DP_T100_MAP Fields*

- **Technical table settings**
  The report configures the table with these technical characteristics: Buffering is set to **Buffering Activated**, with the **Fully Buffered** setting selected for **Buffering Type** (see Figure 5.20). Buffering significantly improves runtime performance. During payment processing, the same error codes are looked up repeatedly. Buffering eliminates database reads for frequently occurring error codes, reducing settlement processing time.

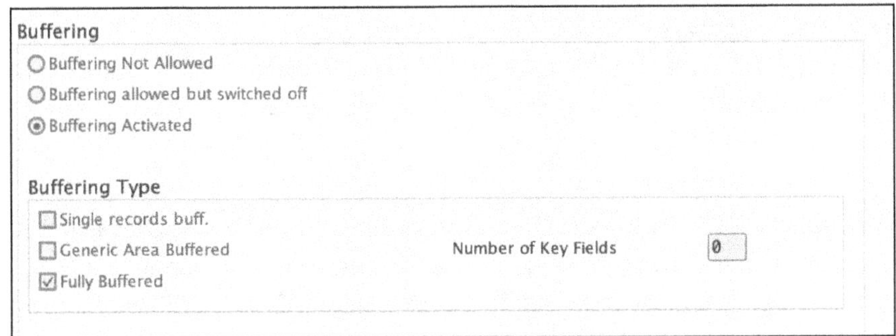


*Figure 5.20: Table FAR_DP_T100_MAP: Technical Characteristics*

Some common implementation issues are as follows:

- **Table `FAR_DP_T100_MAP` already exists from SAP Note 2999922**
  Newer versions of SAP Note 2999922 may include table `FAR_DP_T100_MAP` creation in report NOTE_2999922_II. This is not an error. Run Transaction SE11, display the table,

and verify the structure matches the specification in this SAP Note. If the structure is correct, then this error is not applicable.

- **Cannot view table contents in Transaction SE16N**
  The cause is missing table authorization. Request authorization object S_TABU_DIS with activity 03 (display) for authorization group FC39. Alternatively, use Transaction SE16 with table FAR_DP_T100_MAP, which may have different authorization requirements.
- **Table shows as Not Active in Transaction SE11**
  Report execution was interrupted before activation completed. Execute Transaction SE11, enter "FAR_DP_T100_MAP", click **Change**, make no modifications, and click **Activate**. The system should activate the table successfully.

### SAP Note 3027433

SAP Note 3027433 establishes the centralized error-message-mapping framework for the digital payments add-on integration. When PSPs return error codes (declined transactions, insufficient funds, expired cards, etc.), those PSP-specific codes must be translated into standardized SAP message numbers that can be displayed to users, written to application logs, and processed by exception-handling logic. This SAP Note populates the initial mapping table that connects PSP error codes to message class DIGITAL_PAYMENTS message numbers created by SAP Note 2999922.

Without proper error message mapping, payment failures appear as generic technical errors rather than actionable business messages. Users see cryptic PSP codes like card_declined or insufficient_funds instead of properly formatted SAP messages with meaningful text. This SAP Note ensures consistent, professional error messaging across all payment scenarios.

The SAP Note delivers two attached text files containing ABAP report source code that automates the entire message mapping setup:

- **Report SAP NOTE_3027433**
  Creates message text entries in table T100 (SAP's central message repository) and populates message class DIGITAL_PAYMENTS with actual message texts for all 160-plus message numbers.
- **Report Z_FILL_FAR_DP_T100_MAP**
  Populates table FAR_DP_T100_MAP with mappings between PSP error codes and SAP message numbers and links each PSP-specific error code to an appropriate DIGITAL_PAYMENTS message. You must execute this in each system (DEV, QA, PRD) as the table contents don't transport automatically.

The separation into two reports reflects different data management strategies. Report NOTE_3027433 operates on standard SAP table T100, which transports normally through correction instructions. Report Z_FILL_FAR_DP_T100_MAP populates configuration table FAR_DP_T100_MAP, which requires explicit content transport or reexecution per system.

Understanding the following runtime flow will help you troubleshoot mapping issues:

1. The consumer application (SAP ERP) calls the digital payments add-on API to authorize a payment.
2. The add-on communicates with the PSP (e.g., Stripe), which processes the authorization.
3. The PSP returns a decline with error code `insufficient_funds`.
4. The add-on returns this PSP error code to SAP ERP in the API response.
5. The SAP payment processing code looks up `insufficient_funds` in table `FAR_DP_T100_MAP`.
6. The mapping table returns message class `DIGITAL_PAYMENTS` and message number `101`.
7. SAP retrieves the message text from table `T100` for class `DIGITAL_PAYMENTS`, number `101`.
8. The user sees the formatted message **Insufficient funds on payment card** instead of the raw code, `insufficient_funds`.

This abstraction allows SAP to display localized, user-friendly messages regardless of which PSP processed the transaction. Different PSPs may use different error codes for the same condition, but the mapping ensures consistent messaging to users.

The verification steps are as follows:

1. **Check message class texts**
   Run Transaction SE91 (Message Maintenance), enter message class "DIGITAL_PAYMENTS", and click **Display**. The message overview should show 160-plus messages (see Figure 5.21) with English text populated.

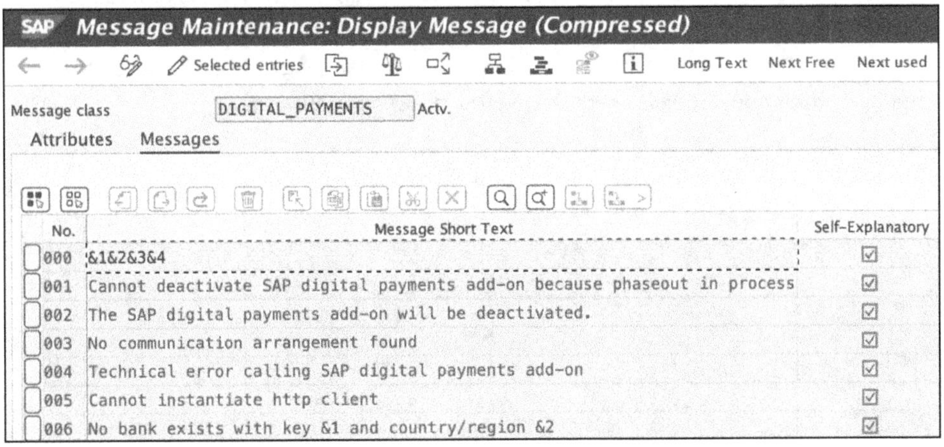

Message Maintenance: Display Message (Compressed)

Selected entries | Long Text | Next Free | Next used

Message class: DIGITAL_PAYMENTS Actv.

Attributes | Messages

| No. | Message Short Text | Self-Explanatory |
|---|---|---|
| 000 | &1&2&3&4 | ☑ |
| 001 | Cannot deactivate SAP digital payments add-on because phaseout in process | ☑ |
| 002 | The SAP digital payments add-on will be deactivated. | ☑ |
| 003 | No communication arrangement found | ☑ |
| 004 | Technical error calling SAP digital payments add-on | ☑ |
| 005 | Cannot instantiate http client | ☑ |
| 006 | No bank exists with key &1 and country/region &2 | ☑ |

*Figure 5.21: Message Class DIGITAL_PAYMENTS Display*

   Open a few messages to verify the text quality and completeness. If message texts are missing or show placeholder text like **&** or **Message not maintained**, report NOTE_3027433 did not execute successfully. Rerun the report and check the log for errors.

2. **Validate mapping table contents**
   Run Transaction SE16N (Table Browser), enter "FAR_DP_T100_MAP" to search for that table, and click **Execute** without filters to display all entries. The table should contain mapping entries linking PSP error codes to `DIGITAL_PAYMENTS` message numbers (see Figure 5.22).

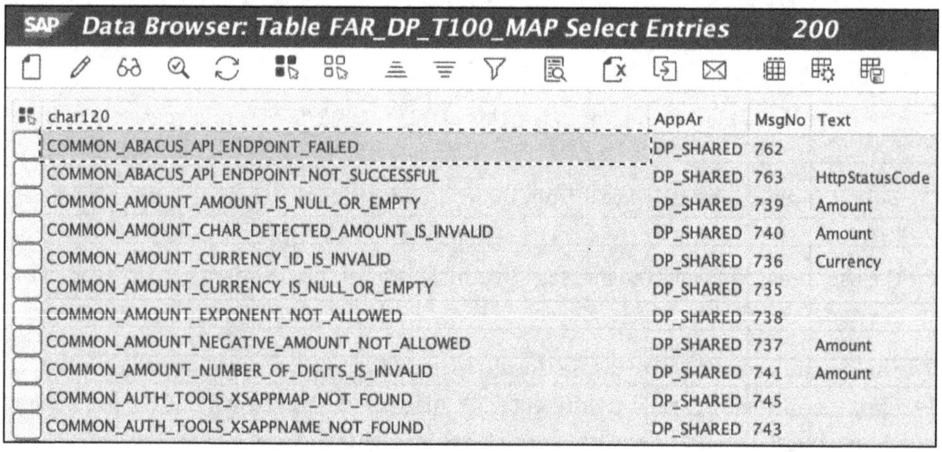

**Data Browser: Table FAR_DP_T100_MAP Select Entries 200**

| char120 | AppAr | MsgNo | Text |
|---|---|---|---|
| COMMON_ABACUS_API_ENDPOINT_FAILED | DP_SHARED | 762 | |
| COMMON_ABACUS_API_ENDPOINT_NOT_SUCCESSFUL | DP_SHARED | 763 | HttpStatusCode |
| COMMON_AMOUNT_AMOUNT_IS_NULL_OR_EMPTY | DP_SHARED | 739 | Amount |
| COMMON_AMOUNT_CHAR_DETECTED_AMOUNT_IS_INVALID | DP_SHARED | 740 | Amount |
| COMMON_AMOUNT_CURRENCY_ID_IS_INVALID | DP_SHARED | 736 | Currency |
| COMMON_AMOUNT_CURRENCY_IS_NULL_OR_EMPTY | DP_SHARED | 735 | |
| COMMON_AMOUNT_EXPONENT_NOT_ALLOWED | DP_SHARED | 738 | |
| COMMON_AMOUNT_NEGATIVE_AMOUNT_NOT_ALLOWED | DP_SHARED | 737 | Amount |
| COMMON_AMOUNT_NUMBER_OF_DIGITS_IS_INVALID | DP_SHARED | 741 | Amount |
| COMMON_AUTH_TOOLS_XSAPPMAP_NOT_FOUND | DP_SHARED | 745 | |
| COMMON_AUTH_TOOLS_XSAPPNAME_NOT_FOUND | DP_SHARED | 743 | |

*Figure 5.22: Table FAR_DP_T100_MAP: Entries*

   If the table is empty, report Z_FILL_FAR_DP_T100_MAP was not executed with parameter `P_UPDATE = 'X'`. Rerun with the checkbox selected.

3. **Test message display**
   Run Transaction SE91, enter "DIGITAL_PAYMENTS" to search for that message class, select **Messages** and enter "101" in the **Number** field, click **Display** (see Figure 5.23), and verify that the message text appears in your logon language. If you require other languages, then translation is needed (covered in SAP Note 3044268).

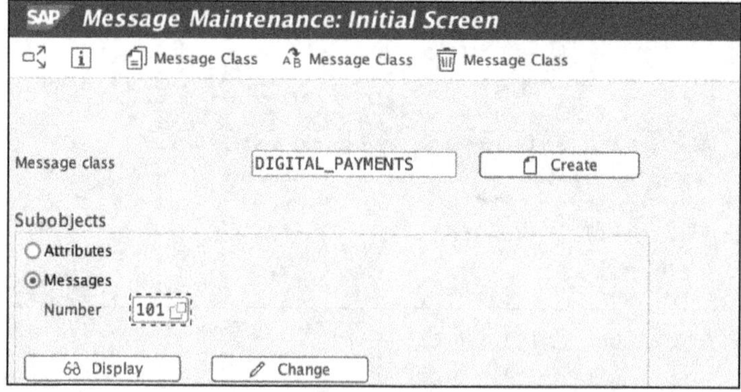


*Figure 5.23: Message Class DIGITAL_PAYMENTS: Transaction SE91 Display*

Table 5.5 lists common implementation issues that can occur when implementing SAP Note 3027433. You can use this table for troubleshooting.

| Issue | Cause and Resolution |
|---|---|
| Report NOTE_3027433 executes but creates no messages | ■ Message class already populated: If messages exist from a previous implementation, the report may skip creation. This is acceptable if messages are complete.<br>■ Authorization issue: The user needs authorization for object S_TABU_DIS with activity 02 (change) for authorization group SS (cross-application). Request authorization from your security team.<br>■ Table T100 is locked: Another process may be updating messages. Wait and retry. |
| Report Z_FILL_FAR_DP_T100_MAP shows **No Update** | Parameter P_UPDATE was not set to 'X'. Reexecute the report, ensure the **Update** checkbox is selected on the selection screen, then execute again. |
| Table FAR_DP_T100_MAP not found | SAP Note 2999922 was not properly implemented. Execute Transaction SE11 and search for table FAR_DP_T100_MAP. If it doesn't exist, return to SAP Note 2999922, execute report NOTE_2999922_II fully, verify the table creation, then retry this SAP Note. |
| Different error codes in production versus test | Report Z_FILL_FAR_DP_T100_MAP was not executed in the production system. Remember that the table contents don't transport automatically. Execute the report in production with P_UPDATE = 'X'. |

*Table 5.5: Common Implementation Issues for SAP Note 3027433*

**SAP Note 3028194**

SAP Note 3028194 delivers the complete business logic for settling (capturing) authorized payment card transactions in SAP ERP. *Settlement* is the critical step in which previously authorized payment amounts are charged to customer payment cards and funds transfer from customer accounts to merchant accounts. This SAP Note collectively handles the entire settlement workflow, including API communication with the digital payments add-on; construction of settlement requests with optional line item detail (Level 2/3 data); processing of settlement responses and updating accounting documents; handling of refund scenarios, including offsetting refunds; comprehensive error management with message mapping; and integration with Transaction FCC1 (Payment Cards: Settlement).

In payment card processing, authorization and settlement are distinct operations separated by time and business logic. *Authorization* occurs during sales order creation and reserves funds on the customer's card without transferring money. *Settlement* occurs later, typically when goods ship or services are delivered; it moves funds from the customer to the merchant. This separation allows businesses to authorize payment before incurring fulfillment costs while ensuring that payment capture only occurs for delivered goods or rendered services.

This SAP Note delivers the technical infrastructure for payment settlement through a combination of manually created ABAP objects and configuration entries. The SAP Note provides complete source code files for all objects but requires manual creation rather than automated deployment through correction instructions. Before creating any ABAP objects, you must extend two core structures to accommodate settlement-specific data fields (these are preimplementation manual activities). These extensions must be completed in every system (development, quality, and production) before importing the SAP Note:

- **Structure `CCDATA` extensions**
  Structure `CCDATA` carries payment card data throughout SAP's payment processing framework. The SAP Note requires adding three new fields: `DP_PAYID`, `DP_TRANS_ID`, and `DP_CHARGE_TRANS_ID` (see Figure 5.24).

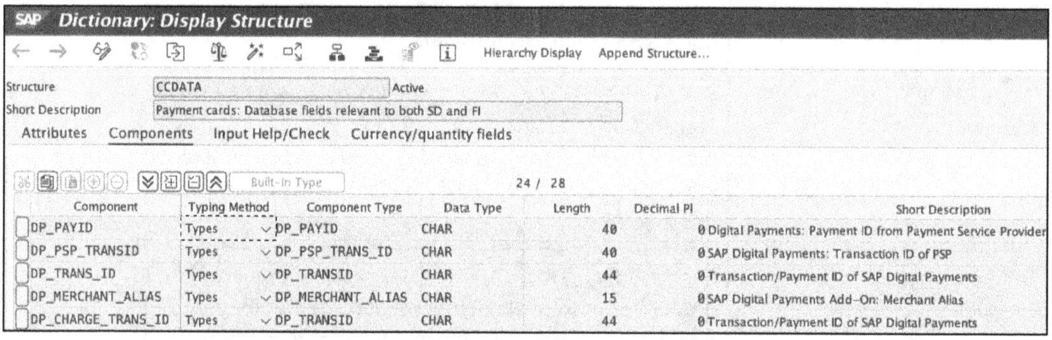

Dictionary: Display Structure

Hierarchy Display Append Structure...

Structure: CCDATA Active
Short Description: Payment cards: Database fields relevant to both SD and FI

Attributes Components Input Help/Check Currency/quantity fields

Built-In Type 24 / 28

| Component | Typing Method | Component Type | Data Type | Length | Decimal Pl | Short Description |
|---|---|---|---|---|---|---|
| DP_PAYID | Types | DP_PAYID | CHAR | 40 | 0 | Digital Payments: Payment ID from Payment Service Provider |
| DP_PSP_TRANSID | Types | DP_PSP_TRANS_ID | CHAR | 40 | 0 | SAP Digital Payments: Transaction ID of PSP |
| DP_TRANS_ID | Types | DP_TRANSID | CHAR | 44 | 0 | Transaction/Payment ID of SAP Digital Payments |
| DP_MERCHANT_ALIAS | Types | DP_MERCHANT_ALIAS | CHAR | 15 | 0 | SAP Digital Payments Add-On: Merchant Alias |
| DP_CHARGE_TRANS_ID | Types | DP_TRANSID | CHAR | 44 | 0 | Transaction/Payment ID of SAP Digital Payments |

*Figure 5.24: Structure CCDATA: Three New Fields Added*

- **Structure `CCSET` extensions**
  Structure `CCSET` defines settlement data passed between Transaction FCC1 and settlement function modules. The SAP Note requires adding two fields at specific positions:
  - In Transaction SE11, for structure `CCSET`, insert field `DP_TOKEN` immediately after existing field `CCNUM` (see Figure 5.25).
  - Append field `DP_CHARGE_TRANS_ID` at the end of the structure (see Figure 5.26).

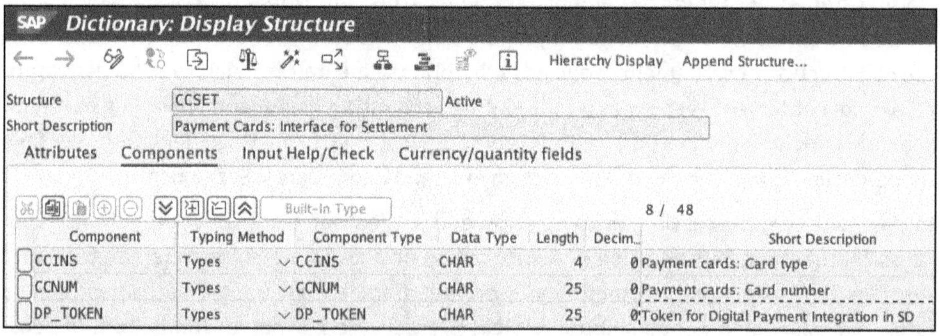

Dictionary: Display Structure

Hierarchy Display Append Structure...

Structure: CCSET Active
Short Description: Payment Cards: Interface for Settlement

Attributes Components Input Help/Check Currency/quantity fields

Built-In Type 8 / 48

| Component | Typing Method | Component Type | Data Type | Length | Decim... | Short Description |
|---|---|---|---|---|---|---|
| CCINS | Types | CCINS | CHAR | 4 | 0 | Payment cards: Card type |
| CCNUM | Types | CCNUM | CHAR | 25 | 0 | Payment cards: Card number |
| DP_TOKEN | Types | DP_TOKEN | CHAR | 25 | 0 | Token for Digital Payment Integration in SD |

*Figure 5.25: Structure CCSET: New Field DP_TOKEN*

**Dictionary: Display Structure**

Hierarchy Display Append Structure...

Structure: CCSET Active

Short Description: Payment Cards: Interface for Settlement

Attributes Components Input Help/Check Currency/quantity fields

Built-In Type 47 / 48

| Component | Typing Method | Component Type | Data Type | Length | Decim... | Short Description |
|---|---|---|---|---|---|---|
| DP_PSP | Types | DP_PAYMENT_SERV... | CHAR | 4 | 0 | Payment Service Provider for Digital Payments |
| DP_CHARGE_TRANS_ID | Types | DP_TRANSID | CHAR | 44 | 0 | Transaction/Payment ID of SAP Digital Payments |

*Figure 5.26: Structure CCSET: New Field DP_CHARGE_TRANS_ID*

Field positioning matters because the settlement processing code reads CCSET fields sequentially. Placing DP_TOKEN after CCNUM maintains the logical grouping of card-related fields and ensures compatibility with code that expects this order.

The SAP Note also provides 10 text files containing complete ABAP source code. Each file corresponds to one object (class, interface, or function module) that must be created manually. The objects must be created in dependency order to avoid activation errors from missing references, as follows:

- **Interfaces (create first)**
  Interfaces define contracts that implementing classes must fulfill. Create all three of the following interfaces before any classes to ensure that class implementations can reference interface types:
  - Interface ZIF_FAR_DP_TYPES (Type Definitions): Run Transaction SE24 (Class Builder) and create interface ZIF_FAR_DP_TYPES (see Figure 5.27). Copy the complete contents from source file *ZIF_FAR_DP_TYPES.txt* into the interface definition to define the necessary type structures. Activate the interface. The interface should activate without errors as it contains only type definitions with no executable logic.

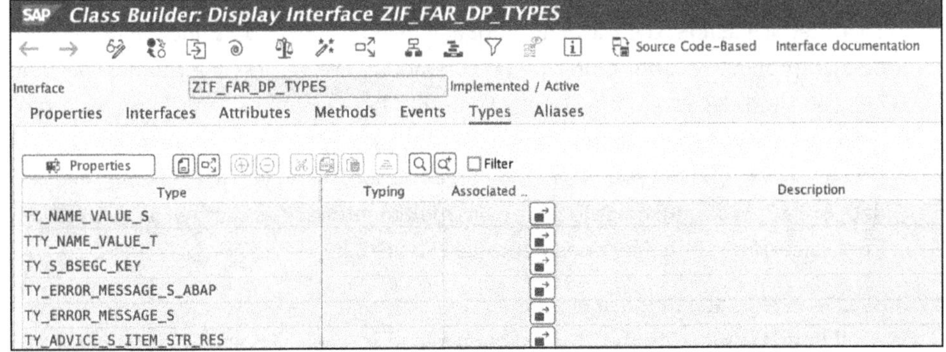


*Figure 5.27: Interface ZIF_FAR_DP_TYPES: Create Type Structures in Transaction SE24*

- Interface ZIF_FAR_DP_PROXY (HTTP Proxy Contract): Run Transaction SE24, create interface ZIF_FAR_DP_PROXY, copy the contents from source file *ZIF_FAR_DP_PROXY.txt* and activate the interface. This interface defines two method signatures: method SEND_POST_REQUEST and method SEND_GET_REQUEST (see Figure 5.28).

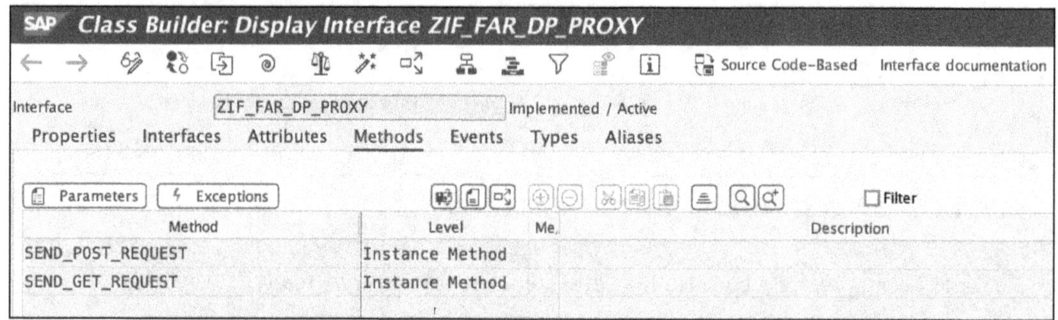


*Figure 5.28: Interface ZIF_FAR_DP_PROXY: Methods*

- Interface ZIF_FAR_DP_GET_L2L3_DATA (Level 2/3 Data Provider): Run Transaction SE24, create interface ZIF_FAR_DP_GET_L2L3_DATA, copy the contents from source file *ZIF_FAR_DP_GET_L2L3_DATA.txt*, and activate the interface (see Figure 5.29).

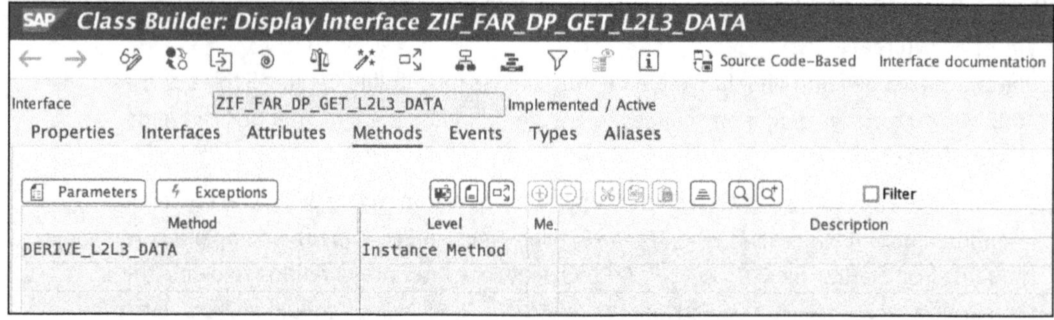


*Figure 5.29: Interface ZIF_FAR_DP_GET_L2L3_DATA Implemented and Active*

This interface defines data structures for enhanced payment data, including line item details (product, quantity, unit price, commodity codes), tax information per line item, and address information (ship-to and ship-from locations). This interface is provided for customer implementation. SAP does not deliver an implementing class because Level 2/3 data sources vary by organization. When you are sending Level 2/3 data to PSPs, you must create your own implementing class that retrieves this data from sales orders or billing documents (see Figure 5.30).

- **Classes (create in dependency order)**
  After all interfaces, create the six classes specified in Table 5.6, in the order shown to avoid missing dependency errors.

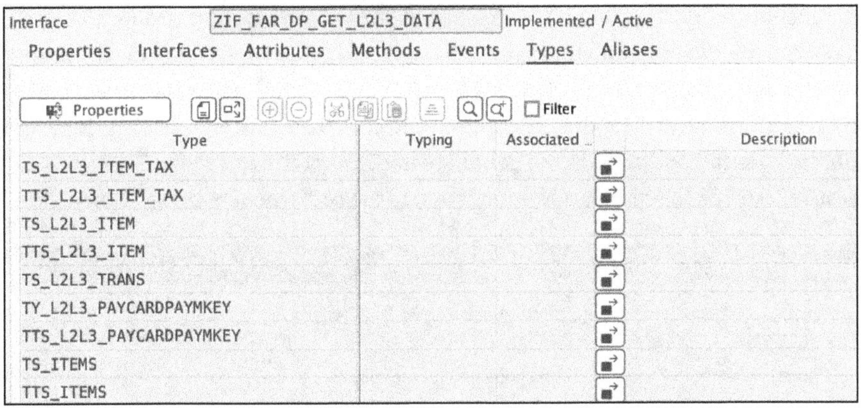

Figure 5.30: Interface ZIF_FAR_DP_GET_L2L3_DATA: Types

| Object Type | Object Name | Source File |
|---|---|---|
| Class | ZCL_FAR_DP_UTIL | *ZCL_FAR_DP_UTIL.txt* |
| Class | ZCL_FAR_DP_COUNTRY_UTIL | *ZCL_FAR_DP_COUNTRY_UTIL.txt* |
| Class | ZCL_FAR_DP_ABAP_JSON | *ZCL_FAR_DP_ABAP_JSON.txt* |
| Class | ZCL_FAR_DP_PROXY_FACTORY | *ZCL_FAR_DP_PROXY_FACTORY.txt* |
| Class | ZCL_FAR_DP_PROXY | *ZCL_FAR_DP_PROXY.txt* |
| Class | ZCL_FAR_DP_SETTLEMENT | *ZCL_FAR_DP_SETTLEMENT.txt* |

*Table 5.6: SAP Note 3028194: Class List*

- **Function module creation**
  Function modules reside within function groups. Create a function group (Transaction SE80) such as ZSAP_DP_SETTLMENT to contain the digital payments settlement function module (see Figure 5.31).

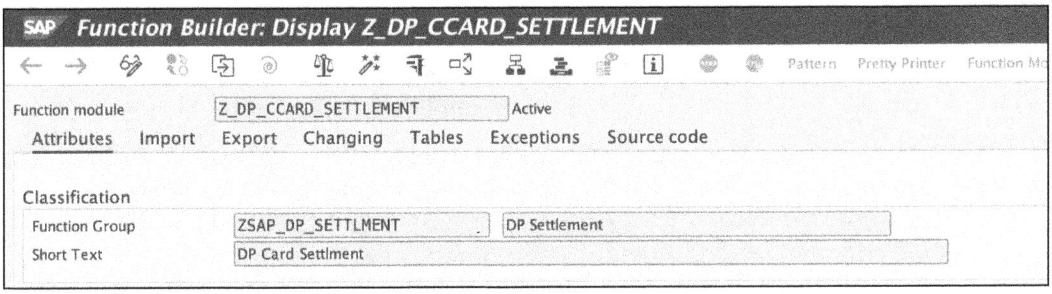

*Figure 5.31: Function Group ZSAP_DP_SETTLMENT*

Create function module Z_DP_CCARD_SETTLEMENT: Within function group ZSAP_DP_SETTLMENT, right-click the function group name in the navigation tree. Select **Create • Function Module**, then enter "Z_DP_CCARD_SETTLEMENT" for the **Name**, enter "Digital Payment Card Settlement" for the Description, and click **Save** (see Figure 5.32). Switch to the **Source Code** tab and copy in the complete implementation from file *Z_DP_CCARD_SETTLEMENT.txt*. The function module implements the settlement orchestration logic.

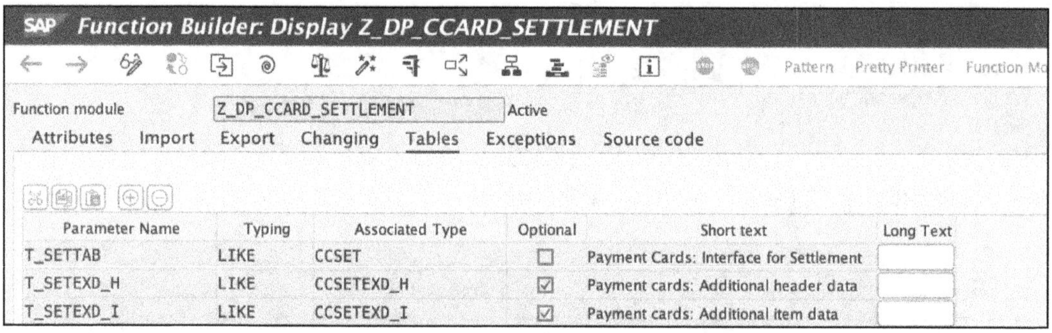

Function Builder: Display Z_DP_CCARD_SETTLEMENT

Pattern Pretty Printer Function Mo

Function module Z_DP_CCARD_SETTLEMENT Active

Attributes Import Export Changing Tables Exceptions Source code

| Parameter Name | Typing | Associated Type | Optional | Short text | Long Text |
|---|---|---|---|---|---|
| T_SETTAB | LIKE | CCSET | ☐ | Payment Cards: Interface for Settlement | |
| T_SETEXD_H | LIKE | CCSETEXD_H | ☑ | Payment cards: Additional header data | |
| T_SETEXD_I | LIKE | CCSETEXD_I | ☑ | Payment cards: Additional item data | |

*Figure 5.32: Function Module Z_DP_CCARD_SETTLEMENT Creation*

After all objects are created and activated, configure table TCCAA to connect Transaction FCC1 with the new settlement function module. Execute Transaction SM30, enter "TCCAA" to search for the table/view, and click **Maintain**. Create a new entry or modify an existing entry with the values shown in Table 5.7.

| Field Name | Technical Field | Description | Example Value | Notes |
|---|---|---|---|---|
| **Chart of Accounts** | KTOTPL | Chart of accounts for company settling payments | 1 | Match your company code's chart of accounts from table T001 |
| **Receivables Account** | HKONT_V | General ledger account for payment card receivables | 146500 | Must be reconciliation account, open item managed |
| **Clearing Account** | HKONT_N | General ledger clearing account for virtual bank | 113150 | Virtual bank clearing account per SAP Note 3003533 |
| **Settlement Function** | FNSET | Function module for settlement | Z_DP_CCARD_SETTLEMENT | The function module created in this SAP Note |

*Table 5.7: Table TCCAA Configuration for Digital Payments Settlement*

The following are some additional details about each field:

- **KTOTPL (Chart of Accounts)**
  Verify your company code's chart of accounts by executing Transaction SE16N on table T001, entering your company code, and noting the KTOPL field value. Multiple company codes may share one chart of accounts. If you have multiple charts of accounts in your system, create separate table TCCAA entries for each.
- **HKONT_V (Receivables Account)**
  This general ledger account must be configured as a reconciliation account in your chart of accounts and must allow open item management. Verify the configuration in Transaction FS00; display the account and check that the **Open Item Management** checkbox is selected and that **Reconciliation Account Type** is set to **D** (customer).
- **HKONT_N (Clearing Account)**
  This account must match your virtual bank clearing account configuration from SAP Note 3003533. The account number typically follows the pattern in which the main virtual bank account ends in 00 (e.g., 113100) and the clearing account ends in a specific suffix such as 50, 70, or 30 (e.g., 113150, 113170, 113130). Verify the account exists in Transaction FS00.
- **FNSET (Settlement Function Module)**
  Enter "Z_DP_CCARD_SETTLEMENT", the function module created in this SAP Note. This field tells Transaction FCC1 which function module to call when executing payment card settlement runs.

The verification steps are as follows:

1. Verify object existence and activation: Run Transaction SE80, navigate to the package where you had created the classes, select **Class Library**, and verify all six classes are present: ZCL_FAR_DP_ABAP_JSON, ZCL_FAR_DP_COUNTRY_UTIL, ZCL_FAR_DP_PROXY, ZCL_FAR_DP_PROXY_FACTORY, ZCL_FAR_DP_SETTLEMENT, and ZCL_FAR_DP_UTIL (see Figure 5.33).

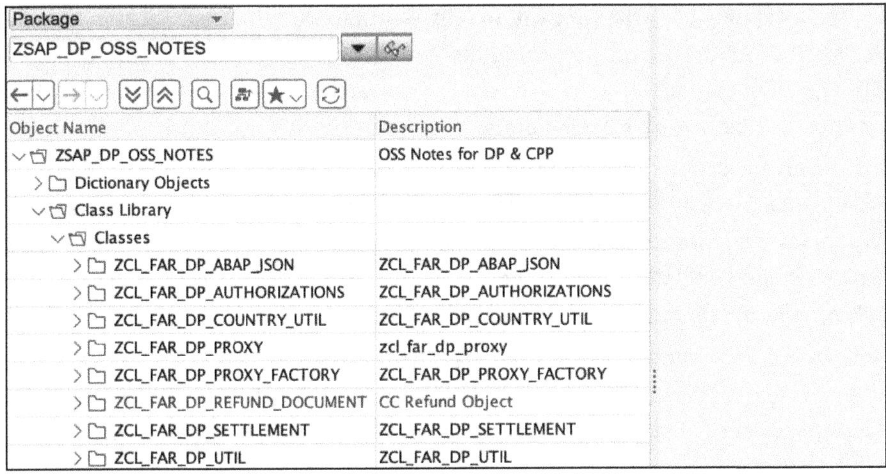


*Figure 5.33: SAP Note 3028194: New Classes Verification Step*

2. Select **Interfaces** and verify that ZIF_FAR_DP_TYPES, ZIF_FAR_DP_PROXY, and ZIF_FAR_DP_GET_L2L3_DATA exist (see Figure 5.34).

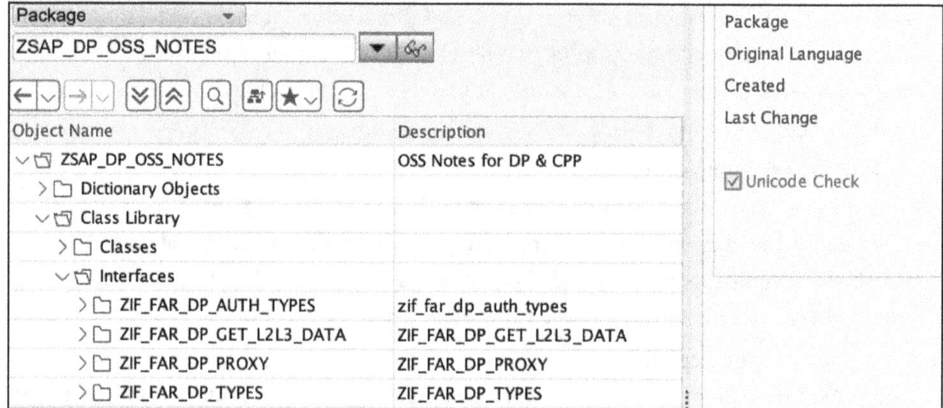


*Figure 5.34: SAP Note 3028194: New Interfaces Verification Step*

3. Configure table TCCAA: Run Transaction SM30, display table TCCAA, and verify an entry exists for your chart of accounts with FNSET = Z_DP_CCARD_SETTLEMENT. The receivables and clearing accounts should match your SAP Note 3003533 virtual bank configuration.
4. Extended program check: Run Transaction SE24, display class ZCL_FAR_DP_SETTLEMENT, and navigate to menu **Class • Check • Extended Program Check**. The system performs comprehensive syntax validation. Zero errors and warnings confirm a clean implementation. Address any errors; most indicate incorrect source file copying or missing prerequisite objects. Run Transaction SE37, display function module Z_DP_CCARD_SETTLEMENT, and navigate to menu **Function Module • Check • Extended Check**. Verify that no errors are reported.

Some common implementation issues are as follows:

- **Class ZCL_FAR_DP_ABAP_JSON activation fails; parent class not found**
  Your SAP_BASIS release doesn't include the /ui2/cl_json framework (requires SAP_BASIS 740+). Check your SAP_BASIS release in Transaction SPAM. If it's below 740, then you cannot use the standard JSON library.
- **Structure extension fails; field already exists**
  SAP Note 3000524 may have already added some of these fields. Execute Transaction SE11, display structure CCDATA, and review all append structures to identify which fields already exist. Add only the missing fields. If all three fields already exist from SAP Note 3000524, you can skip the CCDATA extension for this SAP Note.
- **Function module activation shows Unknown Type Reference**
  Interface ZIF_FAR_DP_TYPES was not created or activated. Return to the interface creation steps, create ZIF_FAR_DP_TYPES, activate it successfully, then retry the function module activation.

- **TCCAA configuration cannot be saved; field FNSET too long**
  Table TCCAA field FNSET has a limited length (typically 30 characters). Function module name Z_DP_CCARD_SETTLEMENT is 23 characters, which should fit. Verify you're not adding extra spaces.
- **Settlement fails with RFC destination DIGITALPAYMENTS not found**
  RFC destinations were not configured per SAP Note 3003533. Complete SAP Note 3003533 configuration, including creating both RFC destinations (DIGITALPAYMENTS and DIGITALPAYMENTS_OAUTH) in Transaction SM59 before testing settlement processing. The settlement code explicitly checks for these destination names; using different names requires code modification.
- **OAuth token retrieval fails**
  RFC destination DIGITALPAYMENTS_OAUTH has incorrect credentials. Verify that the username (client ID) and password (client secret) in Transaction SM59 match exactly the values from your SAP BTP service key. Recopy the credentials carefully, ensuring there are no leading/trailing spaces. The credentials are case-sensitive and must match exactly.

After successfully implementing SAP Note 3028194, the settlement infrastructure exists but cannot be fully tested until the complete configuration per SAP Note 3003533 is finished. The objects are in place and activated, ready for RFC destination configuration, virtual bank setup, and electronic bank statement configuration.

Settlement processing requires the complete configuration, including virtual bank setup (SAP Note 3003533), payment advice processing (multiple SAP Notes), and potentially refund configuration (SAP Note 3212765). Testing the settlement functionality before completing the full configuration typically results in confusing error messages about missing configuration rather than helpful feedback about the implementation quality.

### SAP Note 3041924

SAP Note 3041924 delivers the authorization processing logic for payment cards in SAP ERP Sales and Distribution. Whereas SAP Note 3028194 handled settlement (capture), this SAP Note handles the first step of the payment flow: authorizing payment cards during sales order creation. Authorization reserves funds on customer payment cards without charging them, enabling businesses to verify payment availability before incurring fulfillment costs.

The SAP Note provides three objects that implement payment card authorization processing: one interface defining authorization-specific data types, one class containing authorization business logic, and one function module integrating with SAP ERP Sales and Distribution authorization processing, as follows:

- **Interface ZIF_FAR_DP_AUTH_TYPES**
  Run Transaction SE24, create interface ZIF_FAR_DP_AUTH_TYPES, copy the source code from *ZIF_FAR_DP_AUTH_TYPES.txt*, and activate the interface.

- **Class ZCL_FAR_DP_AUTHORIZATIONS**
  Run Transaction SE24, create class ZCL_FAR_DP_AUTHORIZATIONS, copy the source code from *ZCL_FAR_DP_AUTHORIZATIONS.txt*, and activate the class.
- **Function module Z_DP_CCARD_AUTHORIZATION**
  In the same function group in which Z_DP_CCARD_SETTLEMENT was created, create function module Z_DP_CCARD_AUTHORIZATION, copy the source code from *Z_DP_CCARD_AUTHORIZATION.txt*, and activate the function module.

After creating the objects, update table TCCAA (the same table configured in SAP Note 3028194) to add the authorization function module, as follows:

1. Run Transaction SM30, enter "TCCAA" to search for the table, click **Maintain**, locate your existing entry (created for SAP Note 3028194), and add field FNAUT (Authorization Function) and set **Value** to **Z_DP_CCARD_AUTHORIZATION**.
2. Your TCCAA entry now contains both the settlement function (FNSET = Z_DP_CCARD_SETTLEMENT) and the authorization function (FNAUT = Z_DP_CCARD_AUTHORIZATION).
3. Save the configuration to your transport request.

The verification steps are as follows:

1. Run Transaction SE24, then verify that interface ZIF_FAR_DP_AUTH_TYPES exists with the status set to **Active** (see Figure 5.35).

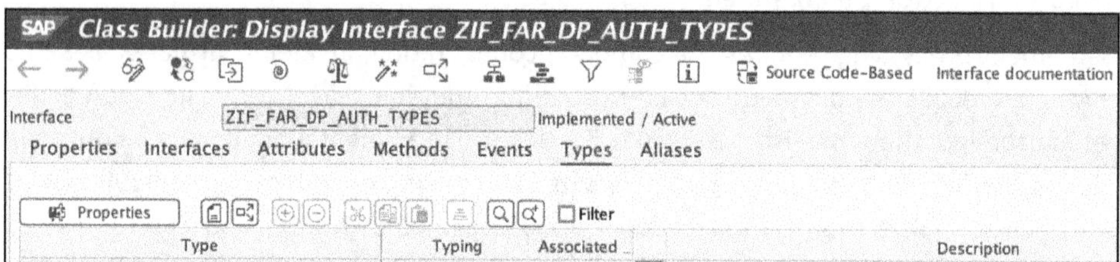


*Figure 5.35: Interface ZIF_FAR_DP_AUTH_TYPES: Active Status*

2. Run Transaction SE24, display class ZCL_FAR_DP_AUTHORIZATIONS, and verify it contains the three methods with status active as shown in Figure 5.36.

*Figure 5.36: Class ZCL_FAR_DP_AUTHORIZATIONS: Methods Verification Step*

3. Run Transaction SE37, display function module `Z_DP_CCARD_AUTHORIZATION`, and verify it exists in your customer function group with all four table parameters properly defined as shown in Figure 5.37.

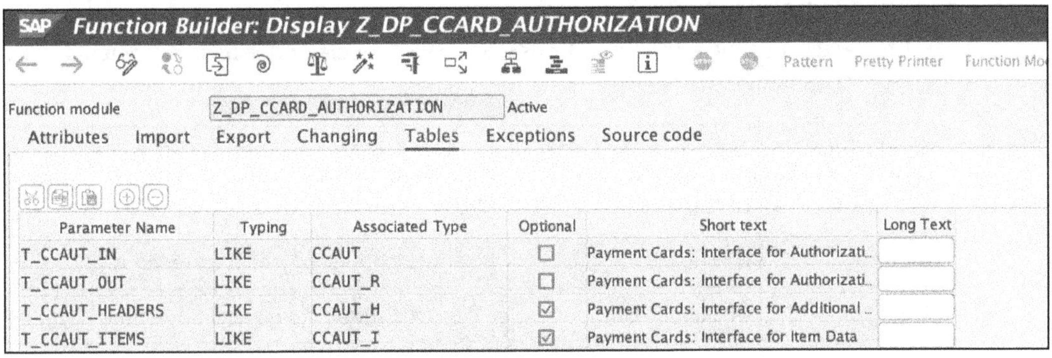

Function Builder: Display Z_DP_CCARD_AUTHORIZATION

Function module Z_DP_CCARD_AUTHORIZATION Active

Attributes Import Export Changing Tables Exceptions Source code

| Parameter Name | Typing | Associated Type | Optional | Short text | Long Text |
|---|---|---|---|---|---|
| T_CCAUT_IN | LIKE | CCAUT | ☐ | Payment Cards: Interface for Authorizati... | |
| T_CCAUT_OUT | LIKE | CCAUT_R | ☐ | Payment Cards: Interface for Authorizati... | |
| T_CCAUT_HEADERS | LIKE | CCAUT_H | ☑ | Payment Cards: Interface for Additional ... | |
| T_CCAUT_ITEMS | LIKE | CCAUT_I | ☑ | Payment Cards: Interface for Item Data | |

*Figure 5.37: Function Module Z_DP_CCARD_AUTHORIZATION: Table Parameters*

4. Execute Transaction SM30, display table `TCCAA`, and verify your entry now shows both `FNSET` and `FNAUT` populated with the respective function module names.

Two common issues are as follows:

- **Cannot add `FNAUT` to `TCCAA` entry**
  The table `TCCAA` entry doesn't exist from SAP Note 3028194. Return to SAP Note 3028194, complete table `TCCAA` configuration with `FNSET` first, then add the `FNAUT` field.
- **Function module fails; class `ZCL_FAR_DP_AUTHORIZATIONS` not found**
  Class was not created or activated. Create and activate the class before the function module.

#### SAP Note 3044268

SAP Note 3044268 enables summarized (compressed) posting for payment advice processing. Without this SAP Note, each individual payment transaction from the digital payments add-on posts as a separate accounting document, creating high posting volumes in systems processing thousands of daily transactions. This SAP Note implements logic to aggregate multiple payment transactions into summary postings, reducing the document count and improving month-end closing performance.

The SAP Note delivers correction instructions that modify electronic bank statement processing to support compressed posting. After implementation, payment advice processing creates summary documents that group multiple transactions rather than show individual postings per transaction.

After applying the correction instructions, configure new posting rules and modify the existing rules to enable summarized posting, as follows:

1. If not already configured, create account symbol `BANK_CHARGES` and assign an appropriate general ledger account for payment processing fees. This configuration typically exists from SAP Note 3003533, but verify the global settings for electronic bank statements in Transaction SPRO.
2. Create new posting rule keys in Transaction SPRO for electronic bank statements using the information from Table 5.8.

| Posting Rule | Text | Purpose |
| --- | --- | --- |
| `DEFE` | DP Ext. Payment Fee | Posts payment processing fees |
| `DPMI` | DP Manual In | Manual inbound payment corrections |
| `DPMO` | DP Manual Out | Manual outbound payment corrections |

*Table 5.8: Electronic Bank Statements: New Posting Rule Keys*

3. For each key created in Step 2, define the posting logic as per the SAP Note instructions, and update posting rules `DEIN` and `DEOU` to use posting area 1. In Transaction SPRO, for electronic bank statements, select **Assign External Transaction Types to Posting Rules**; for transaction category `CAMT053`, add new assignments. Modify the interpretation algorithm for external transaction type `PMNTEXTPPOSC`.

**Impact of Algorithm Change**

Interpretation algorithm 001 posts each transaction individually. Algorithm 000 enables compressed posting, in which multiple transactions aggregate into single document postings. This change is the key enabler for summarized posting.

After this SAP Note is implemented and configured, summarized posting works as follows:

1. Report RFDP_ADVICE_V2 retrieves payment advice from the digital payments add-on.
2. Payment transactions are grouped by posting rule and date.
3. Electronic bank statement processing creates summary documents instead of individual documents.
4. Posting to account DPEXTCLEAR occurs without automatic clearing (items remain open).
5. Run report SAPF124 (Clearing Transaction) against the clearing account to finally clear the open items.

This two-step approach (summarized posting followed by clearing) reduces document volume while maintaining detailed transaction traceability in the clearing document line items.

### SAP Note 3212765

SAP Note 3212765 enables refund processing for settled payment card transactions. This SAP Note creates a dedicated transaction code (Transaction FIN_DP_REFUND_PC) that allows finance users to refund any cleared customer line item that was paid via the digital payments add-on. The SAP Note handles both current-year and prior-year refunds with configurable general ledger account determination and supports parked refund documents for approval workflows.

The SAP Note delivers one correction instruction that creates report RFIN_DP_REFUND_PC and related objects. It also requires extensive manual activities to create data elements, tables, and maintenance views that support refund configuration.

The objects created by correction instruction are as follows:

- Report RFIN_DP_REFUND_PC: Interactive refund processing program
- BAdI implementation `FIN_IMP_DP_REFUND_PC` (implementing class `CL_IM_FIN_DP_REFUND_PC`)
- Messages: DIGITAL_PAYMENTS 800, 801

To create objects manually, follow these steps:

1. Run Transaction SE11, then create the data elements shown in Table 5.9.

| Data Element | Description | Domain | Short Label | Medium Label | Long Label |
|---|---|---|---|---|---|
| `DP_BLART_CR` | Document Type for Credit Memos | `BLART` | DocType CM | Doc. Type for CM | Document Type for Credit Memos |
| `DP_COPY_CY_ACCASS` | Copy Current Year Account Assignment | `XFELD` | CpyCYAcc | Copy CY Acc.Assgnmt | Copy Current Year Account Assignment |
| `DP_COPY_PY_ACCASS` | Copy Previous Year Account Assignment | `XFELD` | CpyPYAcc | Copy PY Acc.Assgnmt | Copy Previous Year Account Assignment |
| `DP_RACCT_PY` | Account Number, Old Fiscal Year | `SAKNR` | AN, Old FY | Acct No, Old FY | Account Number, Old FY |

*Table 5.9: Data Elements*

2. Run Transaction SE11, then create table FAR_DP_REFUND_PC with the fields shown in Table 5.10.

| Field | Key | Data Element | Description |
|---|---|---|---|
| MANDT | X | MANDT | Client |
| BUKRS | X | BUKRS | Company code |
| BLART_CREDIT | | DP_BLART_CR | Document type for credit memo |
| SAKNR | | SAKNR | General ledger account for current year refunds |
| COPY_FUND | | DP_COPY_CY_ACCASS | Copy current year controlling assignment |
| SAKNR_ALT | | DP_RACCT_PY | General ledger account for prior year refunds |
| COPY_FUND_ALT | | DP_COPY_PY_ACCASS | Copy prior year controlling assignment |
| SAKNR_CC | | HKONT_V | General ledger account for card refund posting |
| SGTXT_ADD | | SGTXT | Line item text |

*Table 5.10: Table FAR_DP_REFUND_PC Fields*

3. Run Transaction SE11, then create maintenance view FARV_DP_REFUNDPC for table FAR_DP_REFUND_PC. Include all fields from FAR_DP_REFUND_PC.
4. Run Transaction SE11, then create table FAR_DP_PT_CCINS as per the SAP Note instructions and activate it.
5. Create maintenance view FARV_DP_PT_CCINS. Add tables FAR_DP_PT_CCINS, TB033-CAT, TB033, and TB033T with join conditions as specified in the SAP Note. Include fields MANDT, DP_PAYMENT_TYPE, CCINS, and BEZ30, and activate the table.
6. Run Transaction SE38, change report RFIN_DP_REFUND_PC (created by correction instruction), select **Text Elements**, and add text symbols and selection texts per the SAP Note specification. Upload documentation from attached ITF (Interchange Text Format) files using Transaction SE61.

After implementation, configure refund processing, as follows:

- Refund settings: Run Transaction SM30, enter "FARV_DP_REFUNDPC", and create an entry per company code as per the SAP Note instructions.
- Payment type mapping: Run Transaction SM30, enter "FARV_DP_PT_CCINS", and map payment types to card types as per the SAP Note instructions.

Execute Transaction FIN_DP_REFUND_PC to process refunds:

1. Select cleared customer line items with digital payment data (fields PAYS_PROV, PAYS_TRAN, DP_PAYMENT_TYPE, and DP_TRANS_ID must not be blank).
2. Review selected items and maximum refund amounts.
3. Enter refund amounts (partial or full).
4. Execute. The system creates a credit memo and posts the refund via the digital payments add-on.
5. The refund document must be settled via Transaction FCC1 for PSP processing.

The report also displays parked refund documents referencing payment card payments for approval and posting.

### SAP Note 3273859

SAP Note 3273859 is a technical correction SAP Note that fixes a specific error caused by deimplementing and reimplementing SAP Note 3000524. This SAP Note is only needed if you encounter syntax error **Include LFACGFAR not found** in program SAPLFACG.

This SAP Note applies only to this specific scenario: An older version of SAP Note 3000524 was implemented in your system, and that older version was deimplemented (removed). Then, the current version of SAP Note 3000524 was implemented, and you see syntax error **Include LFACGFAR not found** in program SAPLFACG. If you implemented SAP Note 3000524 directly without deimplementation cycles, then you do not need this SAP Note.

### SAP Note 3276162

SAP Note 3276162 resolves performance issues encountered during payment advice processing in which the system cannot find payment references sent from the digital payments add-on. The SAP Note extends index tables BSIS (General Ledger Account Open Items) and BSAS (General Ledger Account Cleared Items) with digital payments fields and creates extension indexes to prevent full table scans during advice reconciliation.

Payment advice processing fails to match transactions because searches against large tables BSIS and BSAS without proper indexing result in timeouts or missing references. This typically manifests as advice items not being cleared against open receivables even though the payments were processed successfully.

The SAP Note delivers one correction instruction, plus manual activities to extend two critical accounting index tables with digital payments fields and create four extension indexes for performance optimization.

The manual activities you need to perform are as follows:

1. **Extend Table BSIS (General Ledger Account Open Items)**
   In Transaction SE11, for table BSIS, create append structure DIGITAL_PAYMENTS_BSIS (see Figure 5.38) and activate.

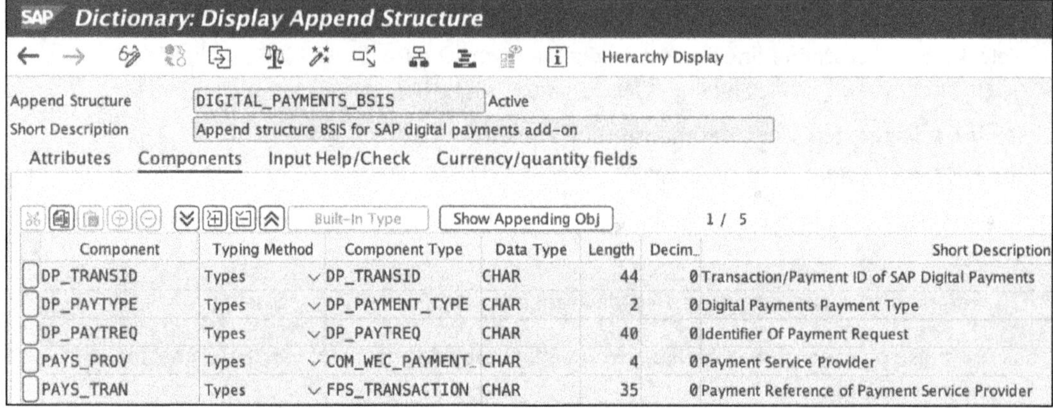


*Figure 5.38: Table BSIS: Append Structure*

2. **Extend Table BSAS (General Ledger Account Cleared Items)**
   Repeat append structure creation for table BSAS and activate (see Figure 5.39).

Dictionary: Display Append Structure

Append Structure: DIGITAL_PAYMENTS_BSAS Active
Short Description: Append structure BSAS for SAP digital payments add-on

Attributes | Components | Input Help/Check | Currency/quantity fields

| Component | Typing Method | Component Type | Data Type | Length | Decim. | Short Description |
|---|---|---|---|---|---|---|
| DP_TRANSID | Types | DP_TRANSID | CHAR | 44 | 0 | Transaction/Payment ID of SAP Digital Payments |
| DP_PAYTYPE | Types | DP_PAYMENT_TYPE | CHAR | 2 | 0 | Digital Payments Payment Type |
| DP_PAYTREQ | Types | DP_PAYTREQ | CHAR | 40 | 0 | Identifier Of Payment Request |
| PAYS_PROV | Types | COM_WEC_PAYMENT... | CHAR | 4 | 0 | Payment Service Provider |
| PAYS_TRAN | Types | FPS_TRANSACTION | CHAR | 35 | 0 | Payment Reference of Payment Service Provider |

*Figure 5.39: Table BSAS: Append Structure*

3. **Create extension indexes**
   For tables BSIS and BSAS, create extension indexes for DP_PAYTREQ, PAYS_TRAN, and HKONT and activate.

After manual activities and correction instruction, configure table T042IPSP to enable the enhanced advice matching logic as per the SAP Note instructions. If processing multiple company codes, create separate T042IPSP entries as per the instructions.

The SAP Note's correction instruction modifies the advice processing logic to use table BSIS (indexed) instead of cluster table BSEG (nonindexed) when searching for external payment references. This change, combined with the extension indexes, improves advice-processing performance in systems with large transaction volumes.

The impact of this SAP Note is as follows:

- Before this note, advice processing searched cluster tables BSEG and RFBLG, causing full table scans to take minutes to hours in large systems.
- After this note, advice processing uses indexed table BSIS with extension indexes, reducing the search time to seconds even with millions of line items.

**SAP Note 3277101**

SAP Note 3277101 enhances search functionality on table BSEGC (Payment Card Accounting Data) to enable searching by digital payment transaction ID (DP_TRANSID). This allows users and programs to quickly locate payment card accounting documents using the transaction ID returned by the digital payments add-on. If you cannot efficiently search payment advice items in table BSEGC using the digital payment transaction ID field, then this SAP Note provides the solution.

This SAP Note applies one correction instruction that modifies the table BSEGC search capabilities to include DP_TRANSID as a searchable field, improving query performance when looking up payment transactions by their digital payments add-on transaction identifier.

For implementation, apply the correction instruction through Transaction SNOTE; no manual activities are required. To verify, execute Transaction SE16N, enter table BSEGC, and verify that DP_TRANS_ID appears in the field selection list and can be used as a search criteria.

**SAP Note 3290490**

SAP Note 3290490 corrects a program error in payment advice processing in which searches for payment requests fail due to incorrect payment type mapping for external payment card payments. When searching for digital payment advice items using payment requests fails because external payment card payments use different payment type values than those expected by the search logic, this SAP Note provides the solution.

This SAP Note applies one correction instruction that fixes the payment type mapping logic to correctly handle external payment card transactions during advice item searches.

For implementation, apply the correction instruction through Transaction SNOTE. No manual activities are required.

**SAP Note 3341226**

SAP Note 3341226 provides control over which payment advice transactions post as summarized accounting documents versus individual postings. Without this SAP Note, the condensed posting behavior is hard-coded in the advice-processing logic. After this SAP Note, you control condensed postings through posting rule configuration. When advice transactions post summarized into single accounting documents while others post individually, with no configuration option to control this behavior, this SAP Note provides the solution.

This SAP Note applies one correction instruction that modifies advice processing (report RFDP_ADVICE_V2) to check the posting type field in posting rules configuration, enabling dynamic control of condensed posting behavior.

The configuration after implementation is as follows:

1. In the IMG (Transaction SPRO), go to **Financial Accounting • Bank Accounting • Business Transactions • Payment Transactions • Electronic Bank Statement • Make Global Settings for Electronic Bank Statement • Define Posting Rules**.
2. For each posting rule, the **Posting Type** field controls the condensed posting behavior:
   - Values **1**, **2**, or **3**: Transactions post condensed (multiple transactions in one document).
   - All other values: Transactions post individually (one document per transaction).
3. SAP recommends condensed posting only for these transaction types:
   - PMNTCCRDPOSC: Payment card positive postings (compressed)
   - PMNTEXTPPOSC: External payment positive postings (compressed)

For all other transaction types, use posting types other than 1, 2, or 3 to maintain individual posting for better traceability.

**SAP Note 3349208**

SAP Note 3349208 fixes a reference field mismatch between settlement documents (from Transaction FCC1) and payment advice documents (from report RFDP_ADVICE_V2) that prevents automated clearing via report SAPF124. When the accounting document references (BKPF-XBLNR and BSEG-ZUONR) from the settlement run differ from the references in the payment advice document, making automated clearing fail, this SAP Note provides the solution.

This SAP Note applies one correction instruction that synchronizes reference field generation between settlement program RFCCSSTT (called by Transaction FCC1) and advice processing program RFDP_ADVICE_V2, ensuring both create identical reference values.

For implementation, apply the correction instruction through Transaction SNOTE. No manual activities are required. After this SAP Note, settlement and advice documents use matching references, enabling automated clearing with report SAPF124 to reconcile payment card clearing accounts.

## 5.2 Payment Card Configuration

Payment card configuration in SAP ERP establishes the master data foundation for processing payment card transactions through the digital payments add-on. This configuration defines which card brands your organization accepts, how card data is validated and stored, the categories used for grouping similar card types, blocking mechanisms for lost or stolen cards, and the general ledger account structure supporting payment processing from authorization through settlement. Unlike SAP S/4HANA, where payment card configuration integrates seamlessly with business partner master data, SAP ERP requires configuration across multiple tables and Transaction SPRO paths, spanning cross-application components and FI.

The configuration completed in this section becomes active immediately for sales order processing, customer master data maintenance, and payment card validation. Incomplete or incorrect configuration causes authorization failures, prevents card entry in sales orders, or allows invalid card data to enter the system. Thorough configuration following SAP Best Practices and alignment with your digital payments add-on PSP configuration ensures smooth payment processing from order creation through settlement and reconciliation.

### 5.2.1 Payment Card Master Data

Payment card master data defines the card types, validation rules, general ledger account assignments, and central settings that control payment card processing throughout SAP ERP. This master data configuration must be completed before any payment card can be entered in customer master records or sales orders. The configuration spans three primary areas: payment card type definition and validation; central payment card settings for FI integration; and general ledger account structure, including virtual bank setup for digital payments add-on processing.

The general ledger account structure setup varies based on your own implementation and organization needs. Let's review some best practices that can be considered in your configuration. You cannot reuse existing payment card general ledger accounts from legacy payment gateway configurations in your SAP ERP system when implementing the digital payments add-on integration. There are a few reasons you need to create a new general ledger account:

- **Incompatible account mappings**
  Existing general ledger account mappings from accounts receivable clearing accounts to PSP clearing accounts contain settings and control functions designed for legacy payment card types. These configurations include field status groups, account assignment logic, and posting key determinations that work for traditional clearing house integrations but are incompatible with digital payments add-on processing requirements.
- **Incorrect account technical settings**
  Existing legacy payment card accounts may not be configured with the technical settings required for digital payment processing. For example, your existing general ledger accounts may not be set up as open item managed, which is required for digital payments.
- **Different posting process**
  The digital payments clearing process uses different posting flows than legacy implementations. Traditional payment card processing posts directly to cash accounts, while the digital payments add-on uses virtual bank infrastructure with electronic bank statement processing for payment advice reconciliation.

The choice of number ranges depends on your chart of accounts structure and account numbering conventions. The critical requirement is that virtual bank accounts follow the 00 ending pattern and clearing accounts follow the matching number pattern with 30, 70, or other suffixes.

The general ledger accounts required for digital payments add-on integration vary depending on your company's financial management and reporting and reconciliation needs. Some general guidance is as follows:

- **Virtual bank main account (account type: bank subaccount)**
  This general ledger account can be applicable for all company codes with payment card processing needs in your landscape.
- **PSP settlement clearing account (account type: bank subaccount)**
  This general ledger account can be applicable for all company codes with payment card processing in your landscape.
- **Accounts receivable clearing accounts for SD invoices and accounts receivable documents**
  These are typically set up per company code and can differentiate by card type, currency, sales org, and more as per your organization needs.

Accounts receivable clearing accounts can differ by card type, currency, or sales org, based on your business requirements. Organizations have flexibility in their account granularity as follows:

- **Single account per company code**
  One accounts receivable clearing account for all payment card transactions regardless of card type or currency; simplest configuration, but provides limited reporting segmentation
- **Separate accounts per card type**
  Different accounts receivable clearing accounts for Visa, Mastercard, Amex, and so on; enables card brand–specific reporting and settlement analysis
- **Separate accounts per currency**
  Different accounts receivable clearing accounts for USD, EUR, GBP, and so on; supports multicurrency operations with clean currency separation in settlement processing
- **Separate accounts per sales organization**
  Different accounts receivable clearing accounts per sales org or business unit; enables profit center reporting and separate settlement monitoring per organizational unit

The choice depends on reporting requirements, settlement complexity, and reconciliation processes. More granular separation provides better visibility but increases configuration and account maintenance effort.

Some organizations may not choose to use the payment advice process with digital payments, in which case additional bank clearing accounts will not be needed. This decision affects account structure requirements:

- Without payment advice processing, you only need a virtual bank main account, a PSP clearing account, and accounts receivable clearing accounts. Settlement posts directly without electronic bank statement processing, and you have relatively simplified configuration with fewer accounts.
- With payment advice processing, you need an additional *payment in* clearing account, potentially a *payment out* account, electronic bank statement processing (Transaction FF_5), additional posting rules, and account symbol configuration—resulting in a more complex setup, but with automated reconciliation enabled.

You should decide on payment advice usage during the project planning phase. Enabling payment advice after the initial go-live may require account structure changes and configuration updates, affecting all company codes.

### 5.2.2 Payment Card Type Configuration

The configuration in this section covers four interconnected areas. First, we define the payment card types in the cross-application component, establishing the card brands and the validation rules applied to card numbers. From there, payment card data synchronization ensures those card type definitions are correctly mapped between the business partner master record and the customer master, which is essential for downstream sales order processing. The SD payment card settings then activate card functionality within the sales process, covering card categories, payment card plan types, blocking reasons, authorization validity periods, checking groups, and authorization/settlement control per account. Finally, SAP Note 3003533 provides the comprehensive configuration guide for completing the technical integration: virtual bank setup, general ledger account structure, electronic bank statement configuration, and RFC destination setup for API connectivity to the add-on.

#### Maintain Payment Card Types

Payment card type configuration occurs through the SAP Implementation Guide (IMG) in the **Cross-Application Components** area, which houses configuration affecting multiple SAP modules simultaneously. Navigate using Transaction SPRO, following this menu path: **Cross-Application Components • Payment Cards • Basic Settings • Maintain Payment Card Type**.

This IMG activity opens a maintenance view for table `TB033` (Payment Cards: Type), containing all configured payment card types in your system. The configuration can also be accessed directly via Transaction SM30 by entering table/view `TB033` or generated maintenance view `V_TB033`.

SAP delivers standard payment card types for major card brands as part of the base system installation. These standard types provide baseline configuration suitable for most implementations and serve as templates when creating custom card types. Table 5.11 lists the standard payment card types delivered by SAP.

| Card Type Code | Card Type Description | Card Category | Standard Validation Rule | Usage |
|---|---|---|---|---|
| VISA | Visa | Credit card | Card number starts with 4 | Global credit/debit cards |
| MC / MAST | Mastercard | Credit card | Card number starts with 5 | Global credit/debit cards |
| AMEX | American Express | Credit card | Card number starts with 3 | Global credit cards |
| DINE | Diners Club | Credit card | Specific numbering pattern | International credit cards |
| DISC | Discover | Credit card | Card number starts with 6 | US-based credit cards |
| JCB | Japan Credit Bureau | Credit card | Specific numbering pattern | Japan-based credit cards |

*Table 5.11: Standard SAP Payment Card Types*

You can create additional custom payment card types for regional card schemes not included in SAP standard delivery, corporate purchasing cards with special processing requirements, store-branded cards or private label cards, gift cards processed through payment infrastructure, or digital wallet payment methods represented as card types for unified processing. To create custom payment card types, execute Transaction SPRO and navigate to **Maintain Payment Card Type**, then click **New Entries**. Finally, configure the fields as listed in Table 5.12 and save the configuration.

| Field Name | Technical Field | Description | Configuration Guidance |
|---|---|---|---|
| **Payment Card Type** | `CCINS` | Four-character type code | Use meaningful codes matching card brands |
| **Description** | `BEZ30` | Card type description | Business-friendly name for user selection |
| **Payment Card Category** | `CCCAT` | Links to table `TB034` category | Assign appropriate category (credit, debit, etc.) |
| **Card Number Check** | | Luhn algorithm validation | Enable for card types requiring validation |
| **Starting Number** | | Expected card number prefix | Used for validation (e.g., Visa starts with 4) |

*Table 5.12: Payment Card Type Configuration Fields*

You can refer to the *SAP Digital Payments Add-On Integration Guide* for the latest set of code lists published by SAP (Figure 5.40 shows an example list). The integration guide is available here: *http://s-prs.co/v629404*.

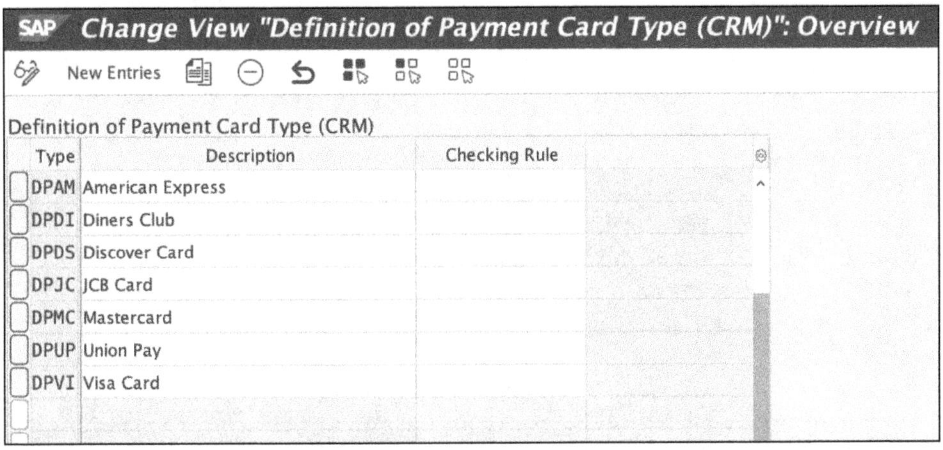

Change View "Definition of Payment Card Type (CRM)": Overview

New Entries

Definition of Payment Card Type (CRM)

| Type | Description | Checking Rule |
|---|---|---|
| DPAM | American Express | |
| DPDI | Diners Club | |
| DPDS | Discover Card | |
| DPJC | JCB Card | |
| DPMC | Mastercard | |
| DPUP | Union Pay | |
| DPVI | Visa Card | |

*Figure 5.40: Payment Card Type Definition*

### Payment Card Data Synchronization

In SAP ERP, the **Assign Payment Cards** setting in Transaction SPRO (found under **Cross application components • Master Data Synchronization • Customer/Vendor Integration • Business Partner Settings • Settings for Customer Integration • Field Assignment for Customer Integration • Assign Attributes**) is used to synchronize payment card data between the business partner and the customer master record. The primary functions of this setting include the following:

- **Data consistency**
  Assigning payment cards ensures that payment card details (such as card type and card number) maintained in the business partner are correctly updated and mapped to the corresponding fields in the customer master.
- **Customer/vendor integration**
  This is a critical step for systems undergoing conversion to SAP S/4HANA or using customer/vendor integration (CVI) in SAP ERP. This mapping allows the system to automatically transfer payment card attributes when creating or updating a customer through Transaction BP.
- **Field mapping**
  This setting specifically handles the assignment of payment card attributes for contact persons associated with the customer, ensuring that their specific payment details are synchronized across both master data objects.
- **Transaction processing**
  Proper assignment is required for subsequent business processes that use payment cards, such as sales order processing or automatic billing, for which the system must identify the correct card information from the integrated master data.

The configuration screen, shown in Figure 5.41, is used to map the internal payment card type codes used in the business partner records to the card type codes used in CVI and subsequent customer master data.

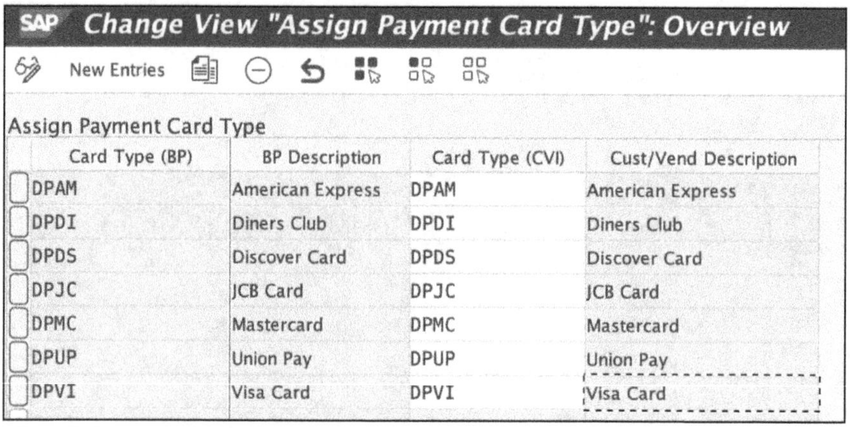

Change View "Assign Payment Card Type": Overview

New Entries

Assign Payment Card Type

| Card Type (BP) | BP Description | Card Type (CVI) | Cust/Vend Description |
|---|---|---|---|
| DPAM | American Express | DPAM | American Express |
| DPDI | Diners Club | DPDI | Diners Club |
| DPDS | Discover Card | DPDS | Discover Card |
| DPJC | JCB Card | DPJC | JCB Card |
| DPMC | Mastercard | DPMC | Mastercard |
| DPUP | Union Pay | DPUP | Union Pay |
| DPVI | Visa Card | DPVI | Visa Card |

*Figure 5.41: Assign Payment Card Types*

This screen ensures data consistency during synchronization processes, particularly when a business is moving toward or already using the business partner as the central master data object (a mandatory step for SAP S/4HANA conversion). The mapping ensures that when payment card information is entered in the business partner record (or entered during order processing for new cards), the system correctly assigns the corresponding values in the underlying customer master record when synchronization occurs. This is vital for downstream processes like payment authorization and settlement.

### Sales and Distribution: Payment Card Settings

In SD, there are a few configuration details that are needed to establish a framework for managing payment cards within the sales process. They ensure the system can correctly identify accepted cards, define processing rules, activate functionality in relevant sales documents, and manage exceptions (like blocking specific cards). In Transaction SPRO, navigate to **Sales and Distribution • Billing • Payment Cards**. You will see the following configuration options:

- **Maintain Card Types** and **Maintain Card Categories**: Card types and card categories define what kind of cards are accepted and how they should be validated internally.
- **Maintain Payment Card Plan Types**: Payment card plan types act as the switches that enable payment card functionality for specific sales transactions (e.g., a standard sales order vs. a free sample delivery).
- **Maintain Blocking Reasons**: Blocking reasons provide a necessary control mechanism to manage exceptions and mitigate risk by preventing specific cards from being used.

We will discuss the configuration of each of these, as well as how to specify the authorization validity period, maintain checking groups, assign checking groups, and assign authorizations, in the following sections.

#### *Maintain Card Types*

Under **Maintain Card Types,** you set the specific types of payment cards accepted by your business, such as Visa or Mastercard. You can assign a unique four-character code (e.g., DPVI for Visa, DPMC for Mastercard) and optionally link it to a standard SAP function module that includes the checking algorithm for the card number format. Figure 5.42 provides an example list of add-on-specific card types that can be configured.

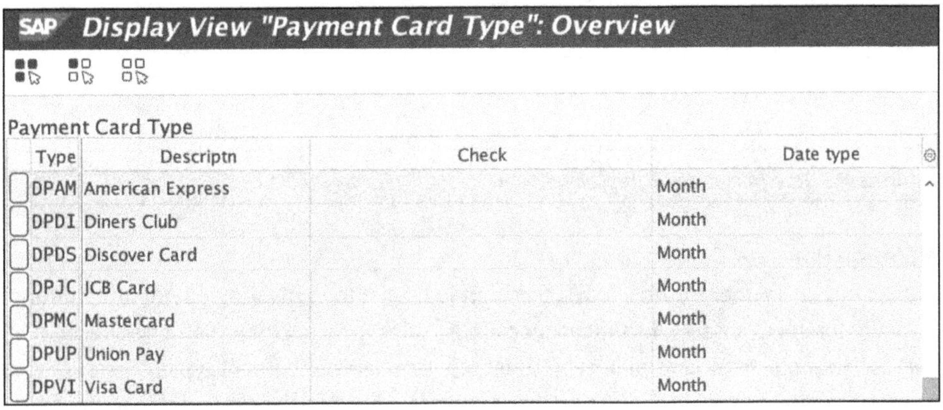

Display View "Payment Card Type": Overview

Payment Card Type

| Type | Descriptn | Check | Date type |
|---|---|---|---|
| DPAM | American Express | | Month |
| DPDI | Diners Club | | Month |
| DPDS | Discover Card | | Month |
| DPJC | JCB Card | | Month |
| DPMC | Mastercard | | Month |
| DPUP | Union Pay | | Month |
| DPVI | Visa Card | | Month |

*Figure 5.42: Add-On Specific Card Types*

#### *Maintain Card Categories*

The **Maintain Card Categories** configuration groups card types into broader categories (e.g., credit cards, procurement cards). Figure 5.43 shows example card categories for your reference. Both card categories shown, credit card and procurement card, are predelivered and available by default in standard SAP systems (both SAP ERP and SAP S/4HANA).

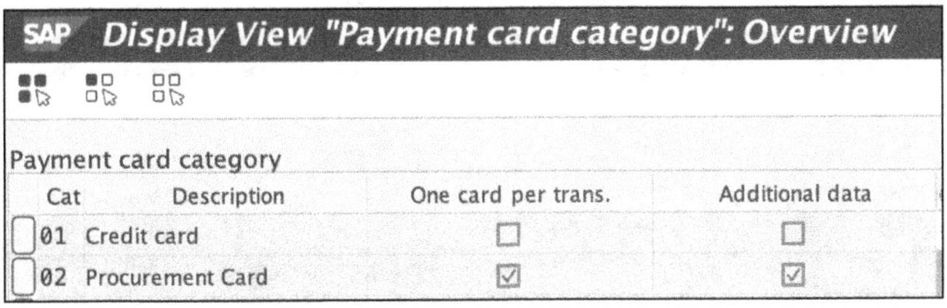

Display View "Payment card category": Overview

Payment card category

| Cat | Description | One card per trans. | Additional data |
|---|---|---|---|
| 01 | Credit card | ☐ | ☐ |
| 02 | Procurement Card | ☑ | ☑ |

*Figure 5.43: Define Card Categories*

You can then assign specific card types to the credit card category. In Transaction SPRO, follow this menu path: **SAP Reference IMG • Sales and Distribution • Billing • Payment Cards • Maintain Card Categories • Determine Card Categories.** In the table that is displayed (see

Figure 5.44), choose a card type entry (e.g., DPAM). Press F4 in the **Cat** (category) field (this field is next to the **Limit** field) for search help, choose **01** (**Credit card**), and save the entry. Repeat this step for the other card types that are in scope for your implementation. This helps in controlling the acceptable number ranges and other specific processing rules, especially for nonstandard card types.

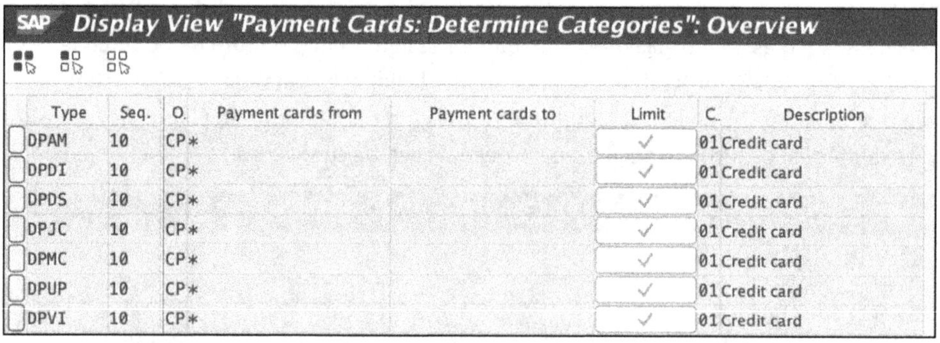

Display View "Payment Cards: Determine Categories": Overview

| Type | Seq. | O. | Payment cards from | Payment cards to | Limit | C. | Description |
|---|---|---|---|---|---|---|---|
| DPAM | 10 | CP* | | | ✓ | 01 | Credit card |
| DPDI | 10 | CP* | | | ✓ | 01 | Credit card |
| DPDS | 10 | CP* | | | ✓ | 01 | Credit card |
| DPJC | 10 | CP* | | | ✓ | 01 | Credit card |
| DPMC | 10 | CP* | | | ✓ | 01 | Credit card |
| DPUP | 10 | CP* | | | ✓ | 01 | Credit card |
| DPVI | 10 | CP* | | | ✓ | 01 | Credit card |

*Figure 5.44: Determine Card Categories*

### *Maintain Payment Card Plan Types*

The **Maintain Payment Card Plan Types** step is crucial for activating payment card functionality for specific sales document types (e.g., sales orders, credit memo requests). You can assign a payment card plan type to all relevant sales document types. The standard SAP system provides plan type 03 for processing payment cards. Without this assignment, the **Payment Cards** tab or field will not be visible during order processing.

Follow this menu path in Transaction SPRO: **SAP Reference IMG • Sales and Distribution • Billing • Payment Cards • Maintain Payment Card Plan Types**. In the table that is displayed (see Figure 5.45), choose the order type entry (e.g., sales document type **OR1** in the **SaTy** field), press F4 in the **PT** field for search help, and choose **03** for **Billing Plan Type**. Repeat this step for all sales document types in your implementation scope.

Figure 5.45 shows an example configuration listing the card types and their assignments.

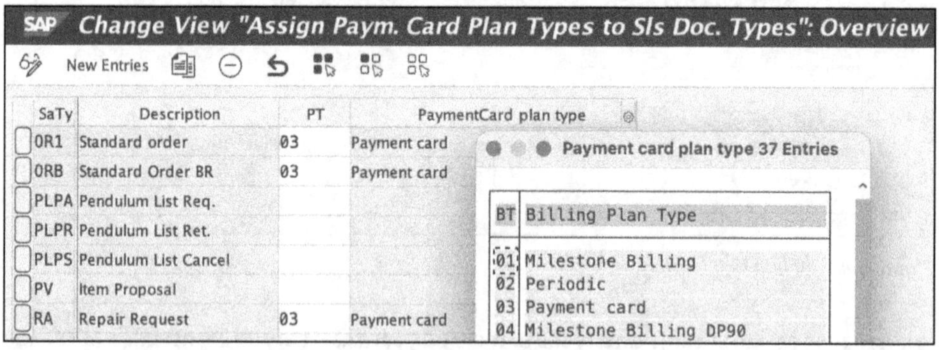

Change View "Assign Paym. Card Plan Types to Sls Doc. Types": Overview

New Entries

| SaTy | Description | PT | PaymentCard plan type |
|---|---|---|---|
| OR1 | Standard order | 03 | Payment card |
| ORB | Standard Order BR | 03 | Payment card |
| PLPA | Pendulum List Req. | | |
| PLPR | Pendulum List Ret. | | |
| PLPS | Pendulum List Cancel | | |
| PV | Item Proposal | | |
| RA | Repair Request | 03 | Payment card |

Payment card plan type 37 Entries

| BT | Billing Plan Type |
|---|---|
| 01 | Milestone Billing |
| 02 | Periodic |
| 03 | Payment card |
| 04 | Milestone Billing DP90 |

*Figure 5.45: Payment Card Plan Type Assignments*

### *Maintain Blocking Reasons*

**Maintain Blocking Reasons** allows you to define a list of reasons for blocking a payment card within the system. These blocking reasons can be entered in the customer (payer) master record to prevent transactions with that specific card. Follow this menu path in the IMG: **Sales and Distribution • Billing • Payment Cards • Maintain Blocking Reasons.** There, you can check the list of blocking reasons configured in your SAP system. The standard system includes blocking reason 01 for lost cards (see Figure 5.46).

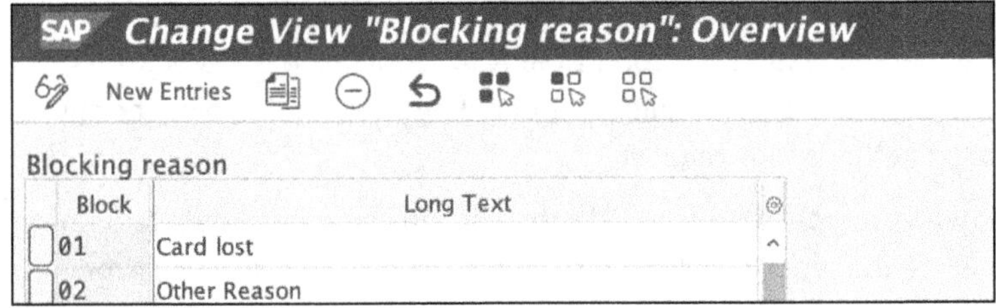

Change View "Blocking reason": Overview

New Entries

Blocking reason

| Block | Long Text |
|---|---|
| 01 | Card lost |
| 02 | Other Reason |

*Figure 5.46: Blocking Reasons*

### *Specify Authorization Validity Period*

This configuration allows you to define the number of days for which an authorization is valid for each payment card type and determines the internal expiration date for the authorization obtained from the clearing house. Navigate in Transaction SPRO to **Sales and Distribution • Billing • Payment Cards • Authorization and Settlement • Specify Authorization Validity Periods.** In the table that is displayed, enter the validity period (as a number of days) in the **Valid** field for each of the card types in your implementation scope. This validity period typically is organization-specific and will mostly be uniform across all card types in your implementation scope.

If the sales and delivery process takes longer than the specified validity period, the system will mark the existing authorization as expired and require reauthorization before allowing subsequent activities, like delivery or billing. Figure 5.47 shows an example in which the validity period is set to 28 days. In this example, the system mandates reauthorization if the delivery is scheduled for day 29 or later.

Change View "Payment Card Type: Validity Period of Auth.": Overview

| Type | Description | Valid |
|---|---|---|
| DPAM | American Express | 28 |
| DPDI | Diners Club | 28 |
| DPDS | Discover Card | 28 |
| DPJC | JCB Card | 28 |
| DPMC | Mastercard | 28 |

*Figure 5.47: Validity Period of Authorization*

### *Maintain Checking Groups: Authorization Horizon*

The authorization horizon and the authorization validity period work in tandem to manage when authorizations are obtained and when they expire. This setting, defined in Customizing for credit management, specifies how many days before a material availability date or billing date the system should initiate a payment card authorization. In Transaction SPRO, navigate to **Sales and Distribution • Billing • Payment Cards • Authorization and Settlement • Risk Management for Payment Cards • Maintain Checking Groups**.

Within the **Maintain Checking Groups** activity, you define checking groups and specify the number of days for the authorization horizon in the **AHorizn** field. The system uses this number of days before the material availability date or billing date to determine the appropriate time to initiate an authorization request.

You also set the **Preau** (preauthorization) indicator here if you want the system to perform a preliminary check when the order date falls outside the main authorization horizon (see Figure 5.48). The goal is to obtain authorization close to the actual shipping date so that it does not expire before the goods are delivered and billed.

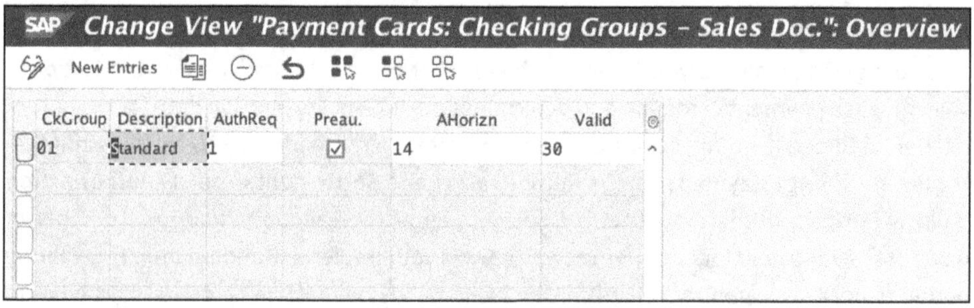

Change View "Payment Cards: Checking Groups - Sales Doc.": Overview

New Entries

| CkGroup | Description | AuthReq | Preau. | AHorizn | Valid |
|---|---|---|---|---|---|
| 01 | Standard | 1 | ☑ | 14 | 30 |

*Figure 5.48: Authorization Horizon and Preauthorization Settings*

During order creation, when a sales order is saved, the system checks if the material availability date falls within the authorization horizon. If the date is within the horizon, authorization is requested immediately for the full amount. If the date is outside the horizon (too far in the future), the system might perform a *preauthorization* (a soft check, often for a minimal amount, like $1) to verify the card details, but it does not request the full authorization.

The authorization obtained (either the initial preauthorization or the full authorization) is only valid for the number of days specified in the **Specify Authorization Validity Periods** setting. If the delivery or billing process extends beyond this validity period, the system will require a fresh reauthorization, which is typically handled by a batch job report run daily. The horizon determines when to authorize initially, and the validity period determines for how long that authorization is good.

### *Define Sales Document Types: Assign Checking Groups*

To assign checking groups to sales document types, you can navigate in Transaction SPRO to **Sales and Distribution • Sales • Sales Documents • Sales Document Header • Define**

**Sales Document Types**. Here you can view existing sales document types (see Figure 5.49; such as **OR1** for a **Standard order**) or create new custom types by copying an existing one.

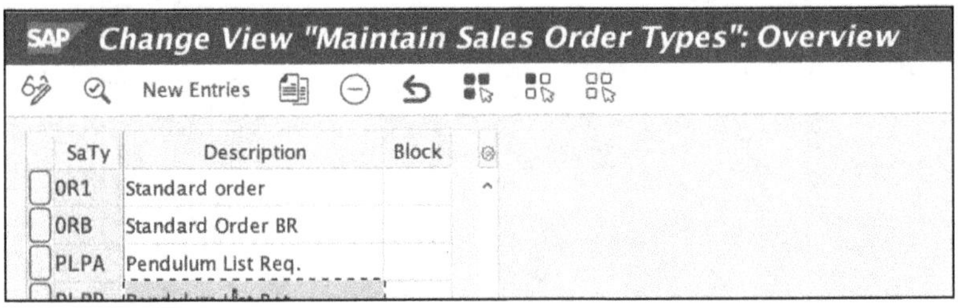

Change View "Maintain Sales Order Types": Overview

New Entries

| SaTy | Description | Block |
|---|---|---|
| OR1 | Standard order | |
| ORB | Standard Order BR | |
| PLPA | Pendulum List Req. | |

*Figure 5.49: Select Order Type Entry*

Select the required sales order type and navigate to the **Billing** section, as shown in Figure 5.50. You can assign the **Checking group** configured earlier to the chosen sales order type and save the configuration.

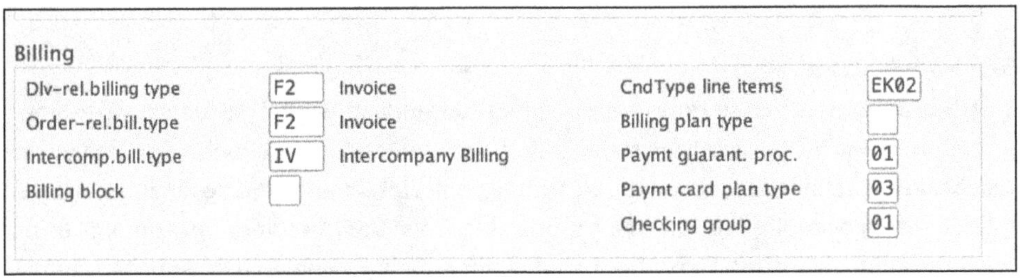

Billing

| | | | | |
|---|---|---|---|---|
| Dlv-rel.billing type | F2 | Invoice | CndType line items | EK02 |
| Order-rel.bill.type | F2 | Invoice | Billing plan type | |
| Intercomp.bill.type | IV | Intercompany Billing | Paymt guarant. proc. | 01 |
| Billing block | | | Paymt card plan type | 03 |
| | | | Checking group | 01 |

*Figure 5.50: Assign Checking Group to Sales Order Type*

### *Set Authorization/Settlement Control Per Account*

This customization determines which general ledger accounts are used for the accounting entries and which technical routines communicate with the clearing house for each step of the payment process. It links financial general ledger accounts to specific technical functions that communicate with external clearing houses for real-time payment card processing. You can access this configuration by navigating in Transaction SPRO to **Sales and Distribution • Billing • Payment Cards • Authorization and Settlement • Maintain Clearing House • Set Authorization/Settlement Control** per account.

This configuration is performed for each chart of accounts and involves linking specific general ledger accounts with function modules and their remote function call (RFC) destinations. As part of the implementation steps for SAP Note 3041924, you created function module `Z_DP_CCARD_AUTHORIZATION`, and as part of the implementation steps for SAP Note 3028194, you created function module `Z_DP_CCARD_SETTLEMENT`. Assign them as shown in Figure 5.51:

- **Authorization**: Custom function module `Z_DP_CCARD_AUTHORIZATION` is used to send an authorization request to an external clearing house when a sales order is saved.
- **Settlement**: Custom function module `Z_DP_CCARD_SETTLEMENT` initiates the process to transfer funds from the customer's bank to the merchant's bank account after billing.

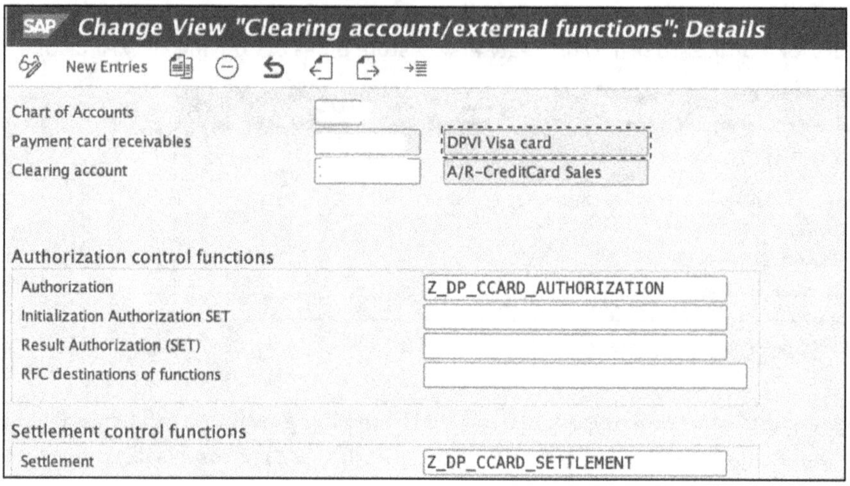

Change View "Clearing account/external functions": Details

New Entries

Chart of Accounts
Payment card receivables DPVI Visa card
Clearing account A/R-CreditCard Sales

Authorization control functions
Authorization Z_DP_CCARD_AUTHORIZATION
Initialization Authorization SET
Result Authorization (SET)
RFC destinations of functions

Settlement control functions
Settlement Z_DP_CCARD_SETTLEMENT

*Figure 5.51: Assign Function Modules*

**SAP Note 3003533**

SAP Note 3003533 is a comprehensive configuration guide for digital payments add-on integration with SAP ERP. It provides step-by-step configuration instructions for setting up virtual banks, general ledger accounts, electronic bank statements, RFC destinations, and all related settings required for payment processing. This SAP Note does not contain correction instructions or manual activities to implement, but it can be used as a master configuration reference. This SAP Note provides configuration guidance across six major areas: central payment card settings in FI, virtual bank and bank account setup, general ledger account creation and assignment, electronic bank statement configuration, RFC destination setup for API connectivity, and PSP synchronization.

Beyond individual card type configuration, SAP requires central settings that control how payment card data integrates between SD and FI. These settings affect all payment card types universally and must be configured before payment processing can occur. Navigate via Transaction SPRO to **Financial Accounting • Accounts Receivable and Accounts Payable • Business Transactions • Payments with Payment Cards • Make Central Settings for Payment Cards**. This configuration screen controls a few integration settings:

1. **Retain cus. item (checkbox)**
   Select this checkbox to retain the customer line item in the accounting document when transferring payment card transaction data from SD to FI. When checked, the FI document generated during billing maintains the customer account posting in addition to the general ledger account postings for payment card receivables and clearing accounts. The checkbox must be selected for the digital payments add-on integration to maintain proper audit trails and enable customer-level reconciliation.
2. **Settings for the Settlement Document**
   Enter the FI document type (from table `T003`) created when payment card settlement successfully completes via Transaction FCC1. Common values include `AB` (accounting

document), SA (settlement document), or custom document types with appropriate number range and posting characteristics.

3. **Settings for Resetting Clearing When Settlement Unsuccessful**
   Enter the FI document type used when settlement fails and previously cleared items must be reset to open status. The most common value is AB (accounting document) for reversal scenarios.

Next, create a virtual bank account with the corresponding account for digital payments in the system. Use Transaction FIPS to create the bank master record of the virtual bank (PSP). Only one house bank may be created for the bank master record of the virtual bank, and only one virtual bank account is allowed per currency. The general ledger account number for the virtual bank account must end with 00. When you create the corresponding bank clearing accounts, apart from the last two digits, the numbers of the bank clearing accounts must match the number of the bank account.

Make sure that a corresponding general ledger account is defined in the configuration under **Financial Accounting (New) • Accounts Receivable and Accounts Payable • Business Transactions • Incoming Payments • Incoming Payments Global Settings • Define Accounts for Bank Charges (Customers)**. The charges of a payment are automatically posted with the payment document to the defined account for bank charges.

When you set up the general ledger account for PSP clearing, ensure that the sort key is set to **External Document Number**. This is configured in Transaction FS00 for the clearing account ending in 70.

SAP Note 3003533 also provides instructions for electronic bank statement configuration. Navigate to **Financial Accounting • Bank Accounting • Business Transaction • Payment Transactions • Electronic Bank Statement • Make Global Settings for Electronic Bank Statement**, and follow the instructions there to create account symbols, assign accounts to account symbols, tax posting requirement, create posting rule keys, define posting rules, create a transaction category, assign external transaction types, and assign a virtual bank to a transaction category.

To integrate the SAP ERP system with the add-on, you must set up the technical connection to the add-on. Note that the relevant certificate (DigiCert Global Root G2) must exist in the Trust Manager (Transaction STRUST) of the SAP system. You can obtain the certificate directly from the certification service provider (i.e., DigiCert). To do this, create the following two RFC connections of type G in Transaction SM59 (see Figure 5.52):

- **RFC destination: DIGITALPAYMENTS_OAUTH**
  This destination is used to retrieve the OAuth 2.0 authentication token required for secure machine-to-machine (M2M) communication. It points to the authentication server URL of your SAP Business Technology Platform (SAP BTP) subaccount. The SAP system sends its client credentials to this destination to receive a time-limited access token. Follow these steps:
  - Create the RFC destination with the name, connection type and description as shown in Figure 5.53.

Configuration of RFC Connections

Generate RFC Callback Positive Lists | Activate Non-Empty Whitelists | Positive List for Dynamic Connections

RFC callback check not secure

| RFC Connections | Type | PL Act. | Comment |
|---|---|---|---|
| ABAP Connections | 3 | | |
| HTTP Connections to External Server | G | | |
| HTTP Connections to ABAP System | H | | |
| Internal Connections | I | | |
| Logical Connections | L | | |

*Figure 5.52: Transaction SM59: Configuration of RFC Destinations to Add-On*

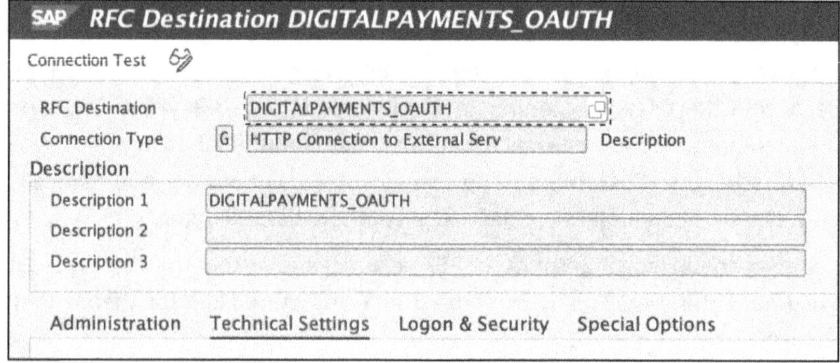


*Figure 5.53: RFC Destination DIGITALPAYMENTS_OAUTH Creation*

- In the **Technical Settings** tab (see Figure 5.54), specify the following:
  - Set **Target Host** as the uaa.url from service key (without https://).
  - Set **Service No.** as **443**.
  - Set **Path Prefix** as **/oauth/token/alias/<your subdomain>.aws-live** (for US10).

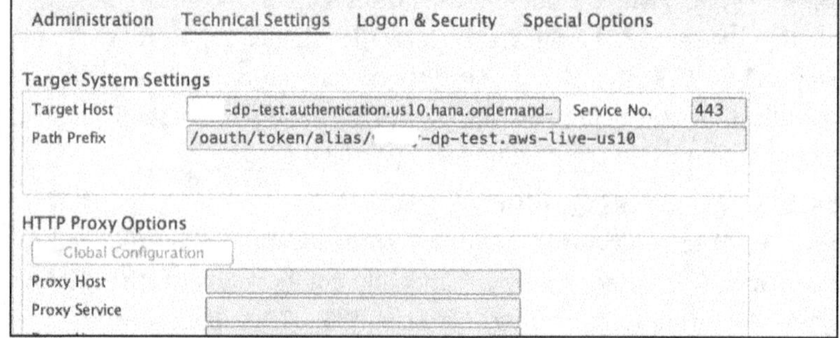


*Figure 5.54: Technical Settings*

- In the **Logon and Security** tab, select **Basic Authentication** and enter the uaa.clientid and uaa.clientsecret from the service key of your subaccount, as shown in Figure 5.55.

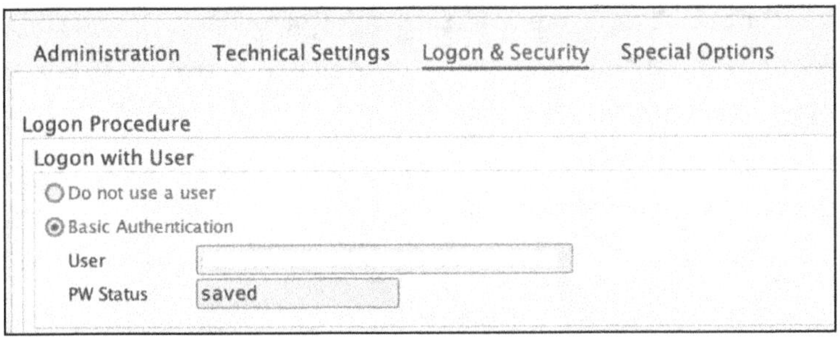

Figure 5.55: RFC Destination: Basic Authentication Credentials

- In the **Security Options** tab (see Figure 5.56), specify the following:
  - Set **SSL** as **Active**.
  - Set **SSL Certificate** as **Default SSL Client (Standard)**.
- Save the RFC destination.
- It is not possible to check the connection using the connection test function.

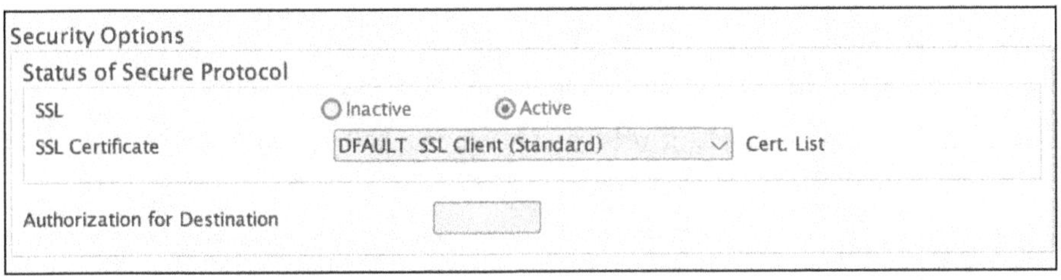

Figure 5.56: RFC Destination: Security Options

- **RFC destination: `DIGITALPAYMENTS`**
  This is the core business API destination used to send actual payment transaction data (e.g., card registration, authorization, or settlement). It points to the actual service endpoint of the digital payments add-on (the */core* URL). Instead of using a user name and password, it is configured to use the OAuth 2.0 configuration tied to the first destination. Follow these steps:
  - Create the RFC destination, with the details shown in Figure 5.57.

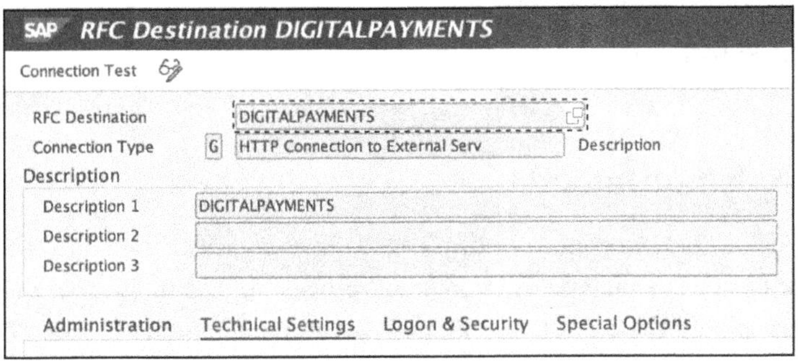

Figure 5.57: RFC Destination DIGITALPAYMENTS: Creation Step

- In the **Technical Settings** tab (see Figure 5.58), specify the following:
  - Set the **Target Host** details based on where your add-on tenant is setup and whether it is a test client or production client.
  - Set **Service No.** as **443**.
  - Set **Path Prefix** based on your SAP version.

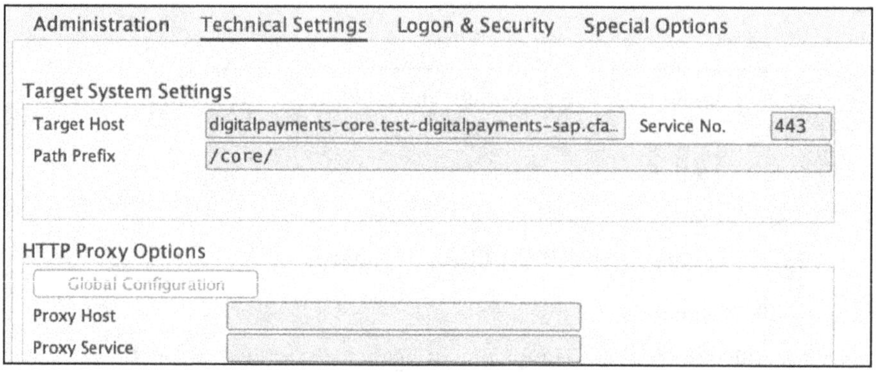

*Figure 5.58: RFC Destination DIGITALPAYMENTS: Technical Settings*

The two destinations form a secure tunnel for processing digital payments:

- Authentication: When a payment process begins, the SAP system uses `DIGITALPAYMENTS_OAUTH` to request a secure token from SAP BTP.
- Authorization: Once the token is received, the system attaches it to the header of the actual payment request.
- Data exchange: The system then uses the `DIGITALPAYMENTS` destination to send the payment details to the cloud add-on.
- Routing: The SAP digital payments add-on receives this secure request and routes it to the specific PSP (e.g., Stripe, Cybersource) configured in your tenant.
- This architecture ensures that sensitive payment card data never enters your SAP system; instead, it is handled via secure tokens (tokenization), reducing your PCI DSS compliance burden.

After completing all the configuration, execute program FAR_DP_UPDATE_PSP_PAYMENT_METH via Transaction SE38. As a result, the current PSP code values are transferred from the digital payments add-on to SAP ERP.

## 5.3 Sales and Distribution Integration

Integration of the SAP digital payments add-on with SAP ERP Sales and Distribution (SD) is the logical next step for business enablement of payment-processing capabilities. In the previous sections, we completed the add-on-specific SAP Notes implementation, worked through the configuration settings in Transaction SPRO, and established a secure tunnel between SAP ERP and SAP BTP. This section focuses on how that architecture facilitates the order-to-cash lifecycle.

In a traditional SAP ERP environment, payment card processing was handled by storing sensitive primary account numbers (PANs) directly in tables like table `VCNUM` or `FPLTC`. The digital payments add-on fundamentally alters this design by enforcing a token-first approach. The SAP ERP system no longer acts as a vault for sensitive data; instead, it becomes a conduit for tokens, exchanging references with the PSP via the add-on. This new design requires changes to standard SD user exits and copy-control routines to ensure the following requirements are upheld:

- **Sales order entry**
  Tokens are determined correctly from the business partner master data or exchanged in real time from external commerce platforms.
- **Authorization**
  Credit holds are placed dynamically against the digital token rather than static credit limits.
- **Billing and settlement**
  The capture of funds occurs seamlessly during the billing run, triggered by the release of the invoice to accounting.
- **Returns**
  Refunds are processed via API calls that reference the original settlement transaction, closing the audit loop.

The following sections detail the specific implementations and process flows required to tokenize the sales lifecycle, beginning with the critical foundation of the sales order and business partner master data before moving on to credit management integration and billing document processing.

### 5.3.1 Order Processing with Payment Cards

This section explains payment card processing within the sales cycle, from initial order entry through to refunds. We begin with manual order processing, walking through how customer service representatives create sales orders, select stored payment cards from the business partner master, and trigger real-time authorization at the point of sale. From there, we explore e-commerce order integration, in which payment authorization or capture may already have occurred externally before the order reaches SAP ERP, and cover both the one-step *direct capture* pattern (common for digital goods) and the two-step, *authorize, then capture* pattern (preferred for physical goods, for which settlement should align with shipment). Finally, we discuss refund processing, explaining both the approval-based credit memo request workflow and the direct refund capability.

### Business Partner Integration

In the SAP digital payments landscape, the business partner plays an active role in driving the processing logic. When a sales order is created, SAP does not simply look up a customer; it executes a partner determination procedure to find four distinct roles. In any standard sales order, the system must identify four specific entities. Although these are often the same entity (e.g., the customer is also the payer), large organizations may split them into the following categories:

- **Sold-to party (SP/AG)**
  This is the entity placing the order. This partner typically triggers the pricing procedure and availability check.
- **Ship-to party (SH/WE)**
  This is the location where goods are delivered. This partner typically determines the tax jurisdiction code (in North America), which calculates the final tax amount sent to the PSP for the capture request.
- **Bill-to party (BP/RE)**
  This is the address where the invoice is sent. This partner affects the print output of the billing document.
- **Payer (PY/RG)**
  This entity is financially responsible for settling the debt and is critical for payments. When you create a sales order, SAP looks at the payer's master data to find the valid payment card (token). If the sold-to party and payer are different (e.g., a franchisee ordering, but headquarters paying), then the token must exist on the headquarters' (payer's) business partner record. The dynamic credit limit check is performed against the payer.

Roles are a critical aspect of the business partner design. They help with data segmentation and act as a context switch. For a digital payment transaction to flow from end to end, the following must be true:

- The business partner must exist in both the sales role—to authorize the order—and the finance role—to settle the cash. During sales order creation, when a user enters an order in Transaction VA01, the system checks if the business partner is extended to the sales role and looks for sales area data; if not, the order creation fails.
- When a user selects a credit card token, the system looks at the **Payment Transactions** tab.
- When an invoice is created and released to accounting to capture the funds, the system switches to the finance role and looks for company code data like reconciliation account, payment method, and so on.

If you have a valid sales customer but never extended them to the finance role, then the sales order will work, but billing will fail. The system cannot post the accounting document to capture the cash because the financial identity of that partner doesn't exist for that company code.

To make these functions work, the business partner must be extended to specific roles. You cannot simply create a business partner; you must assign the correct technical roles to enable these functions. Table 5.13 provides a list of typical roles provided by SAP for your reference.

| Business Partner Role ID | Role Description | Enables Partner Functions | Technical Table |
|---|---|---|---|
| 000000 | Business partner (general) | Basic address, name | BUT000 |
| FLCU01 | Customer (sales view) | Sold-to, ship-to | KNVV |
| FLCU00 | Customer (FI view) | Payer, bill-to | KNB1 |

*Table 5.13: Business Partner Roles for Payment Processing Functions*

The interaction between the sold-to party and the payer is the most common point of failure in digital payments implementations. If you try to use a business partner as a payer (to use their credit card token) but they have not been extended to the finance role for the specific company code, the sales order will fail. In your implementation, you must ensure that a token is attached to the payer business partner, not necessarily the sold-to business partner. A good understanding of the following tables is important when writing code:

- VBPA (Sales Document: Partner): This table links the specific sales order (VBELN) to the partner functions (PARVW) and the partner number (KUNNR).
- KNVP (Customer Master Partner Functions): This defines the default relationships.
- BUT0CC (Payment Cards): This table stores the digital payments token, linked to PARTNER_GUID.

In the digital payments architecture, the business partner master data serves as the secure vault for customer financial information. Unlike legacy systems that might store card data on individual orders or within the sales area data of a customer record, SAP centralizes this sensitive data within the **Payment Transactions** tab of the business partner record.

This centralization is critical for the order-to-cash process because it decouples the payment card data from specific sales transactions. Once a card is tokenized and stored in the business partner master, it becomes reusable and can be called upon by sales orders, service orders, and accounts receivable and payable without requiring reentering the details. Key data elements on this tab include the following:

- **Card Type (CCINS)**: Identifies the card scheme and drives the validation rules, and determines which general ledger clearing account will eventually be posted to during settlement.

- **Card Number (CCNUM)**: In a digital payments scenario, this field does not contain the raw PAN. Instead, it stores the digital payments token.
- **Card Category**: This field allows you to segment cards based on business function, such as distinguishing between a credit card (for immediate settlement) and a purchasing card (which may require Level 2/3 data for corporate compliance).
- **Validity Dates**: The SAP standard availability check in the sales order will reject any token whose validity date in the master data has passed, even if the token itself is still valid at the PSP.

By ensuring tokens are active and obsolete cards are locked or deleted, you can ensure that the downstream sales order process remains automated and error-free. When a sales order is created, the system immediately queries the **Payment Transactions** tab for valid tokens that are found and presents results to the user. The visibility of the **Payment Cards** tab and fields is controlled by a hierarchy of settings. You can configure the **Payment Cards** section to be hidden for standard roles and visible only for a specific role, as follows:

1. Go to Transaction SPRO and follow menu path **Cross-Application Components • SAP Business Partner • Business Partner • Basic Settings • Field Groupings • Configure Field Attributes per BP Role.**
2. Here, you will see the complete list of business partner roles along with their description. In this example, we have enabled payment cards visibility only for one specific role, financial services business partner, and hidden it for the rest of the roles. Now let's select a role for which it is hidden.
3. Select the **Business Partner (General)** role, as shown in Figure 5.59.

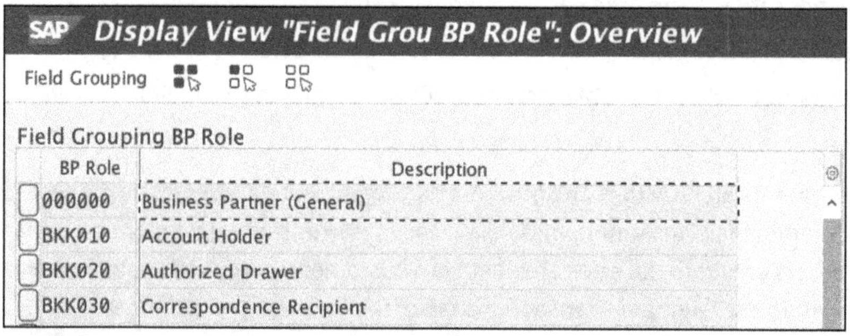

Display View "Field Grou BP Role": Overview

Field Grouping

Field Grouping BP Role

| BP Role | Description |
| --- | --- |
| 000000 | Business Partner (General) |
| BKK010 | Account Holder |
| BKK020 | Authorized Drawer |
| BKK030 | Correspondence Recipient |

*Figure 5.59: Business Partner (General) Role*

4. Select **Payment Cards** in the **Data Set** section (see Figure 5.60). You will see that it is configured with the **Hide** radio button enabled, which means that payment cards will not be visible when this business partner role is chosen in the dropdown in Transaction BP.
5. Now go back one screen and choose a different business partner role, one for which the visibility is enabled; for this example, select **Financial Services Business Partner** (see Figure 5.61).

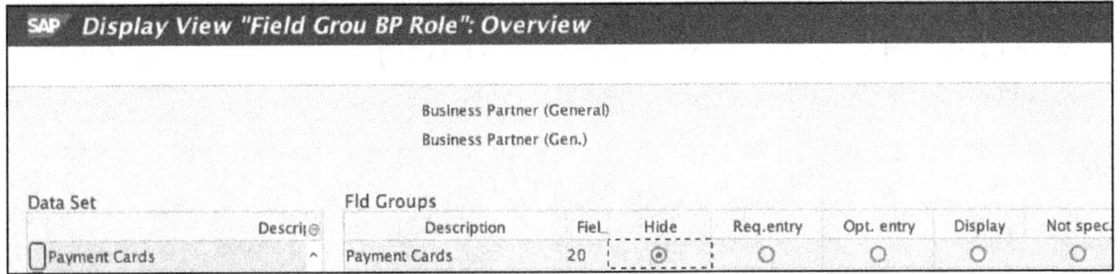


Figure 5.60: Payment Card Visibility: Hide

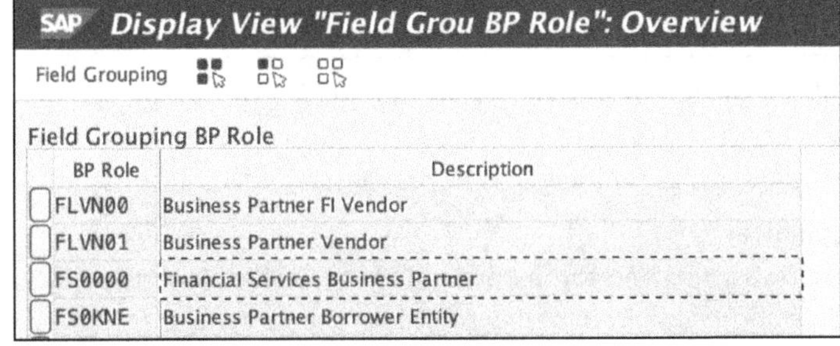


Figure 5.61: Payment Card Visibility: Visible Business Partner Role

6. Select **Payment Cards** in the **Data Set** section. You can see that the visibility is set to **Display** for this role (see Figure 5.62).

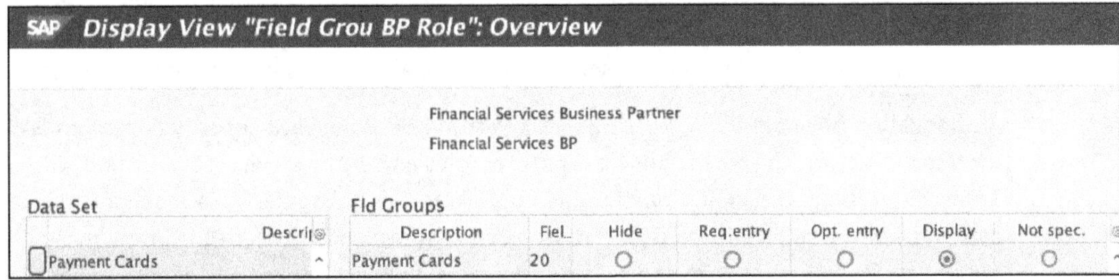


Figure 5.62: Visibility: Display Enabled

7. You can test this by going to Transaction BP and selecting a business partner that has a few cards stored in the **Payment Cards** tab. Select the correct business partner role (see Figure 5.63).

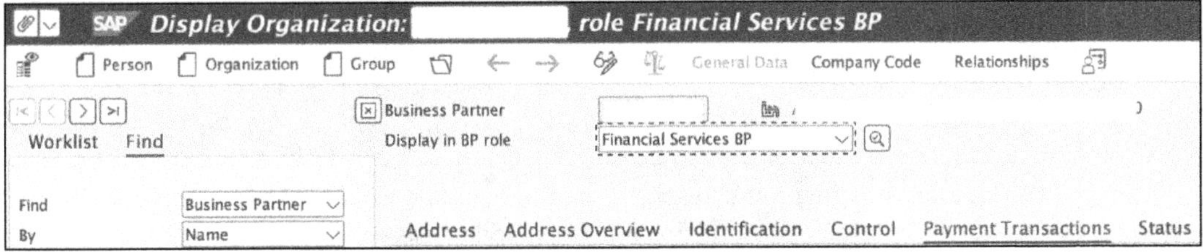


Figure 5.63: Transaction BP: Financial Services Business Partner Role Selection

8. When you navigate to the **Payment Transactions** tab, you will be able to see the **Payment Cards** that are stored in the business partner record. In this example, you can see a list of test cards stored (see Figure 5.64).

Payment Cards

| ID | Type | Description | Card Number | Standard | Description |
|---|---|---|---|---|---|
| 000001 | DPVI | Visa Card | 4HNKCY6YFB4I6SXVRZPPIH77 | ⦿ | 424242xxxxxx4242 |
| 000002 | DPVI | Visa Card | PS5Y2Q6GK4MCP66SRBMNVKNS | O | 400005xxxxxx5556 |
| 000003 | DPVI | Visa Card | JTQX37HMS27QPHMK3PLUVQX4 | O | 424242xxxxxx4242 |
| 000004 | DPAM | American Express | 7CSU6MYZ44QJAHLK7SCI6XUQ | O | 378282xxxxxx0005 |

Card Details ... Change History Entry 1 of 6

*Figure 5.64: Payment Cards: List of Test Cards/Digital Payments Add-On Tokens*

You might want to enable these settings for the following reasons:

- **Data security (PCI compliance)**
  Hiding the **Payment Cards** tab from general roles prevents general users (e.g., those who might only need to update addresses) from seeing sensitive token data or the existence of credit cards. It forces the user to switch to a specific role that can be restricted by SAP authorization objects (like B_BUPA_RLT), ensuring only authorized finance/sales users can access card data.
- **Process discipline**
  Forcing the user to select a specific role ensures that the user is thinking in the context of finance. It prevents the accidental attachment of a personal credit card to a generic business partner role that might be used for nonfinancial purposes (e.g., a contact person).

Authorization object B_BUPA_RLT (Business Partner: BP Roles) is a security control in SAP used to restrict which specific business partner roles a user can view, create, or edit within Transaction BP. In the digital payments implementation, this object is a mechanism that can be used to strictly enforce segmentation of duties, which improves PCI compliance and data hygiene. It ensures that a sales rep processing an order cannot accidentally alter the financial settlement data (the token) established by the credit department. Even if it's only the token-related data, the existence of the card and its metadata (expiration date, cardholder name) is often considered sensitive. By restricting the role via B_BUPA_RLT (see Table 5.14), you restrict access to the data container itself.

| Fields | Description |
|---|---|
| ACTVT | ■ 01 = Create<br>■ 02 = Change<br>■ 03 = Display |
| RLTYP (business partner role category) | Technical identifier of the role (e.g., FS0000 for financial services). |

*Table 5.14: Object B_BUPA_RLT: Controls Access to Business Partner Roles*

Let's look at two example scenarios and how to control access, as follows:

- **Scenario A: The General User (Sales Rep)**
  Permissions assigned: B_BUPA_RLT with ACTVT set as 01, 02, 03 and RLTYP set to FLCU01 (Sales Customer), 000000 (General). This user can open the business partner, change the address, and edit sales area data. They do not have authorization for the financial services role and cannot select it in the dropdown. Because the **Payment Cards** tab is configured to be visible only in that FI role, the sales rep never sees the tab.
- **Scenario B: The Finance User (Payment Specialist)**
  Permissions assigned: B_BUPA_RLT with ACTVT set as 02 (Change) and RLTYP set to FS0000 (Financial Services). They can select the financial services business partner role from the dropdown. Once they select that role, the business data toolset (BDT) configuration triggers, and the **Payment Cards** tab appears, allowing them to enter or token data.

By combining BDT configuration (hiding the tab per role) with B_BUPA_RLT (restricting access to the role), you create a robust, two-layer security model.

SAP S/4HANA provides out-of-the-box integration with the digital payments add-on in the SAP Fiori app user interface to manage payment cards creation and maintenance in the business partner record. In SAP ERP, there are no out-of-the-box SAP Fiori apps or any other integration provided by SAP. You must do custom design and development work to manage payment cards in a business partner record.

Although a business user can manually type a token into Transaction BP, high-volume digital sales require a robust, programmatic integration strategy. There are few ways to get card data into a business partner record:

- **Manual maintenance (SAP GUI)**
  In this option, the SAP user authorized to manage payment cards logs into Transaction BP, navigates to the **Payment Transactions** tab, and manually enters the token and validity dates. This option is viable only for small transaction volumes; it is not scalable and should only serve as a stop-gap arrangement until automated integrations are enabled.
- **Batch input (batch data communication)**
  In this option, screen scraping or a batch input session is used to simulate a user typing into Transaction BP. This option is not recommended because if a screen field moves, the interface breaks.
- **Direct BAPI integration**
  The modern standard is to use Business Application Programming Interfaces (BAPIs). These are stable, upgrade-safe function modules provided by SAP to manipulate master data without touching the UI.

In most SAP customer environments, organizations operate e-commerce platforms—whether SAP or non-SAP—that facilitate customer self-service payment options for product purchases. Beyond these e-commerce systems, SAP customers may also maintain self-service payment portals, enabling users to manage invoices, quotes, or proforma payments as well as subscription renewals via payment cards. Within these portals, customers routinely administer their payment methods by adding new credit cards for subscriptions or removing expired ones. To support an optimal checkout experience, it is essential that such updates are reflected promptly and securely within the business partner master data.

Important elements of this integration's design are as follows:

- When a customer interacts with the web frontend and saves a new card, the portal can be configured or enhanced to send a JSON payload to SAP containing the tokenized data (SAP token) and the PSP metadata (PSP token).
- When a card is added via the portal, instead of storing just the digital payments add-on token, you could consider a twin token-storage strategy. In this approach, the portal application passes the digital payments add-on token (for sales order processing) and the PSP token to SAP. This ensures that if a refund is required months later, the system has the specific external reference needed to process the credit, even if the SAP token has changed.
- You can write custom code in SAP to call the standard BAPIs to create, link, or lock cards in the database. Your design should consider a logical deletion approach.
- When a customer deletes their card on the portal, the SAP integration should not physically delete the record from the database. Deleting a card that was used on a previous invoice breaks the audit trail and can cause settlement errors in historic reporting. Instead, your integration should perform a logical deletion. The card is flagged with a specific lock reason (e.g., **03—Portal Delete**). This effectively hides the card from future sales orders while preserving the historical link for finance.

You might consider using the following standard BAPIs:

- **Card Master Creation (`BAPI_PCA_MASTER_CREATE`)**
  Before a card can be assigned to a partner, it must exist as a master record in the central card table (table `CCARD`). The `BAPI_PCA_MASTER_CREATE` BAPI helps to establish this foundation. If this BAPI is skipped, the system will reject any attempt to link the token to a business partner, as the payment card ID would be considered nonexistent.
- **Business Partner Link (`BAPI_BUPA_PCARD_ADD`)**
  Once the card master exists, it must be linked to a specific business partner. The `BAPI_BUPA_PCARD_ADD` API helps to create this relationship in table `BUT0CC`. This is the specific API that makes the card visible in the **Payment Transactions** tab.
- **Managing Extended Metadata** (`FS_API_BP3100_ADD`)
  Standard SAP BAPIs do not have fields for the PSP token. One way to solve this is to use table `BP3100` (Business Partner: Additional Information) and use the `FS_API_BP3100_ADD` API to store the external tokens as generic text strings linked to the partner. This ensures that when a Refund API is called later, the system can look up the original Stripe reference associated with the SAP token, if needed.

- **Delete Payment Card Details (`BAPI_BUPA_PCARD_REMOVE`)**
  This BAPI removes the link between a card and a business partner (used in deletion).
- **Commit Database Changes (`BAPI_TRANSACTION_COMMIT`)**
  Commits the changes to the database.

### Order Processing with Payment Cards

In this section, we will discuss order processing—from manual sales order entry through e-commerce integration to refund processing via credit memos. We will also discuss how tokens flow through SD transactions, including sales order entry, where tokens are selected from business partner master data or received from external platforms; authorization, where credit validation occurs against tokenized cards rather than static credit limits; billing and settlement, where fund capture occurs seamlessly during invoice release; and refunds, where credits reference the original settlement transactions through API calls.

#### *Manual Order Processing with Payment Cards*

Manual order processing represents the traditional sales scenario in which customer service representatives, inside sales teams, or call center agents create sales orders directly in SAP using Transaction VA01. For payment card transactions, the manual process requires selecting stored payment cards from customer business partner records, triggering automatic authorization during order save, managing authorization failures through sales order blocks, and ultimately settling authorized transactions during billing and settlement processing.

Sales orders with payment card processing use Transaction VA01 (Create Sales Order) with the payment card plan functionality enabled. The payment card plan, accessible through the sales order header, contains all payment card and authorization data for the transaction. Before payment cards can be entered or selected in sales orders, a few prerequisites must be met, as follows:

1. The sales document type must have a payment card plan type assigned. Execute Transaction VOV8 (Sales Document Types), select your document type (e.g., `OR` for standard orders, `DR` for debit memo requests), and verify that the **Paymt card plan type** field (see Figure 5.65) in the **Billing** section contains a value (typically **03**).

Billing

| | | | | |
|---|---|---|---|---|
| Dlv-rel.billing type | | | CndType line items | |
| Order-rel.bill.type | | Invoice | Billing plan type | |
| Intercomp.bill.type | | Intercompany Billing | Paymt guarant. proc. | 01 |
| Billing block | | | Paymt card plan type | 03 |
| | | | Checking group | 01 |

*Figure 5.65: Transaction VOV8: Payment Card Plan Type Verification*

2. If the field is blank, payment card tabs will not appear in sales orders of that document type. Click in the field, press F4 for value help, select the payment card plan type (commonly **03**), and save the configuration. This assignment controls whether users can enter payment cards for specific order types.
3. The payer business partner must have payment cards stored in the **Payment Transactions** tab. Payment card data is stored centrally in table `BUT0CC`, linked to the business partner GUID, not to individual sales areas or company codes. This centralization enables card reuse across multiple sales documents without reentry.
4. To verify one or more payment cards exist for a customer, execute Transaction BP (Business Partner Maintenance), enter the customer number, select the appropriate business partner role for which to enable payment card visibility (e.g., financial services business partner), navigate to the **Payment Transactions** tab, and select the **Payment Cards** subtab. The system displays stored payment cards with the card type, masked card number (token), expiration date, and validity status. New cards can be integrated from either an external e-commerce application or a payment portal application. Or you could enable interactive voice response (IVR) integration to accept customer credit cards securely over IVR, and then integrate IVR-generated tokens into SAP.

Let's create an example sales order for order type `ZOR`, sales organization `US01`, customer `1000000000`, and material `DPMAT01`. We'll walk through the process and explain the add-on integration points and validation steps:

1. Execute Transaction VA01. On **Create Sales Order: Initial Screen** (see Figure 5.66), enter the following information:
   - **Order Type**: Select from dropdown (e.g., **ZOR**)
   - **Sales Organization:** Your sales organization code (e.g., **US01**)
   - **Distribution Channel**: Your distribution channel (e.g., **00** for direct sales)
   - **Division**: Product division (e.g., **00** for cross-division)

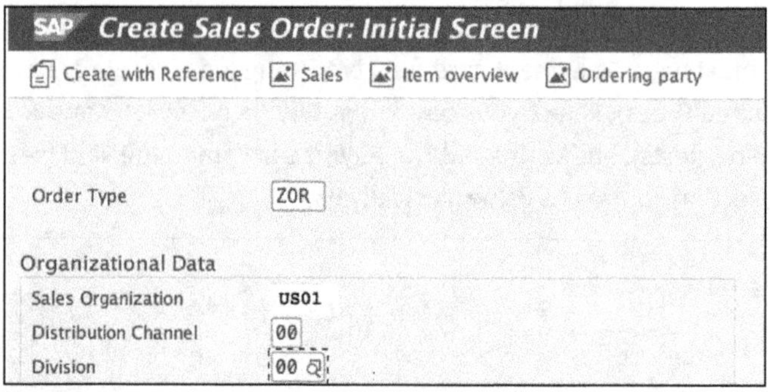

*Figure 5.66: Create Sales Order: Initial Screen*

2. Press Enter to proceed to the order overview screen. On the **Create Sales Order: Overview** screen (see Figure 5.67), enter the following:
   - **Sold-to Party**: Enter customer number "1000000000". The system automatically determines the ship-to, bill-to, and payer information via partner determination.

- **Purchase Order Number**: Enter the customer's PO reference (optional).
- **Requested Delivery Date:** Enter a required delivery date (if not autopopulated).

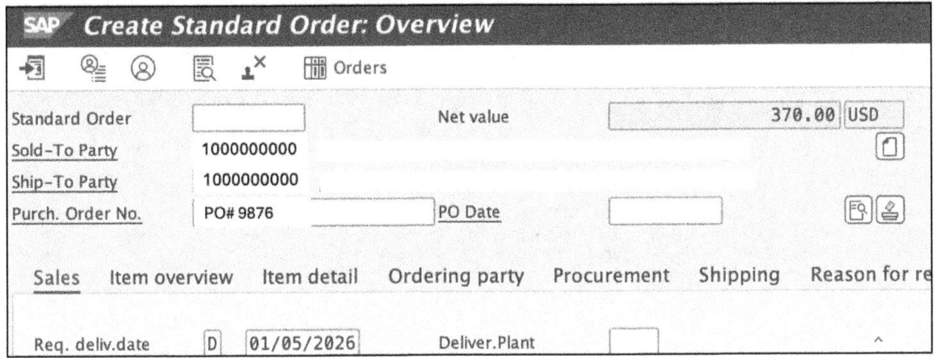


*Figure 5.67: Create Standard Order: Overview*

3. Enter line items. For **Material**, enter the product/service code (e.g., "DPMAT01"), and for **Order Quantity**, enter the quantity ordered (e.g., "1"), as shown in Figure 5.68. The system performs an available-to-promise (ATP) check, determines pricing, and calculates totals including tax.

| Material | Order Quantity | Title |
|---|---|---|
| DPMAT01 | 1 | Demo Material |
| | | |
| | | |

*Figure 5.68: Sales Order: Enter Line Items*

4. Navigate to the sales order header by clicking the **Header** button or selecting **Goto • Header** from the menu. Switch to the **Payment Cards** tab (see Figure 5.69). The payment card plan screen appears, showing sections for card data, authorization information, and settlement details.

Sales | Shipping | Billing Document | Payment cards | Accounting | Conditions | Account assignment | Partners

| | | | |
|---|---|---|---|
| Authorized | 0.00 | Total | 395.90 USD |
| NextDlv/Se | 395.90 | Next date | 01/05/2026 |

Status of last authorization

| | | |
|---|---|---|
| Requirement | ☐ Not Relevant | ☐ Authorization block |
| Call | ☐ Not Relevant | |
| Response | ☐ Not Relevant | |

*Figure 5.69: Sales Order Header: Payment Cards Tab*

5. In the payment card plan overview, in the **Card Number** field, press F4 to display match codes showing all valid payment cards stored in the payer's business partner master. Select the appropriate card from the match code list (see Figure 5.70).

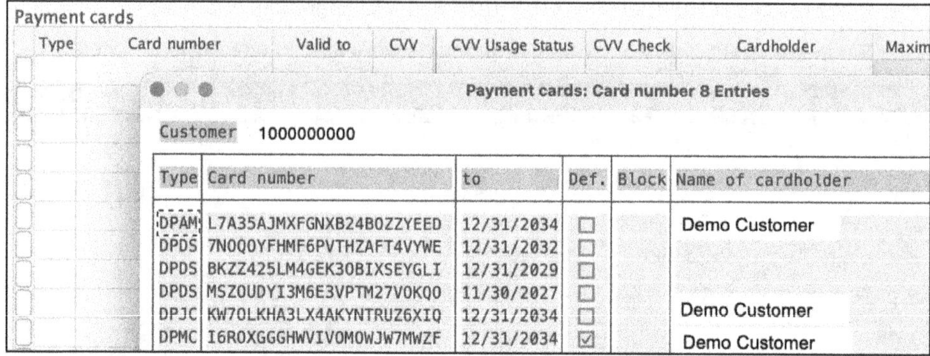


*Figure 5.70: Payment Cards List*

6. Save the sales order using menu option **Sales Document • Save**. During the save process, the system automatically calls function module Z_DP_CCARD_AUTHORIZATION (created by SAP Note 3041924), which executes these operations:
   - Reads payment card data from the sales order (card token, amount, currency)
   - Calls class ZCL_FAR_DP_AUTHORIZATIONS to construct an authorization API request
   - Sends HTTPS POST to the digital payments add-on */core/v1/authorizations* endpoint via RFC destination DIGITALPAYMENTS
   - Add-on routes the request to the appropriate PSP (Stripe, PayPal, etc.) based on routing rules
   - PSP processes authorization and returns approval or decline
   - Function module updates the sales order with authorization results
   - System displays message **Sales order X has been saved** with an authorization status (see Figure 5.71)

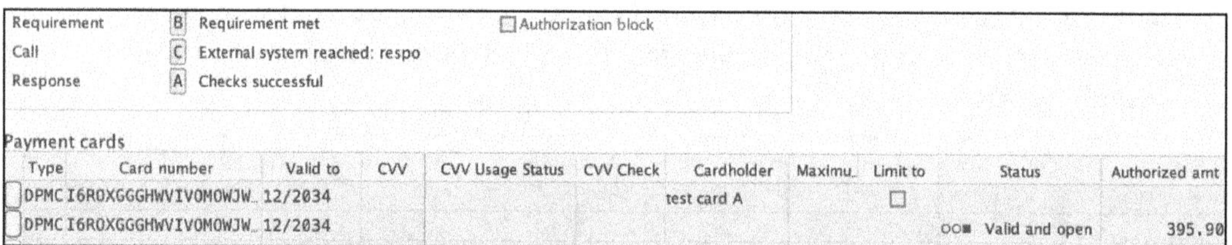


*Figure 5.71: Authorization Status: Traffic Lights*

7. If authorization succeeds, the system saves the authorization details to table FPLTC, including the following:
   - AUNUM: Digital payments authorization ID
   - AUTWR: Authorized amount
   - CCWAE: Authorized currency
   - AUDAT/AUTIM: Authorization date and time (see Figure 5.72)

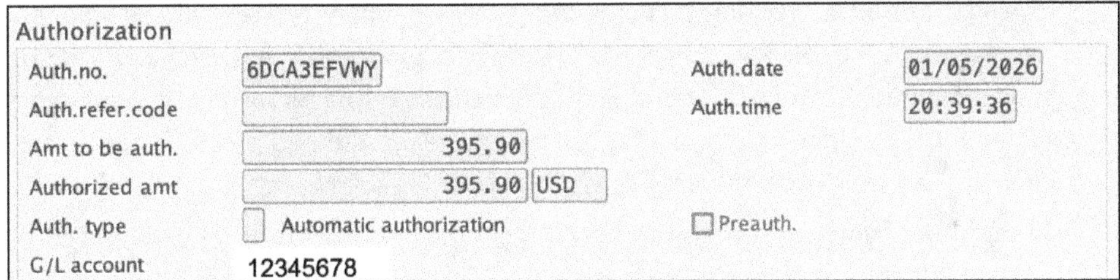


*Figure 5.72: Table FPLTC: Example Values*

8. The sales order is available for further processing (delivery, billing). No blocks are set. If the PSP declines the authorization (insufficient funds, expired card, fraud detection), the system sets a block on the sales order to prevent delivery and billing. The payment card plan shows a red traffic light in the authorization status, displays a decline reason from message class `DIGITAL_PAYMENTS` mapped via table `FAR_DP_T100_MAP` (configured in SAP Notes 3027433 and 3028044), and prevents order progression until the payment issue is resolved.
9. You must either select a different payment card from the customer business partner master, contact the customer for an alternative payment method, or cancel the sales order. The blocked order appears in the payment card worklists for follow-up processing.
10. Delivery processing (for physical goods) works as follows: After the sales order authorization succeeds, standard delivery processing occurs without payment-specific steps. Execute Transaction VL01N (Create Outbound Delivery) with reference to the sales order, perform picking and packing operations, and post goods issue. The delivery document does not interact with payment systems; the authorization from the sales order remains valid until billing occurs.
11. The system checks authorization validity during delivery creation. If significant time has elapsed between order creation and delivery (based on the authorization validity periods configured via Transaction SPRO), the system may warn that the authorization expired and require reauthorization in the sales order before delivery can proceed.

#### *E-Commerce Order Integration with Digital Payments Add-On*

E-commerce integration is a different order-processing pattern compared to manual TransactionVA01 entry. In e-commerce scenarios, customers enter their own orders through web storefronts, payment authorization or capture occurs during checkout before SAP order creation, and the order arrives in SAP ERP as a prepaid transaction requiring different handling than manual orders. There are typically two primary e-commerce integration patterns, as follows:

- **One-step direct capture integration**
  In one-step processing, authorization and settlement occur simultaneously during e-commerce checkout. The customer completes payment before the order reaches SAP

ERP, and the order arrives fully paid, requiring only fulfillment and revenue recognition. This design is common for companies that offer digital products or services with no physical fulfilment involved. The digital fulfilment could mean instantaneous access enabled for the customer or a digital license key generated and provided to the customer (e.g., an anti-virus software license key). Note the following:

- In this flow, if you have obtained consent from the customer to store payment cards on profile in the e-commerce application, then you can enable integration with an SAP ERP business partner to replicate the digital payment add-on tokens. This integration will enable a potential order change or a renewal process to be automated in SAP ERP.
- In the order creation flow, e-commerce applications can post orders to SAP ERP (via IDoc, API, etc.) and, along with the order data (customer, materials, quantities, prices), provide payment reference data for table `FPLTC`, which can be used for a potential order change or cancellation and refund scenarios to be enabled in SAP ERP. The order is created with the payment already captured. No authorization call occurs during order processing because funds are already secured.
- When billing occurs, the system recognizes that the payment was precaptured by checking field `DP_CHARGE_TRANS_ID` in table `FPLTC`. Transaction FCC1 settlement may be bypassed or execute without calling digital payments add-on APIs as capture is already completed. The accounting documents post receivables to the accounts receivable clearing account, but settlement clearing has already occurred externally.

This pattern optimizes e-commerce performance by completing payment processing during checkout without waiting for SAP order creation and billing workflows.

- **Two-step authorize-then-capture integration**
  Two-step processing separates authorization (during e-commerce checkout) from settlement (during SAP billing). This pattern is preferred for physical goods, for which you authorize payment when the customer places the order but only capture funds when goods ship. Consider the following points:
  - In e-commerce authorization (at checkout), first the customer completes checkout. The e-commerce platform then authorizes payment. It can be designed to call a PSP API (e.g., Stripe's API to create a payment intent) and receive authorization, but funds are not captured.
  - Either the e-commerce system or the SAP system can be designed to call the digital payments add-on to validate the PSP payment intent. It creates the authorization record; generates the digital payments card token for the payment method; generates the digital payments authorization ID (`AUNUM`); and saves the PSP payment intent reference, digital payments card token, and digital payments authorization ID, populating table `FPLTC`.
  - The sales order is created with a valid authorization. No authorization API call occurs because authorization already completed during checkout. The order proceeds through standard delivery processing.

- When goods ship and billing occurs, the billing document copies authorization data from the sales order to table BSEGC. Transaction FCC1 settlement processing executes function module Z_DP_CCARD_SETTLEMENT, which calls the digital payments add-on API.
- The settlement function updates table BSEGC, and accounting posts the settlement-clearing entries.

The two-step pattern provides business flexibility, including authorizing payment when a customer commits to a purchase. This ensures the payment method is valid, but captures funds only when goods ship, thus reducing chargebacks for undelivered goods, handling order modifications before capture without refund processing, and canceling authorizations for canceled orders without fund movement. This pattern aligns payment capture with actual fulfillment obligation, improving cash flow accuracy and reducing dispute risk.

For orders processed through e-commerce with the payment already captured, refunds use the digital payments charge ID to reference the original payment. When a refund transaction is initiated in SAP ERP (via a credit memo request or direct refund transaction), the system retrieves the DP_CHARGE_TRANS_ID value from table BSEGC for the original billing document. The refund function calls the digital payments add-on API (/core/v1/refunds/bypaymentbydigitalpaymentservice), passing it the digital payments charge ID, refund amount, and refund currency. The add-on looks up the original charge using DP_CHARGE_TRANS_ID, retrieves the saved Stripe charge ID and merchant ID, and calls the Stripe API to process the refund against the original charge. Stripe processes the refund, creates the refund ID, and returns the refund confirmation. The add-on generates the digital payments transaction ID for the refund and returns both the digital payments transaction ID and Stripe's refund ID to SAP ERP. The refund-by-transaction API eliminates the need to resend card tokens or customer details for refunds, simplifying refund processing and reducing security exposure.

SAP provides two primary mechanisms for credit memo processing: credit memo requests for refunds requiring approval workflows, and direct refund processing using Transaction FIN_DP_REFUND_PC, created by SAP Note 3212765.

For this first mechanism, credit memo requests represent sales documents created to request approval for issuing customer credits. This approval-based approach provides control over refund processing, ensuring proper authorization before payment reversals occur, as follows:

- **Creating a credit memo request**
  - Execute Transaction VA01. Select order type **CR** (Credit Memo Request). Enter the sales organization, distribution channel, and division, then click the **Create with Reference** button.
  - The **Create with Reference** dialog will open. Enter the original sales order number or billing document number in the appropriate tab (**Order** or **Billing Document**) and click the **Copy** button.

- The system displays the **Create Credit Memo Request** screen with data copied from the reference document, including customer information, sold materials and quantities, pricing data, and payment card information (if applicable). Modify the copied quantities if a partial refund is required; change the target quantity to the refund amount and leave items with zero quantity for those not being refunded.
- Navigate to the **Header • Sales** tab, then configure the following:
  - **Billing Block**: The system may automatically set a billing block, requiring approval before a credit memo can be billed.
  - **Order Reason**: Select a reason code (e.g., customer complaint, defective goods, pricing error).
  - **Pricing Date**: This determines which pricing conditions apply (typically, you use the current date).
- Save the credit memo request. The system generates document number and displays a message: **Credit Memo Request has been saved.**

- **Approval and release**
  Credit memo requests typically require approval before billing. The billing block prevents automatic billing until an authorized user releases the document. Approval workflows vary by organization; some use manual release via Transaction VA02 to remove the billing block, others use an automated workflow with electronic approvals, and others rely on integration with external approval systems. After approval, the billing block is removed, enabling credit memo billing.
- **Credit memo billing**
  Execute Transaction VF01, enter the credit memo request number as a reference, and create the credit memo billing document. The credit memo posts these accounting entries:
  - Debit: Revenue account (reverses original revenue)
  - Credit: Customer account (reduces receivables)
  - Debit: Customer account
  - Credit: Accounts receivable clearing account (reverses original accounts receivable posting)

For payment card refunds, the credit memo should trigger refund processing to return funds to the customer's card. This occurs through settlement or dedicated refund processing.

For direct refund processing, SAP Note 3212765 created Transaction FIN_DP_REFUND_PC specifically for processing digital payments add-on refunds against cleared customer line items. This transaction provides direct refund capability without requiring credit memo request workflows, as follows:

- To use Transaction FIN_DP_REFUND_PC, the customer line item to be refunded must have the following digital payments fields populated: `PAYS_PROV` (payment service provider code not initial), `PAYS_TRAN` (PSP transaction reference not initial), `DP_PAYMENT_`

TYPE (payment type not initial), and DP_TRANS_ID (digital payments transaction ID not initial). These fields are populated during original settlement processing via Transaction FCC1.

- A configuration must exist in view FARV_DP_REFUNDPC (per SAP Note 3212765), defining refund general ledger accounts, document types, and cost center assignment rules per company code.
- Payment type to card type mapping must exist in view FARV_DP_PT_CCINS, linking payment type PC to card types (DPVI, DPMC, etc.).
- For refund processing, execute Transaction FIN_DP_REFUND_PC. The selection screen appears with the following criteria:
  - **Company Code**
  - **Customer Number**
  - **Fiscal Year**
  - **Accounting Document Number**
  - Additional filters (posting date, document type, etc.)
- Enter the selection criteria and execute. The system selects cleared customer line items from table BSAD where the digital payments fields are populated, displays selected items with the maximum refund amounts available, and shows pending parked refund documents if any exist.
- Select line items to refund by checking selection boxes, enter the refund amounts (partial or full), and click the **Post** button. The system creates a credit memo document in FI, posts refund entries per the FARV_DP_REFUNDPC configuration and calls the digital payments add-on refund API.
- Report RFIN_DP_REFUND_PC (the underlying program for Transaction FIN_DP_REFUND_PC) executes the following refund logic:
  - Retrieves DP_TRANS_ID or DP_CHARGE_TRANS_ID from table BSEGC for the selected line items
  - Constructs a refund API request with the transaction reference and refund amount
  - Calls the appropriate refund endpoint: either */core/v1/refunds* (if the original card token is available) or */core/v1/refunds/bypaymentbydigitalpaymentservice* (using the digital payments transaction ID)
  - Add-on processes the refund with the PSP
  - System posts accounting documents, creating credit memo and refund-clearing entries

After refund posting via Transaction FIN_DP_REFUND_PC, the refund document appears in Transaction FCC1 for settlement processing. Execute Transaction FCC1 with the refund document selected. The settlement function processes refunds similarly to charges, updating table BSEGC with refund transaction IDs and settlement status. The refund flow completes the payment cycle: original authorization during order creation, settlement (capture) during billing, and refund processing via credit memo or direct refund transaction, all integrated through the digital payments add-on APIs—maintaining a complete audit trail from order through refund.

When integrating e-commerce platforms with SAP ERP for payment card processing, consider these implementation factors:

- **Token lifecycle management**
  E-commerce platforms and SAP must synchronize payment card tokens. When customers add cards in e-commerce portals, tokens should sync to the SAP business partner master (table BUT0CC), enabling use in manual orders and customer service transactions. Token deletion in portals should trigger logical deletion in SAP (card blocked with reason code) rather than physical deletion, maintaining historical references.
- **Duplicate order prevention**
  E-commerce integrations must implement idempotency to prevent duplicate order creation and duplicate payment capture. Use unique transaction IDs from e-commerce as reference keys in SAP, check for existing orders before creating new ones, and implement locking mechanisms to prevent concurrent order creation for the same e-commerce transaction.
- **Authorization expiration handling**
  Two-step patterns with delays between authorization and fulfillment risk authorization expiration. Monitor authorization ages, implement reauthorization workflows for delayed shipments, and communicate expiration risks to warehouse operations, ensuring timely fulfillment within authorization validity.

### 5.3.2 Credit Management Integration

SAP ERP credit management and payment card authorization represent two parallel control mechanisms that work together to minimize payment risk. Credit management evaluates customer creditworthiness based on historical payment behavior, outstanding receivables, and configured credit limits. Payment card authorization validates that the customer's card has sufficient funds and is approved by the issuing bank for the transaction. Both checks can be active simultaneously, with payment card authorization serving as a payment guarantee within the broader credit-management framework.

Credit control areas define the organizational unit where credit management occurs. For payment card integration, the credit control area must be configured with specific update group settings that enable payment card authorization value calculation. Configure the credit control area (in Transaction OB45) with the following information:

- **Update** group: Must be **00012** (see Figure 5.73).
- **Purpose**: This entry is required so the sales order will calculate the value to authorize.

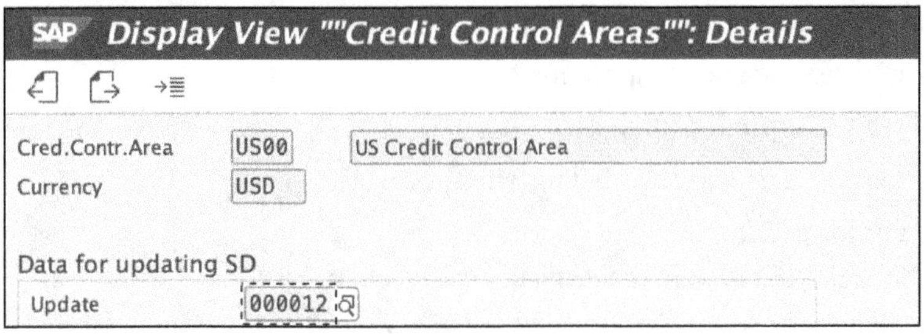

*Figure 5.73: Credit Control Area: Update Group*

Update group 00012 tells the system to include sales order values in authorization amount calculations, ensuring the authorization request sent to the digital payments add-on matches the complete order value, including items, pricing, taxes, and freight charges.

Without the correct update group configuration, authorization amounts may not reflect complete order values, leading to authorization for partial amounts that fail to cover full billing when invoices are created.

Payment guarantee procedures define the sequence of checks executed during sales order processing. You can configure payment guarantee procedures through Transaction SPRO, at menu path **Sales and Distribution • Billing • Payment Cards • Authorization and Settlement • Maintain Payment Guarantee Procedure**. There, define payment cards as a form of payment guarantee (SAP delivers form 02 for payment cards), and maintain the payment guarantee procedure by assigning sequential numbers to different guarantee forms, with payment cards typically assigned sequence 1 (checked first). When multiple payment guarantee forms are configured, the procedure determines their execution order. Payment cards at sequence 1 ensure authorization occurs before checking letters of credit or insurance guarantees.

You can also configure SAP ERP to bypass traditional credit limit checks for customers paying with a payment card as successful payment card authorization provides a stronger payment guarantee than the credit limit evaluation based on historical behavior. You can define credit groups in the IMG under **Sales and Distribution • Basic Functions • Credit Management and Risk Management • Credit Management • Define Credit Groups**. There, select combinations of credit control areas, risk categories, and document types for which credit checking should be bypassed, and mark the **No Credit Check** field with an appropriate value for sales documents, indicating that standard credit limits don't apply when payment cards are used.

This configuration prevents credit blocks on payment card orders even if the customer's receivables exceed the configured credit limit as card authorization provides direct payment confirmation from the issuing bank.

The overall credit status field (CMGST field) in the sales order header (visible in VA03 display under the **Status** tab) reflects the combined results of credit management and payment authorization:

- **Successful payment authorization**
  - Internal risk management status: **Approved**
  - Overall credit status: **Approved** (green)
  - Sales order proceeds to delivery and billing without blocks
- **Failed payment authorization**
  When authorization fails, the system reacts by doing the following:
  - Setting the overall credit status to **Not approved** in the sales order header (status B)
  - Setting an authorization block in the payment card plan
  - Resetting the confirmed quantity to zero
  - Blocking entry in the shipping index to prevent delivery creation
  - Setting requirements in copying control to prevent delivery and goods issue

The blocked order appears in work lists for credit controller review (if the decline reason relates to customer credit) or system administrator review (if the decline relates to technical problems).
To reauthorize after failure, do the following:

- Use report RV21A001 to process blocked sales orders. Remove the authorization block in the payment card plan, correct payment issues (select a different card, update the expiration date, increase the authorization amount), and save the sales order to trigger reauthorization.
- Alternatively, execute Transaction VA02 (Change Sales Order), navigate to the payment card plan, modify payment data, and save. The system automatically reauthorizes when changes are detected in the payment card plan data.

### 5.3.3 Billing Document Processing

Billing triggers the settlement (capture) of previously authorized payment card transactions. When the billing document is created and released to FI, the settlement process executes as follows:

1. **Create billing document**
   Execute Transaction VF01 (Create Billing Document), enter the delivery number or sales order number as a reference, and verify the billing items and amounts. The system copies the payment card data from the sales order into the billing document. Save the billing document, which releases it to FI and triggers receivables posting to the accounts receivable clearing account. The billing document posts the following accounting entries:
   - Debit: Customer account (accounts receivable subledger)
   - Credit: Revenue account
   - Debit: Accounts receivable clearing account configured for the payment card type
   - Credit: Customer account

At this stage, authorization exists, but settlement (fund capture) has not occurred. The receivables remain open on the account receivable clearing account to await settlement processing.

2. **Settlement via Transaction FCC1**
   Execute Transaction FCC1 (Payment Cards: Settlement) to capture authorized funds. The selection screen allows filtering by company code, customer number, billing document number, posting date ranges, and card type. Enter the selection criteria and execute. Transaction FCC1 selects pending settlement items from tables `FPLTC` (authorization data) and `BSEGC` (billing data), groups items by company code and clearing account, reads table `TCCAA` to determine the settlement function module, and calls the settlement function module, passing it the batched settlement items. Function module `Z_DP_CCARD_SETTLEMENT` executes settlement processing.

3. **Settlement success**
   Successfully settled items are marked in `BSEGC` with `SETTL = 'X'`. The open items on AR clearing account are now cleared to the settlement clearing account. Funds have been captured from customer cards and will be transferred to your merchant account by the PSP according to your settlement agreement (typically 2-3 business days).

4. **Settlement failure**
   If settlement fails (expired authorization, PSP communication error, insufficient funds at capture time), Transaction FCC1 displays error messages mapped from PSP error codes via table `FAR_DP_T100_MAP`. Common errors include expired authorization (reauthorization required in sales order), card declined at capture (though previously authorized), and a technical error communicating with the add-on (check RFC destinations). Failed settlements remain open on the accounts receivable clearing account, requiring manual intervention to resolve payment issues with customers or retry settlement after correcting technical problems.

## 5.4 ABAP Development and Enhancements

Implementing the SAP digital payments add-on in an SAP ERP environment is a complex integration project. Unlike SAP S/4HANA, which ships with native hooks for the add-on, SAP ERP requires the developer to intercept standard processing events—sales order saving, delivery creation, billing release—and redirect the logic to the custom digital payments wrapper classes. This section outlines the specific user exits, BAdIs, and custom function modules required to build this bridge.

### 5.4.1 User Exits

In the classic SAP ERP SD architecture, user exits are the primary mechanism for manipulating data during the sales lifecycle.

Most SAP ERP customers use the ORDERS05 IDoc to get orders from upstream customer-facing applications like e-commerce systems or CRM systems. One option is to have these upstream applications integrate with the add-on's APIs. If you have this design in your landscape, the main enhancement you need is VEDA0001. This belongs to the standard IDOC_INPUT_ORDERS function module. Table 5.15 lists the user exits and use cases.

| Component/User Exit | Trigger Point | Common Use Case |
|---|---|---|
| EXIT_SAPLVEDA_001 | Triggers inside the loop as each segment is processed. | Use this for parsing custom segments and moving them into a global memory structure or custom table. This can be used for new digital payments add-on– and PSP-related fields sent from upstream. |
| EXIT_SAPLVEDA_002 | Triggers at the very end, just before the transaction (Transaction VA01) is called. | Use this user exit to manipulate table BDCDATA directly—for example, if you need to force a specific field on the Transaction VA01 screen that isn't mapped by standard logic. |
| EXIT_SAPLVEDA_009 | Triggers after segment looping but before BDC (Batch Data Communication) generation. | Use this for manipulating internal tables. This gives you access to the internal DXVBAK (header) and DXVBAP (item) structures before SAP converts them into screen data. |

*Table 5.15: User Exits and Common Use Cases*

Note that when the IDoc calls Transaction VA01, the standard SD user exits (like MV45AFZZ) will still trigger. If your logic applies to all sales orders (e.g., manual orders), write the code in MV45AFZZ. If your logic applies only to IDocs, write the code in VEDA0001 exits.

For billing documents, use form routine USEREXIT_FILL_VBRK_VBRP within include RV60AFZC to ensure that digital payments data is correctly transported from the sales order header table (VBAK) to the billing document header table (VBRK) during invoice creation. Standard SAP copy control (VTFL) may not automatically map custom or add-on-specific fields like the transaction ID. Common use cases include the following:

- **Mapping payment data**
  Copy the DP_TRANS_ID (digital payments transaction ID) and DP_TOKEN (payment token) values from the sales order payment cards table (FPLTC) into the billing document payment cards table. This ensures that the settlement run has the correct token to capture funds.
- **Authorization validity check**
  Before saving the invoice, verify if the authorization on the token is still valid.

Accounting transfer (EXIT_SAPLV60B_008) executes when the billing document is released to FI to generate the FI invoice. One common use case is to populate the assignment field on the accounts receivable line item within the general ledger, by extracting the digital payments transaction ID (DP_TRANS_ID) from the billing data and mapping it into the assignment field (ZUONR) of the accounting document structure (ACCIT). This helps the electronic bank statement process automatically match the incoming payment to the open receivable, automating the cash application process.

### 5.4.2 BAdI Framework

While user exits modify standard code, BAdIs provide an object-oriented extension method. In the context of the digital payments add-on implementation, one common use case is to use a business partner BAdI (PARTNER_UPDATE) when a business partner is saved in order to validate the token consistency. If a user manually changes a card in Transaction BP, this BAdI could trigger a validation call to the add-on to ensure the token is still active at the PSP. If the PSP responds that the token is deleted, then the BAdI prevents the save in SAP.

### 5.4.3 Add-On-Specific Developments

This section details the custom development objects that can be used when integrating SAP ERP with the digital payments add-on. Best practice is to develop a utility class and implement a few methods. We will present a few use cases for you to consider, including a token exchange, in which a raw PSP token from an external application is converted into a SAP digital payments add-on token that the system can work with; and card registration, which creates the payment card master record and links it to the correct business partner. The final two methods address ongoing card management: retrieving the full set of valid tokens associated with a business partner for use in order processing, and deactivating cards that customers have removed from their profile in a way that preserves historical data integrity rather than physically deleting the record.

#### Token Exchange

In this scenario, assume that the upstream web application was implemented with a direct connection to the PSP (e.g., Stripe), in which the application collects the customer credit card information securely and then receives a token from the PSP. It then sends the PSP token details to SAP. Because SAP cannot use the PSP Stripe token directly, you can develop a method (see Listing 5.1) that takes the raw PSP token and calls the SAP digital payments add-on API (endpoint tokens/getforpaymentcard) to exchange it for an internal SAP digital payments add-on token. This internal token is what you will store in the SAP payment card master.

```
METHOD exchange_token.
IMPORTING
iv_psp_token       TYPE string
```

```
iv_merchant_acct  TYPE string
iv_lifecycle_type TYPE string DEFAULT '01'
EXPORTING
ev_sap_dp_token   TYPE string.
DATA: lv_url          TYPE string,
      lo_http_client  TYPE REF TO if_http_client,
      lv_payload      TYPE string,
      lv_response     TYPE string.

" Constants for JSON construction
CONSTANTS: lc_provider TYPE string VALUE 'STRIPE'.

" 1. Construct the JSON payload for the SAP digital payments add-on API
" This maps the external PSP token to your SAP merchant account
lv_payload = |{{ | &
             | "PaymentCards": [ {{ | &
             |   "PaymentServiceProvider": "{ lc_provider }", | &
             |   "PaytCardByPaytServiceProvider": "{ iv_psp_token }", | &
             |   "MerchantAccount": "{ iv_merchant_acct }", | &
             |   "PaytCardRegnLifeCycleType": "{ iv_lifecycle_type }" | &
             | }} ] | &
             |}}|.

" 2. Create HTTP client (assuming a configured RFC destination for the
add-on)
cl_http_client=>create_by_destination(
  EXPORTING
    destination              = 'SAP_DIGITAL_PAYMENTS'
  IMPORTING
    client                   = lo_http_client
  EXCEPTIONS
    argument_not_found       = 1
    destination_not_found    = 2
    destination_no_authority = 3
    plugin_not_active        = 4
    others                   = 5 ).

IF sy-subrc <> 0.
  " Handle connection error
  RETURN.
ENDIF.

" 3. Configure request
lo_http_client->request->set_method( 'POST' ).
```

```
lo_http_client->request->set_header_field( name = 'Content-Type' value =
'application/json' ).
lo_http_client->request->set_cdata( lv_payload ).

" 4. Send and receive
lo_http_client->send( ).
lo_http_client->receive(
  EXCEPTIONS
    http_communication_failure = 1
    http_invalid_state         = 2
    http_processing_failed     = 3
    others                     = 4 ).

IF sy-subrc <> 0.
  lo_http_client->close( ).
  RETURN.
ENDIF.

" 5. Parse response (simple transformation or regex can be used here)
lv_response = lo_http_client->response->get_cdata( ).

" Extract the SAP token from the JSON response
" (For brevity, assuming a simple regex or parser class is available)
FIND REGEX '"PaytCardByDigitalPaymentSrvc":"(.*?)"' IN lv_response
SUBMATCHES ev_sap_dp_token.

lo_http_client->close( ).
ENDMETHOD.
```

*Listing 5.1: Token Exchange Example ABAP Code*

### Registering a Card

Once you have the token, you must create the master data in SAP. You must create the payment card master (CCARD) that holds the token and expiration date and then link this card to the correct business partner so it can be used in sales orders. The method in Listing 5.2 wraps both BAPIs into a single transaction. It also handles the storage of the original PSP token (Stripe token) in a custom or extension table, as standard SAP CCARD tables often do not have a dedicated field for the raw external token.

```
METHOD register_card.
IMPORTING
iv_partner   TYPE bu_partner
iv_psp_token TYPE string
iv_card_mask TYPE string  " e.g. ************1234
iv_card_type TYPE ccins   " e.g. VI, MC
```

```
iv_valid_to  TYPE datum
EXPORTING
et_return    TYPE bapiret2_t.
DATA: lv_sap_dp_token TYPE string,
      ls_master_data  TYPE bapibus1186_master_data,
      ls_bp_card_link TYPE bapibus1006_pcard_data.

" 1. Exchange the external token for an SAP token
exchange_token(
  EXPORTING
    iv_psp_token      = iv_psp_token
    iv_merchant_acct = 'MY_MERCHANT_ID'
  IMPORTING
    ev_sap_dp_token  = lv_sap_dp_token ).

IF lv_sap_dp_token IS INITIAL.
  " Add error message to et_return
  RETURN.
ENDIF.

" 2. Create SAP Payment Card Master
ls_master_data-card_type   = iv_card_type.
ls_master_data-card_number = lv_sap_dp_token. " SAP Token is the Card
Number
ls_master_data-valid_to    = iv_valid_to.
ls_master_data-mask_number = iv_card_mask.

CALL FUNCTION 'BAPI_PCA_MASTER_CREATE'
  EXPORTING
    card_type   = iv_card_type
    card_number = lv_sap_dp_token
    data        = ls_master_data
    save_direct = abap_true
  TABLES
    return      = et_return.

" Check for errors
IF line_exists( et_return[ type = 'E' ] ).
  RETURN.
ENDIF.

" 3. Link card to business partner
ls_bp_card_link-card_type      = iv_card_type.
ls_bp_card_link-card_number    = lv_sap_dp_token.
```

```
ls_bp_card_link-creditcardname = iv_card_mask. " Store mask as name for
easy ID

CALL FUNCTION 'BAPI_BUPA_PCARD_ADD'
  EXPORTING
    businesspartner = iv_partner
    data            = ls_bp_card_link
  TABLES
    return          = et_return.

" 4. Store external token reference (custom table logic)
" Store the Stripe token (iv_psp_token) in a custom table linked to the
SAP token
" This is vital for reconciliation/search later
INSERT zqtc_token_map FROM @( VALUE #(
  partner     = iv_partner
  sap_token   = lv_sap_dp_token
  stripe_token = iv_psp_token
  created_on  = sy-datum ) ).

" 5. Commit
CALL FUNCTION 'BAPI_TRANSACTION_COMMIT'
  EXPORTING wait = abap_true.
ENDMETHOD.
```

*Listing 5.2: Example ABAP Code to Register Card*

#### Retrieving Tokens

When a sales order is created via an interface (IDoc or API), the system knows the customer's business partner number. It needs to find which card tokens are valid for that customer to attach to the order. The method in Listing 5.3 joins the standard business partner card table (BUT0CC) with the custom mapping table to return a full view of the customer's wallet (masked card, SAP token, and Stripe token).

```
METHOD get_bp_tokens.
IMPORTING
iv_partner TYPE bu_partner
EXPORTING
et_tokens  TYPE ztt_customer_tokens. " Table type of token structure
" Select standard card links joined with custom token data
SELECT a~partner,
       a~ccins AS card_type,
       a~ccnum AS sap_token,
       c~ccname AS masked_number,
       z~stripe_token
```

```
    FROM but0cc AS a
    INNER JOIN ccard AS c
      ON  a~ccins = c~ccins
      AND a~ccnum = c~ccnum
    LEFT OUTER JOIN zqtc_token_map AS z
      ON  a~partner = z~partner
      AND a~ccnum   = z~sap_token
    INTO CORRESPONDING FIELDS OF TABLE @et_tokens
    WHERE a~partner = @iv_partner
      AND c~datbi  >= @sy-datum. " Only valid cards
ENDMETHOD.
```

*Listing 5.3: Example ABAP Code to Retrieve Tokens*

### Deactivating a Card

When a customer removes a card from their profile on the website, you should not physically delete the data from SAP immediately as it may be linked to historical sales orders or invoices. Instead, the best practice is to lock the card. This prevents it from being used on new orders but preserves the data integrity for audit purposes. The method in Listing 5.4 is an example implementation that you can adapt to your specific SAP landscape configuration.

```
METHOD deactivate_card.
IMPORTING
iv_partner   TYPE bu_partner
iv_sap_token TYPE ccnum
EXPORTING
et_return    TYPE bapiret2_t.
DATA: lt_card_details TYPE TABLE OF bapibus1006_pcard_details,
      ls_card_data    TYPE bapibus1006_pcard_data,
      ls_card_x       TYPE bapibus1006_pcard_data_x.
" 1. Fetch current card details to get the card ID
CALL FUNCTION 'BAPI_BUPA_PCARD_GETDETAILS'
  EXPORTING
    businesspartner = iv_partner
  TABLES
    data            = lt_card_details.

READ TABLE lt_card_details INTO DATA(ls_target_card)
  WITH KEY card_number = iv_sap_token.

IF sy-subrc = 0.
  " 2. Prepare the lock flag
  ls_card_data-card_locked = abap_true.
  ls_card_x-card_locked    = abap_true.
```

```
    " 3. Update the card link
    CALL FUNCTION 'BAPI_BUPA_PCARD_CHANGE'
      EXPORTING
        businesspartner  = iv_partner
        card_id          = ls_target_card-card_id
        carddetaildata   = ls_card_data
        carddetaildata_x = ls_card_x
      TABLES
        return           = et_return.
    " 4. Commit if successful
    IF NOT line_exists( et_return[ type = 'E' ] ).
      CALL FUNCTION 'BAPI_TRANSACTION_COMMIT'
        EXPORTING wait = abap_true.
    ENDIF.
  ELSE.
    " Return error: Card not found
  ENDIF.
  ENDMETHOD.
```

*Listing 5.4: Example ABAP Code to Deactivate Card Token*

## 5.5 Authorization and Capture Process

In the order-to-cash lifecycle, the handling of digital payments is split into two distinct financial events: authorization and capture. The SAP digital payments add-on serves as the centralized gateway for these events, abstracting the complexity of communicating with various PSPs.

*Authorization* is the process of verifying that the customer's card is valid and has sufficient funds to cover the transaction and then holding those funds for a specific period. The funds are merely reserved, reducing the cardholder's *open to buy* limit.

The technical execution of an authorization request involves a synchronous round trip between the SAP ERP backend and the SAP digital payments add-on on SAP BTP, as follows:

- **Sales order save**
  Saving an order triggers the SD pricing procedure to calculate the net value plus tax.
- **Function module call**
  The SAP authorization routine (via tables TVCIN and TCCAA) calls a wrapper function module.
- **API payload construction**
  SAP ERP builds a JSON payload with the SAP payment card token (from table FPLTC), amount, currency, and routing parameters like company code and customer country. These determine the correct PSP merchant account (e.g., US Stripe for US customers).

- **REST call**
  The system sends a POST request to /core/v1/authorizations.
- **Response handling**
  On success, the add-on returns a digital payments authorization ID and PSP transaction ID (e.g., Stripe payment intent ID), which are saved in table FPLTC, in the AUNUM and DP_TRANS_ID fields. The sales order credit status is then set to **Approved**.

*Capture*, often called *settlement*, is the action of converting the valid authorization into a finalized charge, causing the funds to move from the customer's bank to the merchant's account. In SAP ERP, this process is decoupled from the sales order and occurs after billing as follows:

- **Billing (Transaction VF01)**
  Authorization data is transferred from the sales order to the billing document when billing occurs.
- **Accounting release**
  The billing document generates an open item in accounts receivable.
- **Settlement run (Transaction FCC1)**
  Transaction FCC1, a nightly batch job, processes payment card settlements.
- **Capture API Call**
  Transaction FCC1 selects open accounts receivable items with valid card data and triggers POST /core/v1/charges via the settlement function module.

In distributed systems, failures occur. The SAP digital payments add-on integration must handle exceptions gracefully to prevent stuck orders or lost revenue—for example:

- **Authorization failures (synchronous)**
  If the API call during the sales order save returns a nonsuccess HTTP code or a business decline (e.g., insufficient funds):
  - User feedback: The error message from the PSP is mapped and displayed directly in the Transaction VA01 status bar.
  - Credit status: The sales order is saved, but the credit status (VBUK-CMGST) is set to B (**Not Approved**).
  - Delivery block: The system automatically applies a delivery block, preventing the warehouse from shipping goods until a valid payment is secured.
- **Capture failures (asynchronous)**
  Settlement happens in the background. If the capture API fails due to an expired authorization or PSP system error:
  - Settlement log: The error is recorded in the settlement log, viewable via Transaction FCC1.
  - Financial impact: The customer accounts receivable item remains open. The system does not clear the item to the PSP clearing account.

- Retry mechanism: Standard SD configuration allows for repetition of settlement. The job can be scheduled to retry failed items the next day.

- **Error code mapping**
PSPs often return cryptic error codes. To make these intelligible to SAP users, the implementation uses a mapping table (`FAR_DP_T100_MAP`). This maps the external PSP error code to a standard SAP table `T100` message, ensuring that logs and user messages are readable and actionable.

## 5.6 Batch Jobs and Background Processing

Although payment card authorization during sales order creation is a real-time, synchronous event, the bulk of financial settlement and reconciliation in a digital payments add-on landscape occurs through background processing. To ensure proper cash flow management, accurate accounting, and system synchronization, a specific sequence of batch jobs must be scheduled and monitored. The following are the standard SAP payment programs:

- **Collective billing processing (Transaction VF04/program SDBILLDL)**
Batch-create billing documents for multiple deliveries or orders. Transaction VF04 processes the billing due list, generating invoices for completed deliveries (post goods issue) or order-related billing. It can combine invoices by customer, billing type, and sales organization, or create separate ones as needed. For payment card processing, Transaction VF04 generates receivables that Transaction FCC1 settles; run the former before the latter to ensure that receivables are ready for settlement.
- **Payment card settlement (Transaction FCC1/program RFCCSSTT)**
The primary settlement processing for payment card transactions occurs through standard SAP Transaction FCC1, which executes in background mode for batch processing of multiple transactions simultaneously.

The digital payments add-on integration introduces specific maintenance programs that keep SAP ERP synchronized with the cloud-based add-on configuration and enable payment advice processing for reconciliation. PSP synchronization program FAR_DP_UPDATE_PSP_PAYMENT_METH synchronizes the PSP list and payment methods from the digital payments add-on to SAP ERP. The recommended schedule is weekly or after any PSP configuration changes in the digital payments add-on tenant. Run the program manually after adding new PSPs or activating new payment methods.

Payment advice processing program RFDP_ADVICE_V2 retrieves payment settlement advice files from the digital payments add-on and posts clearing entries. This program only functions if payment advice processing is enabled in your digital payments add-on configuration. If you are enabling advice processing, you must configure electronic bank statement settings per SAP Note 3003533.

The final background processing step reconciles payment settlements with actual bank deposits, ensuring that digital payments clearing accounts balance to zero after all transactions clear. Electronic bank statement processing (Transaction FF.5) processes electronic bank statement files, clearing PSP accounts to bank accounts. When the PSP deposits net settlement amounts into your bank account (typically two to three business days after capture), it provides electronic bank statement files in formats like CAMT.053, MT940, or BAI. Transaction FF.5 processes these files for posting clearing entries.

Table 5.16 provides a consolidated list of jobs along with their recommended schedules.

| Sequence | Job/Transaction | Frequency | Timing | Purpose |
|---|---|---|---|---|
| 1 | Transaction VF04 (Billing Due List) | Hourly or daily | Business hours | Create billing documents from deliveries |
| 2 | Transaction FCC1 (Settlement) | Daily | After last billing run | Capture funds for billed transactions |
| 3 | Program FAR_DP_UPDATE_PSP_PAYMENT_METH | Weekly | Weekend | Sync PSP list from add-on |
| 4 | Report RFDP_ADVICE_V2 | Daily | After settlement | Retrieve payment advice for reconciliation |
| 5 | Transaction FF.5 or FF67 (Bank Statement) | Daily | After advice processing | Clear PSP clearing accounts |

*Table 5.16: Checklist of Jobs for Payment Processing*

## 5.7 Custom Reports and Interfaces

Standard SAP reports (e.g., report FBL5N for customer line items or Transaction VA05N for sales orders) lack visibility into the specific token IDs, PSP transaction IDs, and authorization codes provided by the digital payments add-on. It is recommended to build a custom digital payments dashboard that joins sales, billing, and token data into a single view. There are several options to build a dashboard; which one you choose depends on your organization's reporting tools preferences. Some generic options include the following:

- Develop a custom transaction by replicating a standard SAP reporting transaction, such as Transaction VA05N, and enhance the selection screen and report layout to incorporate add-on-relevant fields introduced through various SAP Notes. The specific fields to be added should be determined based on the reporting requirements of your organization.

- You can build a CDS view by merging different tables and fields, then use a reporting tool such as Microsoft's Power BI to create an interactive dashboard. Power BI can connect with CDS views that have been exposed as OData services.
- Alternatively, you can develop an SAP HANA view by bringing together the necessary tables and fields and then link it to Power BI through the Power BI SAP HANA connector.

Effective operational monitoring involves examining both functional logs and technical connections. SAP application logs (Transaction SLG1) are provided, and support teams are encouraged to review these logs daily to identify any token exchange failures.

The digital payments add-on also comes equipped with features for logging and monitoring. Its components record comprehensive details about transaction processing, configuration updates, and errors through the SAP Application Logging service for SAP BTP.

In the Cloud Foundry environment, built-in monitoring tools oversee application health, track instance uptime, and measure resource consumption. Administrators can further enhance this oversight by forwarding metrics and alerts to enterprise solutions like SAP Cloud ALM.

## 5.8 Summary

This chapter provided details on the effort required to enable SAP ERP for the digital payments add-on. We began by establishing the master data foundation, converting the business partner into a secure vault for tokenized payment cards. We then moved to the transactional layer, implementing the authorization logic during sales order entry and the capture logic during billing. With the SAP ERP backend now fully enabled to handle secure digital transactions, the following chapters will explore the specifics of migrating this architecture to SAP S/4HANA and the testing methodologies.

Chapter 6
# Implementation in SAP S/4HANA

*With the add-on's architecture established and its platform components configured, we now turn our focus to the SAP S/4HANA system, where payment processing executes within your business workflows. When implementing the digital payments add-on in SAP S/4HANA, architectural concepts become operational reality: activating the integration, configuring payment methods, enabling SAP Fiori apps, and embedding digital payments into the order-to-cash process that drives daily revenue collection.*

This chapter focuses on implementing the SAP digital payments add-on in SAP S/4HANA, where the integration is more streamlined and out of the box. We begin with integration activation steps that establish the connection between SAP S/4HANA and the add-on, followed by payment method configuration across cross-application, sales and distribution, and financial accounting layers. We then cover the standard SAP Fiori applications that provide a modern user interface for payment monitoring and built-in payment processing capabilities. Then we walk you through the complete order-to-cash process integration, from sales order authorization through delivery, billing, settlement, and bank reconciliation. Finally, the chapter contrasts cloud and on-premise deployment options and explains migration considerations when moving from SAP ERP to SAP S/4HANA.

## 6.1 Integration Activation in SAP S4HANA

Unlike SAP ERP, which requires extensive custom development, SAP S/4HANA offers native integration capabilities that significantly reduce implementation complexity. The activation process establishes secure communication between the SAP S/4HANA system and the add-on service running on SAP BTP, enabling seamless payment processing across the business partner master data, sales orders, billing documents, and accounts receivable functions.

The integration approach differs between SAP S/4HANA Cloud Public Edition and SAP S/4HANA on-premise or SAP S/4HANA Cloud Private Edition deployments. Cloud environments use communication arrangements configured through SAP Fiori applications, while on-premise systems rely on RFC destinations managed through traditional SAP transactions. Despite these technical differences, both deployment models achieve the same functional outcome: a secure, standards-based connection that enables digital payment processing without storing sensitive payment card data within the ERP system.

Activating the integration begins with verifying prerequisites and system requirements to confirm your SAP S/4HANA release is ready; then it identifies the specific SAP Notes needed to enable add-on connectivity. From there, you obtain the SAP BTP subscription and service key that provides the credentials SAP S/4HANA needs to authenticate with the add-on, before finally using those credentials to configure the technical RFC destinations that establish the live connection for on-premise and private cloud SAP S/4HANA systems.

### 6.1.1 Prerequisites and System Requirements

Before beginning the integration activation, verify that all necessary prerequisites are in place. Missing prerequisites can cause configuration failures or an incomplete integration that may not become apparent until payment processing is attempted in production scenarios. SAP S/4HANA support for the digital payments add-on began with release 1709, with continuous functional enhancements delivered in subsequent releases and through additional SAP Notes. Before implementing the add-on, you should verify the SAP S/4HANA release level and review the specific capabilities available for your version.

If you are running releases earlier than 1709, you must first upgrade the SAP S/4HANA system before implementing the digital payments add-on. While basic payment card functionality is available from release 1709, later releases provide important capabilities such as external payment methods, payment by link, and enhanced reconciliation features that many organizations require for their digital payment strategies.

### 6.1.2 SAP Note Identification

Integration activation necessitates the implementation of several SAP Notes that provide the technical basis for add-on connectivity. As the add-on is updated through an ongoing feature release cycle, it is advisable to review the most recent SAP Notes relevant to your specific implementation scope. The steps ahead can assist in identifying the appropriate SAP Notes and their respective versions. It is also important to consult with a qualified expert who has prior experience with the add-on to ensure the correct list of SAP Notes is finalized.

To identify SAP Notes, follow these steps:

1. Open your browser and go to the SAP Note search URL: *https://me.sap.com/mynotes/search*. The page will open on the **Expert Search** tab, as shown in Figure 6.1.

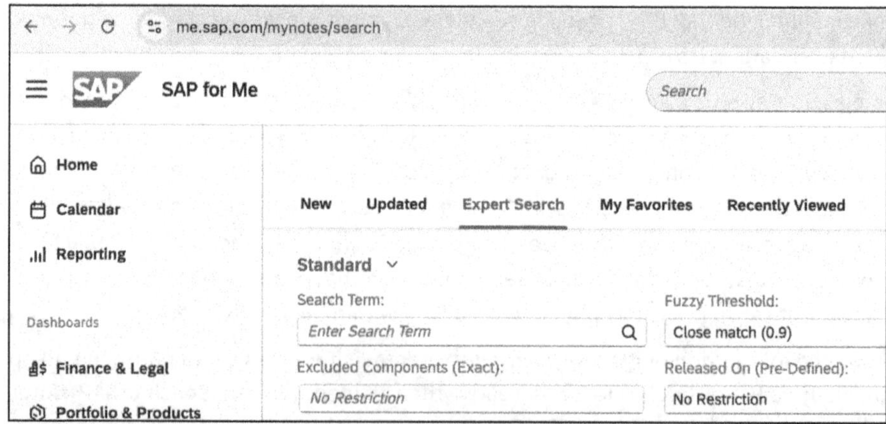


*Figure 6.1: Add-On SAP Note Search*

2. Enter "FIN-FSCM-DP-DP" in the **Component** search field and select the matching result, as shown in Figure 6.2.

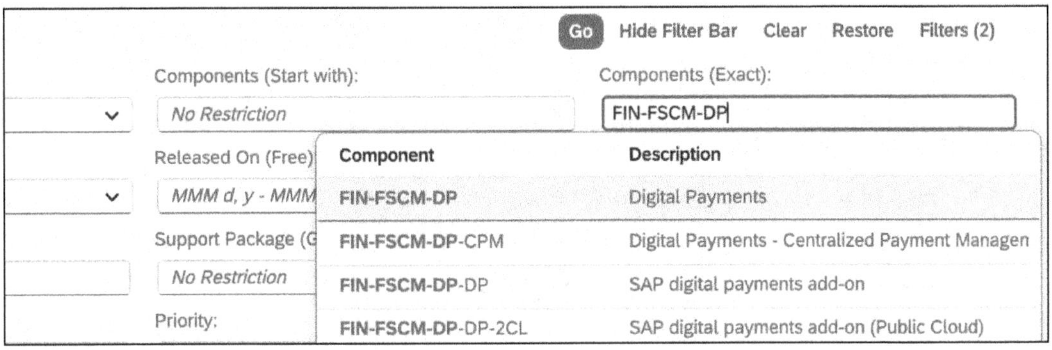

| Component | Description |
|---|---|
| FIN-FSCM-DP | Digital Payments |
| FIN-FSCM-DP-CPM | Digital Payments - Centralized Payment Managen |
| FIN-FSCM-DP-DP | SAP digital payments add-on |
| FIN-FSCM-DP-DP-2CL | SAP digital payments add-on (Public Cloud) |

*Figure 6.2: Add-On Component Search Term*

3. From the **Category** dropdown menu, select **Consulting**, then click **Go**, as shown in Figure 6.3.

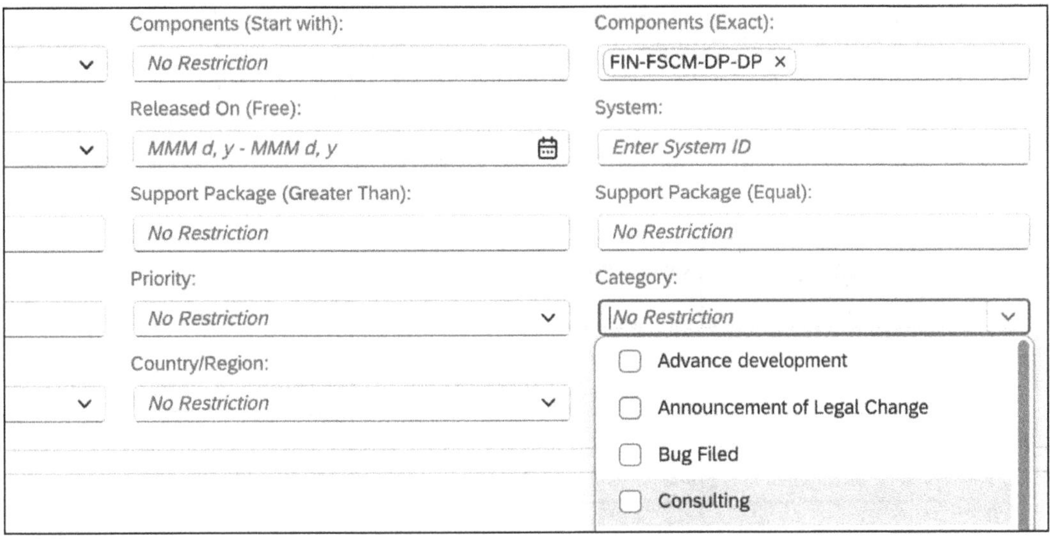

*Figure 6.3: Add-On SAP Notes Category Selection*

4. Review the search results thoroughly and select the SAP Notes relevant to your implementation scope. You may find SAP Notes related to the `FIN-FSCM-DP-DP-2CL` component, which can be disregarded unless you are working with an SAP S/4HANA Cloud Public Edition system. Enhancements for SAP S/4HANA Cloud Public Edition systems are automatically delivered through quarterly SAP release updates. Additional SAP Note categories, such as *program error*, *problem*, and *how-to*, should also be considered and included based on the SAP S/4HANA release in your environment. Among the search results (see Figure 6.4), SAP Note 2524512 is the most crucial and should be reviewed first.

| SAP Component | Number | Version | Title | Category |
|---|---|---|---|---|
| **FIN-FSCM-DP-DP** | **3392691** | **3** | **SAP Digital Payments Add-on: Performing Downtime-free Key Rotation of M2M Credentials** | **Consulting** |
| FIN-FSCM-DP-DP | 3439629 | 4 | Custom Process Handling of SAP Digital Payments responses | Consulting |
| FIN-FSCM-DP-DP | 3602070 | 4 | Reverse Payment Card payment posting of open receivables | Consulting |
| FIN-FSCM-DP-DP | 2730920 | 45 | Enable external payments (alternative payments) with "SAP digital payments add-on" in SAP S/4HANA (on-premise, extended cloud, private cloud edition) | Consulting |
| FIN-FSCM-DP-DP | 2524512 | 59 | Integration of SAP S/4HANA (on-premise extended cloud, private cloud edition) with the application "SAP digital payments add-on" | Consulting |

*Figure 6.4: Add-On SAP Notes Results Review*

### 6.1.3 SAP BTP Subscription and Service Key

SAP S/4HANA integration requires an active SAP digital payments add-on subscription within your SAP BTP global account. This subscription was established during the SAP BTP setup activities described in Chapter 3. For integration activation, you need to generate a service key that contains the connection credentials and endpoint URLs required for SAP S/4HANA to authenticate and communicate with the add-on service.

Service key generation follows the process documented in Chapter 3, Section 3.3. Navigate to your SAP BTP subaccount, access the service instance for the digital payments add-on, and create a new service key with a descriptive name such as *S4HANA-PROD-KEY* or *S4HANA-QAS-KEY*, depending on your environment. The generated service key contains several critical parameters that will be used during RFC destination or communication arrangement configuration. Table 6.1 describes the key service key parameters and their purposes.

| Parameter | Description | Use in SAP S/4HANA Configuration |
|---|---|---|
| `uaa.url` | Authorization server endpoint for OAuth 2.0 token requests | Host and path prefix for `OAUTH` RFC destination |
| `uaa.clientid` | OAuth 2.0 client identifier for the SAP S/4HANA system | User name for `OAUTH` RFC destination authentication |
| `uaa.clientsecret` | OAuth 2.0 client secret for authentication | Password for `OAUTH` RFC destination authentication |
| `url` | Base API endpoint for the add-on core service | Host for main API RFC destination |
| `Subdomain` | Tenant-specific subdomain identifier | Path construction for token endpoint |

*Table 6.1: Service Key Parameters for SAP S/4HANA Integration*

The service key JSON structure includes additional parameters such as certificate information, API version indicators, and landscape-specific settings. Not all parameters are used during SAP S/4HANA configuration, but you can retain the complete service key for reference during troubleshooting or when implementing additional integration scenarios. You can also verify that the service key URLs match the expected data center location (e.g., US10 for the North American data center).

### 6.1.4 Technical Integration Configuration for On-Premise Systems

SAP S/4HANA on-premise and SAP S/4HANA Cloud Private Edition systems establish connectivity to the digital payments add-on through RFC destinations configured in Transaction SM59. The integration requires two distinct RFC destinations serving different purposes in the bidirectional communication flow between SAP S/4HANA and the add-on service running on SAP BTP. Completing the technical integration involves four configuration steps, each of which depends on the one before it. You start by importing the SSL certificate into SAP S/4HANA's trust store to enable encrypted HTTPS communication with SAP BTP. With that secure channel in place, you configure the OAuth RFC destination, which handles authentication by obtaining access tokens from the SAP BTP authorization server. Those tokens are then used by the core API RFC destination, which routes all payment transaction traffic to the add-on service. Finally, integration activation in Customizing switches on the digital payments functionality within SAP S/4HANA so that the RFC connections you have built are used by the system.

#### SSL Certificate Import

Before creating RFC destinations, you must verify that the SAP S/4HANA system's SSL client trust store contains the necessary certificates for encrypted HTTPS communication with SAP BTP. The digital payments add-on service uses certificates issued by DigiCert—specifically, the DigiCert TLS RSA4096 Root G5 certificate. Organizations already using other SAP BTP services may have imported this certificate previously, but verification is essential to avoid connection failures during testing. To proceed, follow these steps:

1. Access Transaction STRUST to open Trust Manager. Navigate to the **SSL Client (Standard)** node in the certificate tree on the left side of the screen. Double-click the node to display the current certificate list. Verify whether a certificate with subject `CN=DigiCert TLS RSA SHA256 2020 CA1` or `CN=DigiCert TLS RSA4096 Root G5` exists in the certificate list (see Figure 6.5). If these certificates are not present, proceed with manual import.

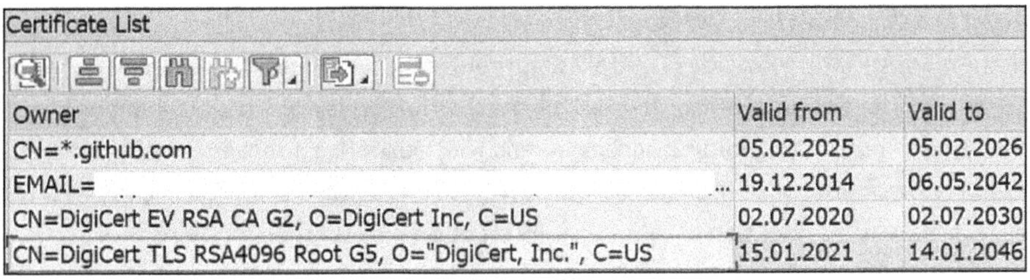

Certificate List

| Owner | Valid from | Valid to |
|---|---|---|
| CN=*.github.com | 05.02.2025 | 05.02.2026 |
| EMAIL= ... | 19.12.2014 | 06.05.2042 |
| CN=DigiCert EV RSA CA G2, O=DigiCert Inc, C=US | 02.07.2020 | 02.07.2030 |
| CN=DigiCert TLS RSA4096 Root G5, O="DigiCert, Inc.", C=US | 15.01.2021 | 14.01.2046 |

*Figure 6.5: DigiCert TLS Certificate*

2. Download the DigiCert root certificate from the official DigiCert repository at *https://www.digicert.com/kb/digicert-root-certificates.htm*. Locate the DigiCert TLS RSA4096 Root G5 certificate in the root certificates section and download it in Base64-encoded format. Save the certificate file to your local workstation with a recognizable filename, such as *DigiCertTLSRSA4096RootG5.crt*.
3. Return to Transaction STRUST and ensure the **SSL Client (Standard)** node is selected. Click the **Import Certificate** button in the toolbar, navigate to the downloaded certificate file, and select it for import. The system displays the certificate details, including the issuer, subject, validity period (see Figure 6.6), and fingerprint. Verify that the certificate information matches the expected DigiCert root certificate, then click the **Add to Certificate List** button to import the certificate into the trust store.

| Certificate | |
|---|---|
| Subject | CN=DigiCert TLS RSA4096 Root G5, O="DigiCert, Inc.", C=US |
| Subject (Alt.) | |
| Issuer | CN=DigiCert TLS RSA4096 Root G5, O="DigiCert, Inc.", C=US |
| Serial Number (Hex.) | 08:F9:B4:78:A8:FA:7E:DA:6A:33:37:89:DE:7C:CF:8A |
| Serial Number (Dec.) | 11930366277458970227240571539258396554 |
| Valid From | 15.01.2021 00:00:00 to 14.01.2046 23:59:59 |
| Algorithm | RSA |
| Key Strength | 4096 |
| Signature Algorithm | RSA+SHA384 |

*Figure 6.6: DigiCert TLS Certificate Details*

4. After importing the certificate, click the **Save** button to persist the changes to the trust store. Transaction STRUST may prompt you for confirmation before saving, as trust store modifications affect all HTTPS communications from the SAP S/4HANA system.

The certificate import is a one-time activity for each SAP S/4HANA system. Once imported, the certificate remains valid until its expiration date, which is typically several years in the future for root certificates. DigiCert periodically issues updated root certificates, so organizations should monitor certificate expiration dates and plan for certificate rotation before expiration to avoid service disruptions.

#### OAuth Authentication RFC Destination

The first RFC destination provides OAuth 2.0 authentication services for API calls from SAP S/4HANA to the digital payments add-on. This destination does not handle payment transactions directly but rather obtains access tokens that authorize subsequent API requests. The OAuth flow follows the industry-standard client credentials grant type designed for server-to-server authentication without user interaction. Follow these steps:

1. Open Transaction SM59 to access the RFC destination configuration. Click the **Create** button to begin defining a new destination. The system presents a dialog requesting the

destination name. Enter "DIGITALPAYMENTS_OAUTH" as the RFC destination name (see Figure 6.7), ensuring the name exactly matches this value; it is referenced by name in the payment-processing programs. Select **connection type G**, which represents the HTTP connection to an external server, enabling HTTPS communication with the SAP BTP authorization server.

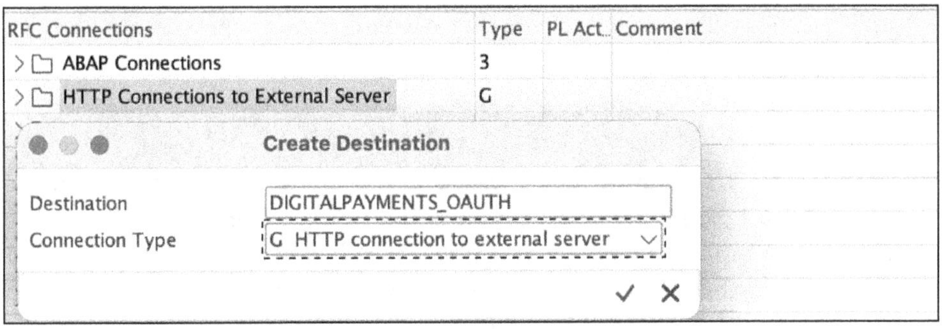

*Figure 6.7: DIGITALPAYMENTS_OAUTH RFC Destination Creation*

2. On the **Technical Settings** tab, configure the connection parameters using values from the service key generated during SAP BTP setup. In the target **Host** field, enter the hostname portion of the *uaa.url* value from the service key. Extract only the hostname component, removing the *https://* protocol prefix. Set the **Port** field to 443, the standard HTTPS port number (see Figure 6.8). The **Path Prefix** field requires careful construction based on your SAP BTP landscape location and subdomain. The path must include the OAuth token endpoint and subdomain-specific routing information.

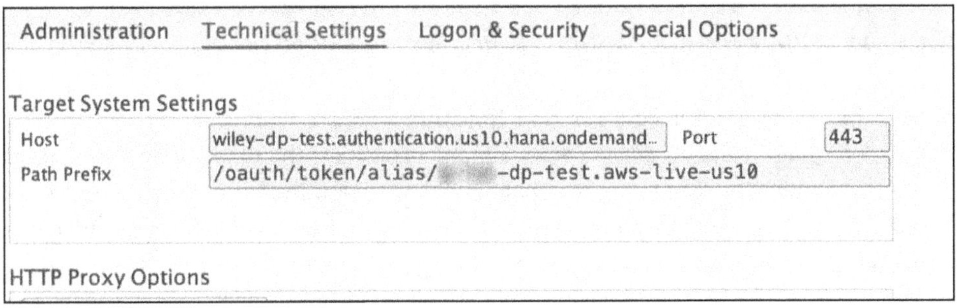

*Figure 6.8: RFC Destination: Technical Settings*

3. Navigate to the **Logon and Security** tab to configure authentication credentials. In the **Logon Procedure** section, select the **Basic Authentication** option (see Figure 6.9). This authentication method sends the client ID and secret as HTTP basic auth credentials during token requests, which is the standard approach for OAuth client credentials flow. Enter the `uaa.clientid` value from the service key into the **User** field. In the **Password** field, enter the `uaa.clientsecret` value from the service key. This is a long random string that must be entered precisely without any modifications or truncation.

Administration Technical Settings Logon & Security Special Options

Logon Procedure
Logon with User
Do Not Use a User
OAuth Settings
Basic Authentication
User
PW Status saved

*Figure 6.9: DIGITALPAYMENTS_OAUTH: Logon Credentials*

4. Configure the SSL security options in the **Security Options** section of the **Logon and Security** tab. Set the **SSL** option to **Active** to enable HTTPS encryption for all communication through this destination. For the **SSL Certificate** field, select **DEFAULT SSL Client (Standard)** from the dropdown list (see Figure 6.10). This selection instructs the system to use the certificates imported into the SSL client (standard) trust store managed through Transaction STRUST, including the DigiCert certificate imported earlier.

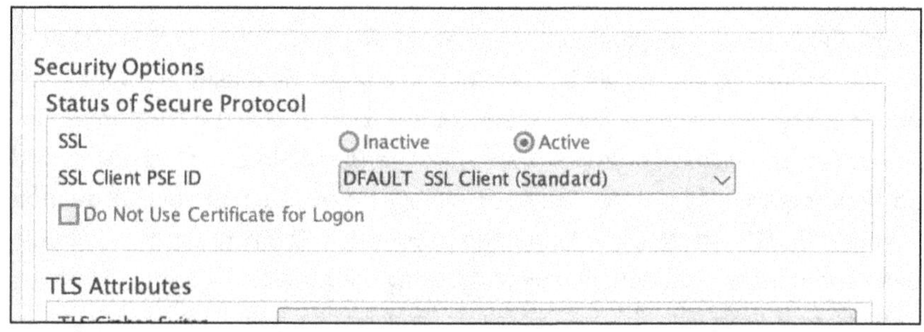


*Figure 6.10: DIGITALPAYMENTS_OAUTH RFC Destination: Security Options*

5. Save the RFC destination configuration. Note that the Transaction SM59 connection test function cannot validate OAuth RFC destinations because the token endpoint requires `POST` requests with specific OAuth parameters that the standard connection test does not support. The OAuth destination will be validated indirectly when the main API destination successfully obtains access tokens during payment processing.

### Core API RFC Destination

The second RFC destination handles all operational API communication from SAP S/4HANA to the digital payments add-on, including payment card registration, authorization requests, settlement processing, refund execution, and payment advice retrieval. This destination routes payment transaction payloads to the add-on's core service API endpoints using the OAuth access tokens obtained through the first destination. Follow these steps:

1. Create a new RFC destination in Transaction SM59 with the name "DIGITALPAYMENTS". This exact naming is required as payment-processing programs reference the destination

by this specific name. Select **G** (**HTTP Connection to External Server**) for **Connection Type**, matching the type used for the OAuth destination (see Figure 6.11).

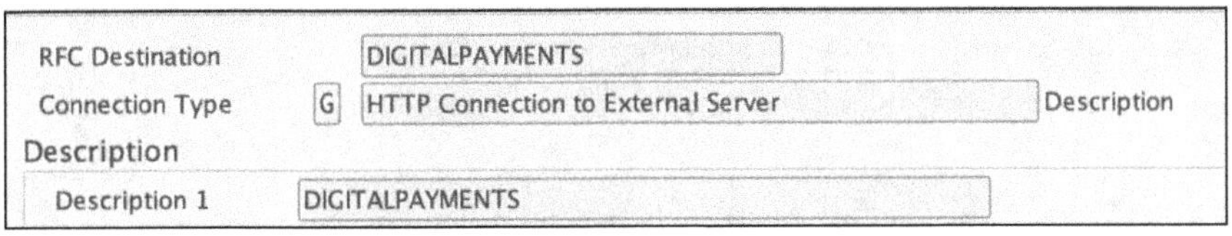

RFC Destination DIGITALPAYMENTS
Connection Type G HTTP Connection to External Server Description
Description
Description 1 DIGITALPAYMENTS

*Figure 6.11: Core API RFC Destination Creation*

2. The **Technical Settings** tab requires different configuration values than those for the OAuth destination, as this destination connects to the add-on's core API service rather than the authorization server. The target **Host** field under **Target System Settings** must contain the fully qualified domain name of the add-on API endpoint (see Figure 6.12), which varies based on your SAP BTP landscape and whether the environment is for testing or production use.

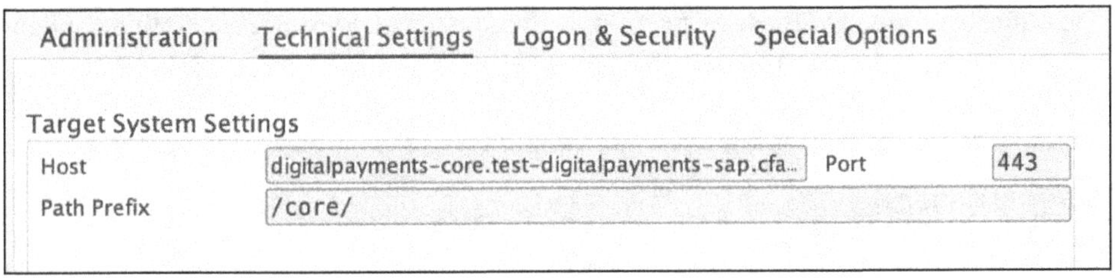

Administration Technical Settings Logon & Security Special Options
Target System Settings
Host digitalpayments-core.test-digitalpayments-sap.cfa... Port 443
Path Prefix /core/

*Figure 6.12: Technical Settings*

3. Select the appropriate **Host** value based on your SAP BTP landscape and deployment environment. Development and quality assurance SAP S/4HANA systems should connect to test add-on tenants, while production SAP S/4HANA systems must connect to production add-on tenants. This will ensure proper separation of environments and prevent test transactions from affecting production payment processing or financial reporting.
4. Set the **Port** field to "443" for HTTPS communication. The **Path Prefix** field depends on your SAP S/4HANA release level and must be configured correctly to ensure that API requests route to the appropriate service version. For SAP S/4HANA release 1709, enter "/core/v1/" as the path prefix. For SAP S/4HANA release 1809 and all subsequent releases, enter "/core/" as the path prefix. The version-specific path ensures compatibility between the SAP S/4HANA client code and the add-on API service contract.
5. The **Logon & Security** tab (see Figure 6.13) configuration differs significantly from the OAuth destination, as this destination does not send user credentials but rather relies on OAuth tokens for authentication. In the **Logon Procedure** section, select the **Do Not Use a User** option. This setting instructs the HTTP client to not send any HTTP basic authentication headers. Instead, the payment-processing programs will programmatically retrieve OAuth tokens from the `DIGITALPAYMENTS_OAUTH` destination and include them

as bearer tokens in the HTTP `Authorization` header of API requests sent through this destination.

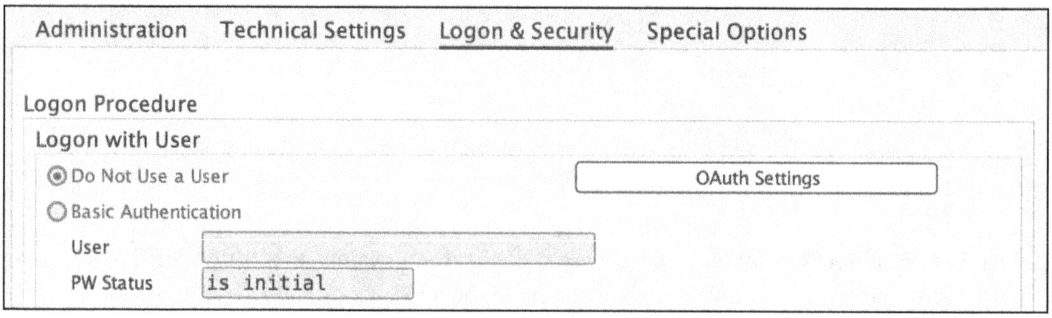

*Figure 6.13: Logon & Security Tab*

6. Configure SSL security identically to the settings for the OAuth destination. Set the **SSL** option to **Active** (see Figure 6.14) and select **DEFAULT SSL Client (Standard)** for the SSL certificate dropdown. These settings apply the same trust store and encryption configuration used for the OAuth connection, ensuring consistent security across both RFC destinations.

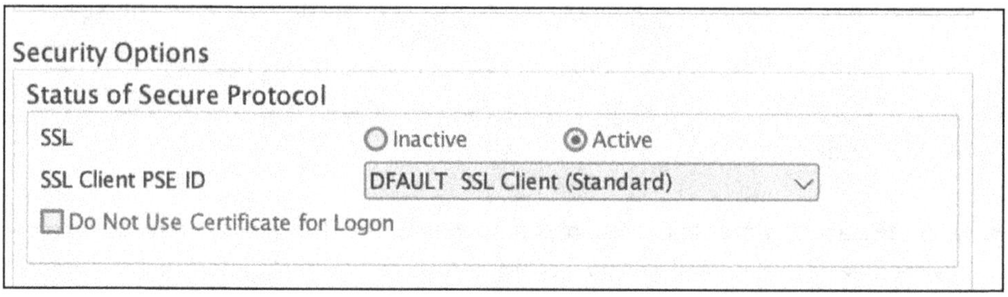

*Figure 6.14: Security Options*

7. Save the RFC destination. As with the OAuth destination, the standard connection test function in Transaction SM59 cannot validate this destination because the add-on API endpoints require OAuth bearer tokens in the request headers, which the connection test does not provide. The destination will be validated during the integration activation testing described later in this section.

The RFC destination configuration represents the technical foundation for all SAP S/4HANA and add-on communication. Any errors in the destination setup will prevent payment processing from functioning correctly. Common configuration mistakes include incorrect path prefixes, mismatched landscape URLs between test and production environments, missing SSL certificates, and typographical errors in host values. Careful attention during RFC destination setup prevents troubleshooting delays during later implementation phases.

#### Integration Activation in Customizing

With the technical connectivity established through RFC destinations, the next step activates the digital payments add-on integration within SAP S/4HANA Customizing. This activation enables payment-processing features across the business partner management, sales and distribution, and financial accounting functionalities.

For SAP S/4HANA on-premise and SAP S/4HANA Cloud Private Edition systems, access the Customizing configuration through Transaction SPRO. Navigate to **SAP Reference IMG** and follow the path **Integration with Other SAP Components • Integration with the SAP Digital Payments Add-On**. This Customizing section contains all configuration activities required to activate and configure the integration.

The integration configuration is organized into logical groupings that cover activation settings, scenario enablement, tenant determination, accounts receivable settings, and payment method configurations. Work through these configuration activities systematically to ensure a complete integration setup. The following subsections describe each configuration activity, discussed in the sequence in which they should be performed.

**Activate SAP Digital Payments Add-On** is a fundamental configuration activity that controls whether the digital payments add-on integration is enabled systemwide. Locate the **Activate SAP Digital Payments Add-On** configuration activity within the integration configuration and execute the activity to display the activation settings screen.

The screen presents a checkbox labeled **Active**. Select this checkbox to enable the integration (see Figure 6.15). When activated, the system unlocks all digital payment processing capabilities, including payment card management in business partners, electronic payment fields in sales documents, authorization request handling, settlement processing, and payment advice reconciliation.

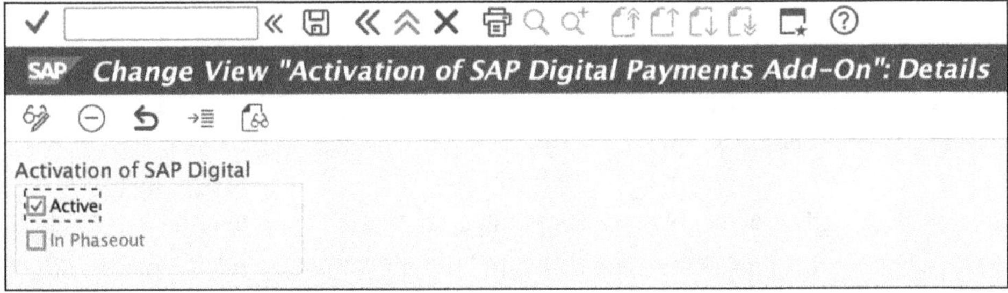


*Figure 6.15: Activation of Add-On in Transaction SPRO*

Save the configuration setting. Deactivating the integration by deselecting the **Active** checkbox disables all payment-processing functions but does not delete existing payment data or configuration settings. You can temporarily deactivate the integration during troubleshooting or when switching between PSPs, then reactivate it without reconfiguring all settings. However, avoid deactivating the integration in production environments during active business hours as in-progress payment transactions may fail if the integration is disabled.

The digital payments add-on supports multiple business scenarios that can be activated independently based on implementation requirements. Access the **Activate SAP Digital Payments Scenarios** configuration activity to define which scenarios your implementation will use. The configuration screen presents a list of available scenarios (see Figure 6.16).

SAP Digital Payments Add-On Scenario 6 Entries

| Scenario | SAP Digital Payments Scenario Description |
|---|---|
| 01 | Digital Payments for Accounts Receivable only |
| 02 | Digital Payments for Accounts Receivable with SD Billing |
| 03 | Digital Payments for Accounts Receivable with SD Order |
| 04 | Digital Payments for Contract Accounting and Convergent Invoicing |
| 05 | Digital Payments for Accounts Receivable with SD Order and Service Processing |
| 06 | Digital Payments for Accounts Receivable with SD Billing and Service Processing |

*Figure 6.16: Add-On: Available Scenarios List*

Add new rows to the configuration table for each scenario you want to activate. Figure 6.17 shows an example in which five scenarios have been activated based on the implementation scope.

SAP Change View "Activation of SAP Digital Paym..": Overview

New Entries

Activation of SAP Digital Payments Scenarios

| Scenario | SAP Digital Payts Scenario Description |
|---|---|
| 01 | Digital Payments for Accounts Receivable only |
| 02 | Digital Payments for Accounts Receivable with SD Billing |
| 03 | Digital Payments for Accounts Receivable with SD Order |
| 05 | Digital Payments for Accounts Receivable with SD Order and Service Processing |
| 06 | Digital Payments for Accounts Receivable with SD Billing and Service Processing |

*Figure 6.17: Active Scenarios*

Each scenario activation enables specific configuration activities within the SAP S/4HANA system. Activating unnecessary scenarios does not harm system functionality, but it may present unused applications and configuration options that could confuse users. Conversely, failing to activate required scenarios will result in missing functionality and error messages when users attempt to access disabled features.

Save the scenario configuration after adding all required entries. The system validates that each scenario is compatible with your SAP S/4HANA release level and that no scenario conflicts exist. Some scenarios have prerequisite dependencies on other scenarios, which the system enforces during validation. If dependencies are not met, error messages indicate which additional scenarios must be activated.

Scenario activation is an incremental process. Organizations can activate additional scenarios after the initial implementation without disrupting existing payment processing. The additional scenario activation requires revalidation of the overall configuration, but it does not necessitate changes to existing payment card processing.

The tenant determination configuration maps company codes to specific add-on tenants and RFC destinations. This configuration enables multitenant scenarios in which different company codes connect to different add-on instances, supporting use cases such as separate PSP relationships for different business units or geographic regions. Execute the **Determine SAP Digital Payments Add-On Tenant** configuration activity.

The configuration screen presents a table with columns for company code (**CoCd**), OAuth RFC Destination (**Digital Payment Destination for OAUTH**), and Core API RFC Destination (**Digital Payment Destination**). For single-tenant implementations in which one add-on instance serves all company codes, create a single table entry with the **CoCd** field left blank. This default entry applies to all company codes in the system. For on-premise systems, enter **DIGITALPAYMENTS_OAUTH** in the **Digital Payment Destination for OAUTH** column and **DIGITALPAYMENTS** in the **Digital Payment Destination** column (see Figure 6.18).

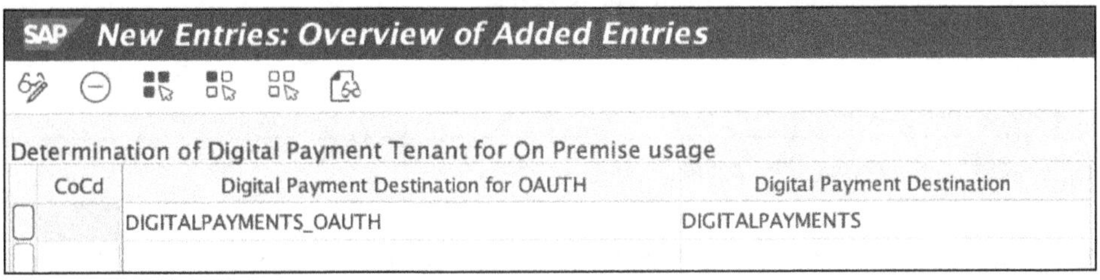

SAP New Entries: Overview of Added Entries

Determination of Digital Payment Tenant for On Premise usage

| CoCd | Digital Payment Destination for OAUTH | Digital Payment Destination |
|---|---|---|
| | DIGITALPAYMENTS_OAUTH | DIGITALPAYMENTS |
| | | |

*Figure 6.18: Add-On Setup for Single Tenant*

Multitenant scenarios require separate table entries for each company code or group of company codes connecting to different add-on instances. For each company code requiring a dedicated tenant, create a table entry with the specific company code value populated. Configure the corresponding RFC destinations or communication arrangements for that company code's add-on tenant. The system evaluates tenant determination entries in sequence, applying the first match found. Entries with specific company codes take precedence over the default, blank company code entry.

Multitenant configurations may require creating additional RFC destination pairs for each tenant, following the same configuration steps described in previous sections but connecting to different add-on service instances. Coordinate with your SAP BTP administrator to obtain separate service keys for each tenant and ensure proper tenant isolation in the SAP BTP cockpit. Save the tenant determination configuration. The system validates that all referenced RFC destinations exist in Transaction SM59. Validation failures indicate missing technical configuration that must be completed before the integration can function.

Most organizations typically implement single-tenant configurations unless specific business requirements necessitate separate PSP relationships or data isolation for different company codes. Multitenant configurations increase administrative complexity and require careful coordination during testing and go-live activities to ensure each company code connects to its intended add-on tenant.

After completing configuration activities, validate the technical connection between SAP S/4HANA and the digital payments add-on to confirm proper integration setup before proceeding with the business configuration. The validation process verifies RFC connectivity, OAuth authentication, API endpoint accessibility, and data synchronization.

SAP provides a standard program, FAR_DP_UPDATE_PSP_PAYMENT_METH (see Figure 6.19), specifically designed to test the digital payments add-on connection and synchronize payment method data from the add-on to SAP S/4HANA. Execute this program through Transaction SE38 by entering the program name and clicking the **Execute** button.

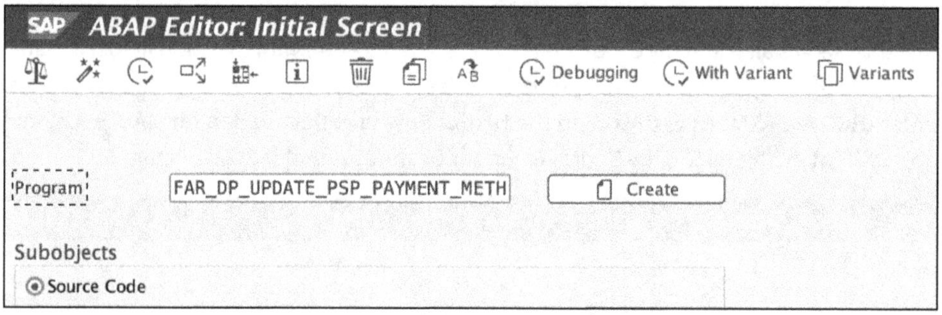

*Figure 6.19: Connection Validation Report*

On the program selection screen, select the **Display log** checkbox (see Figure 6.20) to view detailed logging output during program execution. Click the **Execute** button to run the program.

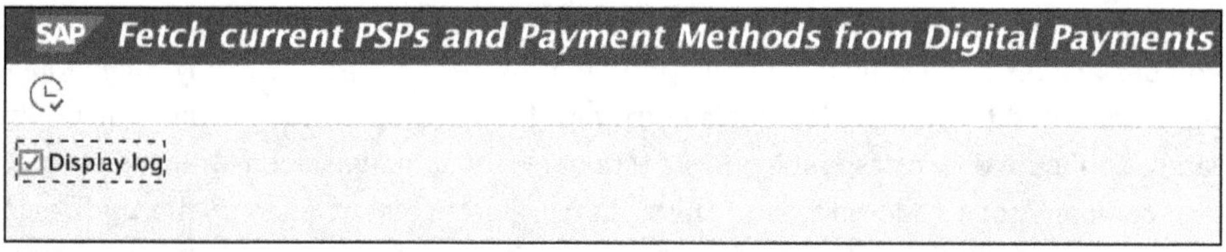

*Figure 6.20: Display Log Checkbox*

The program performs the following validation steps:

1. It validates the tenant determination configuration, verifying that the RFC destinations or communication arrangements are properly configured for the executing user's company code or the default configuration. Any errors in tenant determination halt the program execution, producing error messages that indicate the missing or incorrect configuration entries.
2. The program attempts OAuth authentication using the `OAUTH` RFC destination. It sends an access token request to the SAP BTP authorization server using the client credentials from the RFC destination configuration. Successful authentication confirms that the

OAuth destination parameters are correct, SSL certificates are properly configured, and network connectivity to SAP BTP is functional. Authentication failures produce detailed error messages that include HTTP status codes and OAuth error descriptions.

3. The program calls the add-on's API to retrieve the list of supported payment methods configured in your add-on tenant. This API call validates the core API RFC destination configuration, tests bearer token transmission, and verifies that the add-on service is accessible and responsive. The program logs each API request and response for troubleshooting purposes.
4. Finally, if all the validation steps succeed, then the program synchronizes payment method metadata from the add-on with the SAP S/4HANA system tables. This synchronization ensures that the payment method dropdown lists in SAP Fiori applications and configuration screens display all PSP payment methods available in your add-on tenant configuration. The synchronization updates SAP S/4HANA's internal cache of available payment methods.

The program execution log displays the status of each validation step with clear success or error indicators. Figure 6.21 illustrates a successful program execution log showing that all validation steps completed without errors.

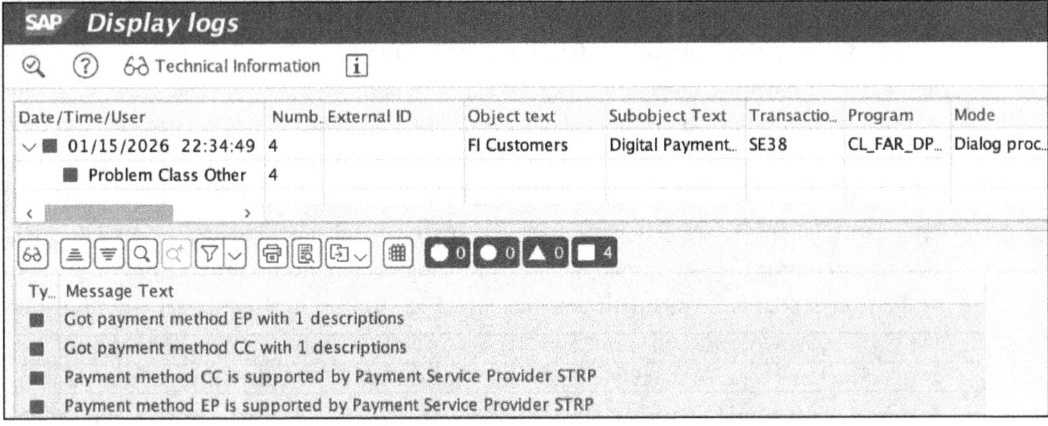


*Figure 6.21: Successful Program Execution*

Common errors encountered during connection testing include the following:

- *HTTP 404 Not Found* errors indicate incorrect path prefix configuration in RFC destinations. Verify that the path prefix matches your SAP S/4HANA release requirements and that no extra characters or spaces were inadvertently included.
- *NIECONN_REFUSED* errors indicate network connectivity problems between SAP S/4HANA and SAP BTP. Check your firewall rules to ensure that outbound HTTPS traffic on port 443 is permitted from SAP S/4HANA application servers to SAP BTP endpoints. Contact your network security team to verify if proxy servers or SSL inspection devices are blocking the connection or if proxy parameters were not entered or configured properly.

- *OAuth authentication errors* with messages about an invalid client or unauthorized access indicate incorrect client ID or client secret values in the OAuth RFC destination. Regenerate the service key in SAP BTP and reconfigure the RFC destination with the new credentials, ensuring that the exact values are copied without modifications.
- *SSL handshake failures* or *certificate validation errors* indicate missing or incorrect SSL certificates in Transaction STRUST. Reimport the DigiCert root certificate and verify that the certificate appears in the SSL client (standard) trust store.
- Empty payment method lists after successful synchronization may indicate incomplete add-on tenant configuration. Verify in the add-on configuration UI that PSPs are activated and routing rules are defined. The SAP S/4HANA system can only display payment methods that exist in the connected add-on tenant.

Resolve all connection test errors before proceeding with additional configuration activities. A functioning connection between SAP S/4HANA and the add-on is essential for all subsequent payment processing operations.

## 6.2 Payment Method Configuration

After establishing technical connectivity between SAP S/4HANA and the digital payments add-on, the next step involves configuring payment methods that control how credit cards and external payment types are processed throughout the order-to-cash cycle. Payment method configuration spans multiple Customizing areas, including cross-application payment card settings, sales and distribution billing parameters, master data synchronization, and financial accounting integration points. The configuration defines which payment types are available for selection, how they are categorized and validated, when authorizations occur during the sales process, and how billing documents determine the correct general ledger accounts for payment postings. SAP S/4HANA's out-of-the-box integration with the digital payments add-on eliminates the need for custom checking routines and authorization function modules that were required in earlier SAP ERP implementations. In addition to payment card and sales document configuration, this section also covers electronic bank statement setup for automated PSP disbursement reconciliation, and tax configuration to ensure that payment card transactions generate the correct tax postings in financial accounting.

### 6.2.1 Cross-Application Payment Card Type Configuration

Payment card types define the specific credit card brands and external payment methods that can be used throughout SAP S/4HANA business processes. These types are configured in the cross-application components area, making them available across sales and distribution, financial accounting, and business partner management. Configuration in this section follows a logical sequence of steps. We begin by maintaining the individual payment card types that represent each card brand or digital wallet your organization will accept, then group those types into payment card categories that simplify downstream configuration. The categories are then assigned back to the card types, linking the two together so that category-based rules apply correctly during processing. Finally, blocking reason configuration gives administrators a controlled way to temporarily suspend specific payment cards without removing them from the system entirely.

**Maintaining Payment Card Types**

Access the payment card type configuration through Transaction SPRO by navigating to **Cross-Application Components • Payment Cards • Basic Settings • Maintain Payment Card Type**. Refer to Table 6.2 for the list of supported card payment types. For the latest list, see SAP Note 2524512.

| Payment Type | Description | Typical Usage |
|---|---|---|
| DPVI | Visa Card | Most common credit card type globally |
| DPMC | Master-/Euro Card | Second most common credit card type |
| DPAM | American Express | Premium credit card; higher merchant fees |
| DPDS | Discover Card | Primarily used in the US market |
| DPJC | JCB Card | Popular in Japanese and Asian markets |
| DPDI | Diners Club | Business travel and corporate cards |
| DPUP | UnionPay | Dominant card network in China |
| DPLB | Local Brand Card | Regional card networks and local issuers |

*Table 6.2: Add-On-Supported Card Payment Types*

Select **New Entries** (see Figure 6.22) and configure the card payment types based on your implementation scope. For example, if you have Visa cards in your implementation scope, you will add a new entry, select DPVI as the **Type** from the search help, and then click the checkbox in the **Dig. Pymts** field. Repeat this step for all card types in your implementation scope.

SAP Change View "Definition of Payment Card Type (CRM)": Overview

New Entries

Definition of Payment Card Type (CRM)

| Type | Description | Rule | Encryption | Dig. Pymts |
|---|---|---|---|---|

*Figure 6.22: Create New Entries*

Figure 6.23 shows an example configuration table of payment card types in the system.

| | | |
|---|---|---|
| DPAM | American Express | ☑ |
| DPDI | Diners Club | ☑ |
| DPDS | Discover Card | ☑ |
| DPJC | JCB Card | ☑ |
| DPMC | Master-/Euro Card | ☑ |
| DPUP | Union Pay | ☑ |
| DPVI | Visa Card | ☑ |

*Figure 6.23: Configuration of Card Payment Types*

External payment methods require additional configuration beyond credit cards as each payment method may support different processing capabilities. Some external payment methods use two-step authorization and capture workflows like credit cards, while others use one-step direct capture processing, in which funds are immediately debited from the payer's account. Table 6.3 defines common external payment types. For the full list, see SAP Note 2730920.

| Payment Type | Description | Processing Type | Regional Focus |
|---|---|---|---|
| DP1P | PayPal Direct Capture | One-step | Global digital wallet |
| DP2P | PayPal with Authorization | Two-step | Global digital wallet with auth |
| ALI1 | Alipay Direct Capture | One-step | China market focus |
| ALI2 | Alipay with Authorization | Two-step | China with preauth support |
| GOP1 | Google Pay Direct Capture | One-step | Global mobile wallet |
| GOP2 | Google Pay with Authorization | Two-step | Global mobile wallet with auth |
| APP1 | Apple Pay Direct Capture | One-step | iOS ecosystem payments |
| APP2 | Apple Pay with Authorization | Two-step | iOS with preauthorization |
| AMA1 | Amazon Pay Direct Capture | One-step | Amazon payments |
| AMA2 | Amazon Pay with Authorization | Two-step | Amazon with auth workflow |
| IDE1 | iDeal Payment | One-step | Netherlands bank transfer |

*Table 6.3: External Payment Types*

| Payment Type | Description | Processing Type | Regional Focus |
|---|---|---|---|
| GIR1 | Giro Pay | One-step | Germany bank transfer |
| SEPA | SEPA Direct Debit | One-step | Europe direct debit |
| EPS1 | E-Payment Standard | One-step | Austria bank transfer |
| KLA1 | Klarna Direct Capture | One-step | Buy now, pay later, Europe |
| KLA2 | Klarna with Authorization | Two-step | Klarna with preauth |
| WEC1 | WeChat Pay Direct Capture | One-step | China mobile payment |

*Table 6.3: External Payment Types (Cont.)*

The payment types only become active when assigned to card categories and sales document types in subsequent configuration steps.

### Defining Payment Card Categories

Payment card categories group related payment types for organizational and processing purposes. Categories enable bulk assignment of properties and simplify configuration by allowing administrators to work with logical groupings rather than individual payment types. Access the card category configuration through **Cross-Application Components • Payment Cards • Basic Settings • Maintain Payment Card Category** in Transaction SPRO.

The standard implementation uses a single category for all digital payments add-on payment types. Define category **01** with the **Description** as **Credit Card** (see Figure 6.24). This category encompasses both traditional credit cards and external payment methods; all digital payment types are processed through similar workflows involving authorization, settlement, and reconciliation.

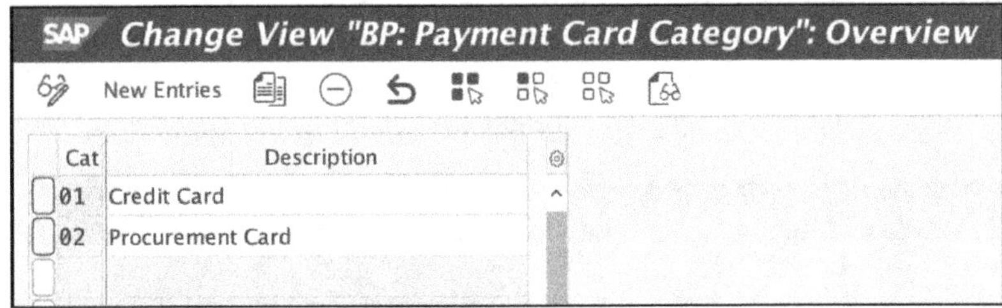


*Figure 6.24: Standard Payment Card Categories*

### Assign Payment Card Category to Payment Card Type

After defining the card category, assign payment card types to the category through the **Cross-Application Components • Payment Cards • Basic Settings • Assign Payment Card Category to Payment Card Type** configuration activity.

Create table entries assigning each payment type in your scope to **Payment card cat.** 01, **Credit Card** (see Figure 6.25). The assignment links the technical payment type code to the logical category, enabling category-based configuration in subsequent activities.

Change View "Assign Payment Card Category to Payment Car..

New Entries

Assign Payment Card Category to Payment Card Type

| Type | Payment card cat. |
|---|---|
| Diners Club Card | Credit Card |
| American Express | Credit Card |
| Diners Club | Credit Card |
| Discover Card | Credit Card |
| JCB Card | Credit Card |
| Master-/Euro Card | Credit Card |
| Union Pay | Credit Card |
| Visa Card | Credit Card |

*Figure 6.25: Configuration Table*

Most implementations benefit from the simplicity of a single category that treats all payment types consistently. Multiple categories increase configuration complexity without providing functional benefits unless distinctly different processing rules are required.

### Blocking Reason Configuration

Define blocking reasons that can be assigned to payment cards in the business partner master data to temporarily prevent their use without deleting the card registration. Access the configuration through **Cross-Application Components • Payment Cards • Basic Settings • Maintain Payment Card Blocks**.

Create blocking reason **01** with description **Card Lost** for situations in which customers report lost or stolen cards. Create blocking reason **02** with description **Other Reason** (see Figure 6.26) for general-purpose blocking, such as suspected fraud, expired cards pending replacement, or temporary payment disputes. Additional blocking reasons can be defined based on your requirements for tracking and reporting on blocked payment methods.

Change View "BP: Blocking Reason for Payment Cards": Overview

New Entries

| Lock | Description |
|---|---|
| 01 | Card Lost |
| 02 | Other Reason |

*Figure 6.26: Maintain Blocking Reasons*

When a payment card is blocked in the business partner master record, the block prevents the card from being selected during sales order creation or payment processing. Existing authorizations on blocked cards remain valid and can be captured if the authorization was obtained before the card was blocked. The blocking functionality provides a safer alternative to deleting payment cards, as blocks can be removed when issues are resolved, whereas deleted cards require complete reregistration through the PSP.

### 6.2.2 Sales Payment Card Configuration

The SAP S/4HANA Sales configuration controls how payment cards are integrated into sales documents, including orders, deliveries, and billing documents. This configuration determines which sales document types support payment card entry, establishes authorization timing rules, and defines account determination logic for payment-related postings. This section works through each layer of the sales and distribution configuration in sequence. We begin with card types and card categories for sales documents, which mirror the cross-application setup but add sales-specific attributes, before covering payment card plan type assignment, which activates the **Electronic Payments** tab on sales document types. We then configure the authorization horizon and checking group settings, which control when authorizations are triggered relative to delivery dates, alongside authorization validity periods, which govern how long an approved authorization remains usable before reauthorization is required. The account determination configuration then establishes which general ledger accounts receive payment card postings at billing, including the classification of direct capture payment types that bypass the standard authorization workflow. Finally, we set up the business partner payment card mapping, which ensures that cards registered in the digital payments add-on are visible and usable across all sales and finance processes.

#### Card Types for Sales Documents

While the cross-application card type configuration establishes the available payment methods systemwide, sales requires its own card type definitions that add functionality-specific attributes. Access the sales card type configuration through **Sales and Distribution • Billing • Payment Cards • Maintain Card Types** in Transaction SPRO.

The sales card type table includes additional columns beyond the basic type and description fields (see Figure 6.27). The **Check** column allows specification of a checking rule for payment card validation; however, this field should be left blank when using the digital payments add-on as all validation occurs through the add-on service. The **Date Type** column controls the format of payment card validity date entry, typically set to **Month** to indicate the month and year format. The **Virtual Card** indicator (found further to the right), when selected, designates payment types that are one-time use cards and not stored in business partner master data.

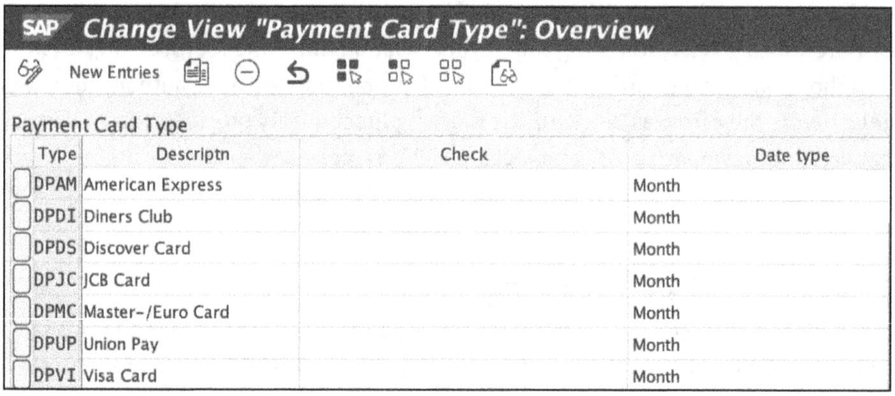

**Change View "Payment Card Type": Overview**

New Entries

Payment Card Type

| Type | Descriptn | Check | Date type |
|---|---|---|---|
| DPAM | American Express | | Month |
| DPDI | Diners Club | | Month |
| DPDS | Discover Card | | Month |
| DPJC | JCB Card | | Month |
| DPMC | Master-/Euro Card | | Month |
| DPUP | Union Pay | | Month |
| DPVI | Visa Card | | Month |

*Figure 6.27: Sales Payment Card Types*

Create sales card type entries for all payment types that will be used in sales documents. The payment type codes must match exactly the codes defined in the cross-application configuration. Refer to SAP Notes 2524512 and 2730920 for the sales card type configuration for credit card and external payment types.

The **Data type** setting of **Month** indicates that payment card expiration dates are entered and stored in MM/YYYY format. This is the standard format for credit card expiration dates globally and aligns with PSP expectations for card validation. Alternative data type values exist for different validity period formats, but this setting is universally appropriate for credit card and digital payment processing.

Organizations accepting one-time payment cards entered directly during sales order creation without prior registration in business partner master data should define additional card type entries with the **Virtual Card** indicator selected. However, the digital payments add-on integration typically processes all cards through business partner registration to maintain token references, making virtual card configuration less common in SAP S/4HANA implementations compared to legacy SAP ERP scenarios.

### Card Category Configuration for Sales Documents

Sales maintains its own card category definitions separate from the cross-application categories configured earlier. These sales-specific categories control how payment cards behave within sales documents and billing processes. Access the configuration through **Sales and Distribution • Billing • Payment Cards • Maintain Card Categories.**

Execute the **Define Card Categories** activity to create category definitions. For the digital payments add-on integration, define category (**Cat**) **01** with **Description** set to **Credit Card** for credit card payments (see Figure 6.28) and, if needed, category **03** with **Description** set to **Payment Service** for external payment types. This category designation aligns with SAP's delivered configuration for payment methods processed through external service providers rather than traditional bank-based card processing. The **One card per trans.** indicator should be left unchecked to allow split payment scenarios. The **Additional data** indicator should also remain unchecked.

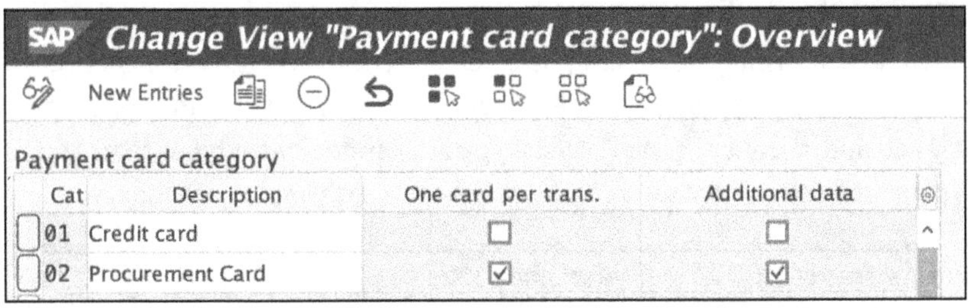

SAP Change View "Payment card category": Overview

New Entries

Payment card category

| Cat | Description | One card per trans. | Additional data |
|---|---|---|---|
| 01 | Credit card | ☐ | ☐ |
| 02 | Procurement Card | ☑ | ☑ |

*Figure 6.28: Define Card Categories*

After defining the categories, assign card types to categories through the **Determine Card Categories** activity within the same configuration node. Create assignment entries (see Figure 6.29) linking each sales card type to category **01**. The assignment table includes additional columns for sequence number, card number range from/to values, and limit amount. For standard digital payments add-on implementations, populate only the **Type** and **Cat** (category) columns, leaving other fields blank.

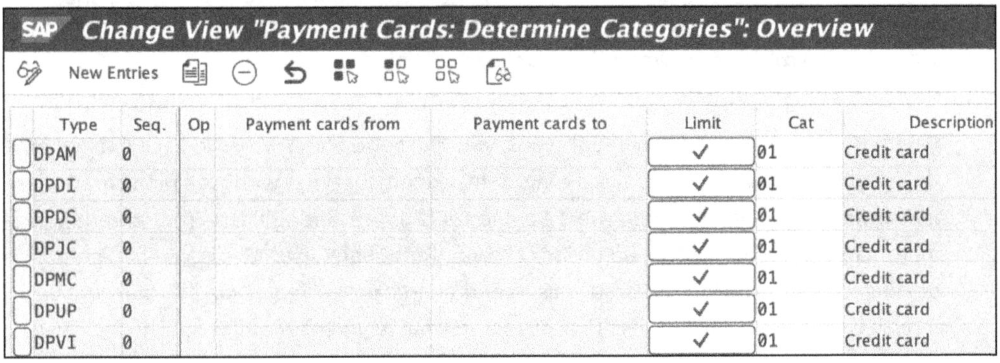

SAP Change View "Payment Cards: Determine Categories": Overview

New Entries

| Type | Seq. | Op | Payment cards from | Payment cards to | Limit | Cat | Description |
|---|---|---|---|---|---|---|---|
| DPAM | 0 | | | | ✓ | 01 | Credit card |
| DPDI | 0 | | | | ✓ | 01 | Credit card |
| DPDS | 0 | | | | ✓ | 01 | Credit card |
| DPJC | 0 | | | | ✓ | 01 | Credit card |
| DPMC | 0 | | | | ✓ | 01 | Credit card |
| DPUP | 0 | | | | ✓ | 01 | Credit card |
| DPVI | 0 | | | | ✓ | 01 | Credit card |

*Figure 6.29: Determine Card Categories: Table Entries*

The sequence number (**Seq.**) column enables prioritization when multiple card categories could apply to a card type, though this is rarely needed in digital payments add-on implementations in which each payment type maps to a single category. The card number range columns allow restricting specific card categories to certain card number patterns, which is useful in legacy implementations with complex card routing logic but not necessary when the add-on handles all PSP routing.

#### Payment Card Plan Type Assignment

Payment card plan types control which sales document types support payment card processing. Assigning plan type **03**, the standard delivered value for payment cards, to a sales document type activates the **Electronic Payments** tab in the document header and enables payment card selection during document creation. Access the configuration through **Sales and Distribution • Sales • Sales Documents • Sales Document Header • Assign Payment Card Plan Types to Sales Document Types**.

The configuration screen displays a table of sales document types with a **PaymentCard Plan Type** column. Add plan type **03** to all sales document types that should support payment card entry and authorization. Table 6.4 lists typical sales document types that require payment card plan type assignment for a complete order-to-cash implementation.

| Document Type | Description | Rationale |
|---|---|---|
| OR | Standard Order | Primary order type for B2C and B2B sales |
| RE | Returns | Enables refund processing to original payment card |
| CR | Credit Memo Request | Credit processing with payment card refund capability |
| DR | Debit Memo Request | Additional charges with payment card settlement |
| CS | Cash Sales | Point-of-sale or immediate payment scenarios |

*Table 6.4: Typical Sales Document Types for Payment Cards*

When using custom sales document types, you must add plan type **03** to each custom type that should support payment processing. Sales document types without a plan type assignment will not display the **Electronic Payments** tab, preventing users from entering payment card information. This behavior can be intentionally used to disable payment processing for specific order types, such as free-of-charge orders (e.g., type **ZFD** in Figure 6.30), internal transfers, or rush orders that bypass normal payment workflows.

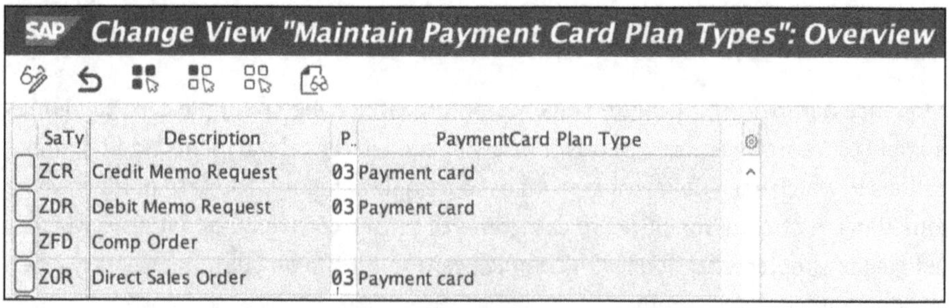

Change View "Maintain Payment Card Plan Types": Overview

| SaTy | Description | P.. | PaymentCard Plan Type |
|---|---|---|---|
| ZCR | Credit Memo Request | 03 | Payment card |
| ZDR | Debit Memo Request | 03 | Payment card |
| ZFD | Comp Order | | |
| ZOR | Direct Sales Order | 03 | Payment card |

*Figure 6.30: Custom Sales Document Types*

The payment card plan type assignment is a prerequisite for payment card authorization to occur during sales order processing. Even with a plan type assigned, authorization requires additional configuration of payment guarantee procedures and checking groups, described in subsequent sections. The plan type assignment simply makes payment card fields available in the user interface without automatically enabling authorization processing.

For returns and credit memo document types, plan type 03 assignment enables the system to copy payment card data from the original sales order or billing document, facilitating automatic refund processing to the customer's original payment method. This copying behavior can be further controlled through additional configuration settings for specific business scenarios that require different refund handling.

### Authorization Horizon and Checking Group Definition

Authorization horizon settings control when payment card authorizations occur relative to delivery or billing dates. These settings prevent premature authorization that could consume customer credit limits for orders with distant delivery dates while ensuring timely authorization before fulfillment activities begin. The configuration balances business requirements for credit risk management against customer experience considerations related to payment card authorization timing.

Checking groups bundle authorization control parameters that can be assigned to sales document types. Each checking group specifies an authorization requirement routine, an authorization horizon in days, a preauthorization validity period, and reauthorization settings. Access the checking group configuration through **Sales and Distribution • Billing • Payment Cards • Authorization and Settlement • Maintain Checking Groups**.

Execute the **Define Checking Groups** activity (see Figure 6.31) to create or modify checking group definitions. The standard configuration uses checking group **01** for all payment card processing.

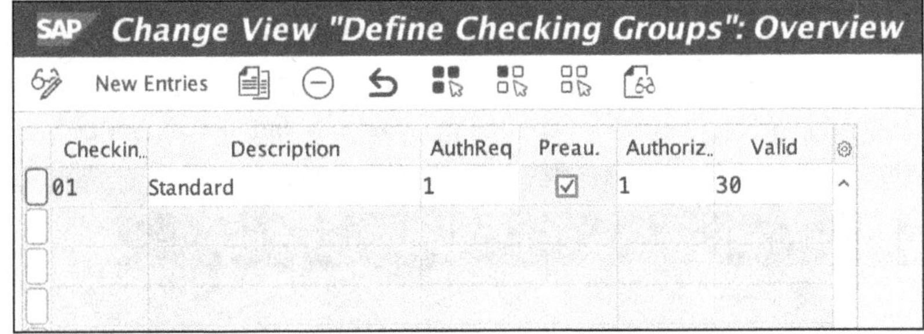


*Figure 6.31: Define Checking Groups Table*

Configure the following parameters for checking group **01**:

- The **AuthReq** (authorization requirement) field specifies which ABAP routine validates whether a document is ready for authorization. Enter "1"; this is the standard delivered routine that checks whether all required payment card data is present and the document status permits authorization. This routine verifies that a payment card is assigned, the card expiration date is valid, the authorization amount can be calculated, and no blocking conditions prevent authorization.

- The **Authoriz...** (authorization horizon) field defines how many days before the material availability date the system triggers a full authorization request. Set this value based on your organization's typical fulfillment cycle and PSP authorization validity periods. A common setting is **1**, indicating that authorization occurs when the delivery date is one day away or closer. For orders with delivery dates beyond the horizon, the system performs preauthorization instead of full authorization.
- The **Valid** (validity period) specifies for how many days a preauthorization remains valid before requiring a refresh. Preauthorizations use nominal amounts to validate card details without reserving the full transaction value. Set this value to **30** days for most implementations, allowing sufficient time for delivery date changes or order modifications without excessive preauthorization refreshes.

If you have complex authorization requirements, you can define multiple checking groups with different horizon and validity settings for different sales document types. For example, express orders requiring immediate fulfillment might use checking group 02 with an authorization horizon of zero days to force immediate authorization, while standard orders use checking group 01 with a one-day horizon. Most implementations use a single checking group supplemented by sales document type configuration.

After defining checking groups, assign them to sales document types through the **Assign Checking Groups** activity (see Figure 6.32) in the same configuration area. The assignment table links sales document types to checking groups, activating the authorization control parameters for those document types.

Create assignment entries for all sales document types that have payment card plan type 03 assigned from the previous configuration activity. The checking group assignment works in conjunction with the payment card plan type assignment to fully enable payment card processing. Sales document types must have both plan type 03 and a checking group assigned to support payment authorization. Missing either assignment results in incomplete payment processing capabilities, such as payment cards being enterable but not authorized, or authorization being attempted but failing due to missing validation parameters.

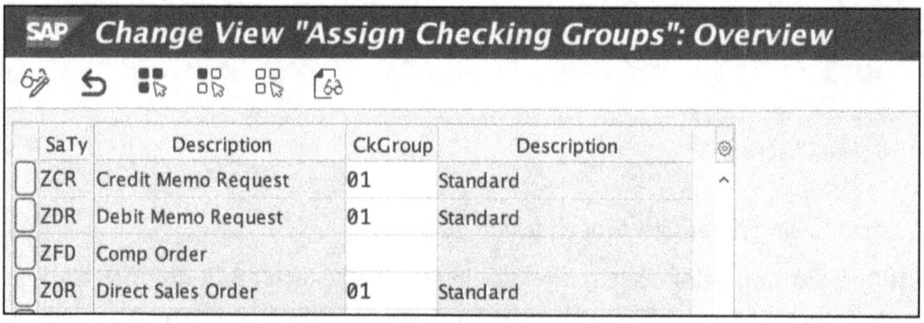

SAP Change View "Assign Checking Groups": Overview

| SaTy | Description | CkGroup | Description |
|---|---|---|---|
| ZCR | Credit Memo Request | 01 | Standard |
| ZDR | Debit Memo Request | 01 | Standard |
| ZFD | Comp Order | | |
| ZOR | Direct Sales Order | 01 | Standard |

*Figure 6.32: Assign Checking Groups Table*

### Authorization Validity Period Configuration

Authorization validity periods define how long a payment authorization remains valid before requiring reauthorization. PSPs typically limit authorization validity to prevent extended holds on customer credit limits and reduce fraud risk from delayed captures. Configure validity periods through **Sales and Distribution • Billing • Payment Cards • Authorization and Settlement • Specify Authorization Validity Periods**.

The configuration screen presents a table (see Figure 6.33) in which each payment card type can have a specific validity period defined. The validity period is expressed in days from the authorization timestamp. Create entries for payment types based on PSP capabilities and business requirements.

Direct capture payment methods do not require validity period configuration because they do not use authorization and capture workflows. The payment is captured immediately when the transaction is initiated, eliminating the concept of authorization validity. Leave validity period entries blank for all direct capture payment types.

The validity period must align with PSP capabilities and merchant agreement terms. PSPs enforce maximum authorization validity periods as part of their risk-management policies. Configuring validity periods longer than PSP limits will result in authorization rejections or automatic authorization expirations before the configured SAP S/4HANA validity period expires. Consult the PSP's documentation or merchant services agreements to verify the appropriate validity settings for each payment type.

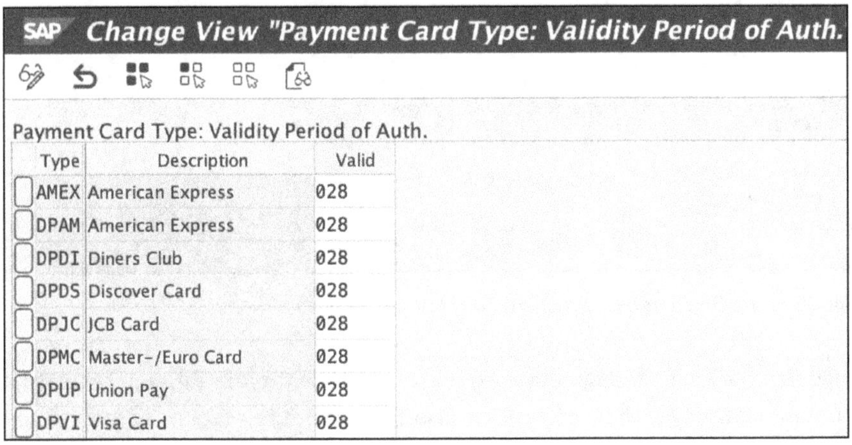

Change View "Payment Card Type: Validity Period of Auth.

Payment Card Type: Validity Period of Auth.

| Type | Description | Valid |
|---|---|---|
| AMEX | American Express | 028 |
| DPAM | American Express | 028 |
| DPDI | Diners Club | 028 |
| DPDS | Discover Card | 028 |
| DPJC | JCB Card | 028 |
| DPMC | Master-/Euro Card | 028 |
| DPUP | Union Pay | 028 |
| DPVI | Visa Card | 028 |

*Figure 6.33: Authorization Validity Period Table*

When an authorization approaches expiration, SAP S/4HANA can automatically trigger reauthorization if the **Checking Group Reauthorization** indicator is active. The system compares the authorization timestamp plus the validity period against the current date during sales document processing and delivery creation. If expiration is imminent or has occurred, the system requests a new authorization from the add-on, updating the authorization data with the refreshed authorization validity period.

### Account Determination for Payment Card Postings

Account determination configuration establishes which general ledger accounts receive payment-related postings when billing documents are released to financial accounting.

The configuration uses the *condition technique*, the same flexible determination logic employed for revenue account determination and pricing, enabling precise control over account assignment based on multiple organizational and transaction characteristics.

Navigate to **Sales and Distribution • Billing • Payment Cards • Authorization and Settlement • Maintain Clearing House • Account Determination • Maintain Procedures to Define Account Determination Procedures**. The procedure acts as a container for condition types that control account determination logic.

SAP delivers standard procedure A00001 for payment card account determination, which includes condition type CC01 for general account assignment. Verify that this procedure exists in your system configuration (see Figure 6.34). For most implementations, the delivered A00001 procedure requires no modifications. Organizations with complex account structures requiring multiple determination steps can define additional condition types and incorporate them into the procedure, though this level of complexity is rarely necessary for digital payments add-on implementations in which the PSP routing handles most payment-specific logic.

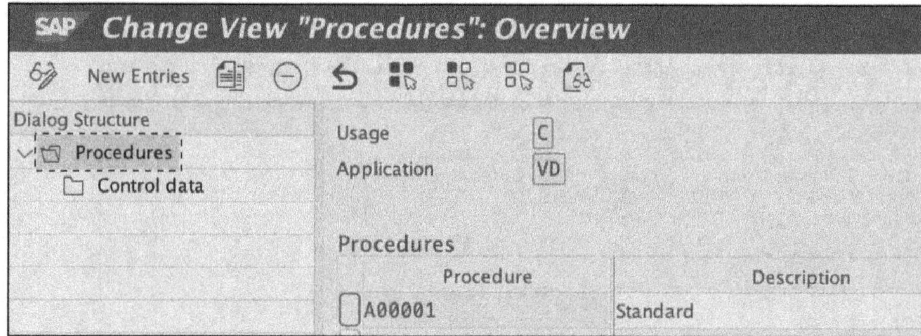


*Figure 6.34: Standard Procedure for Payment Card Account Determination*

After procedures are defined, assign them to billing document types through the **Assign Account Determination Procedures** activity, also located in the account determination configuration area. Execute Transaction OV88 or access the configuration activity through Transaction SPRO navigation (see Figure 6.35).

The assignment table displays billing document types and the account determination procedures. Create entries linking each billing document type that processes payment cards to procedure **A00001**.

Billing document types not listed in the assignment table will not perform payment card-specific account determination, resulting in payment transactions posting to standard revenue and receivables accounts rather than the designated payment card accounts. This could cause reconciliation issues and incorrect financial reporting.

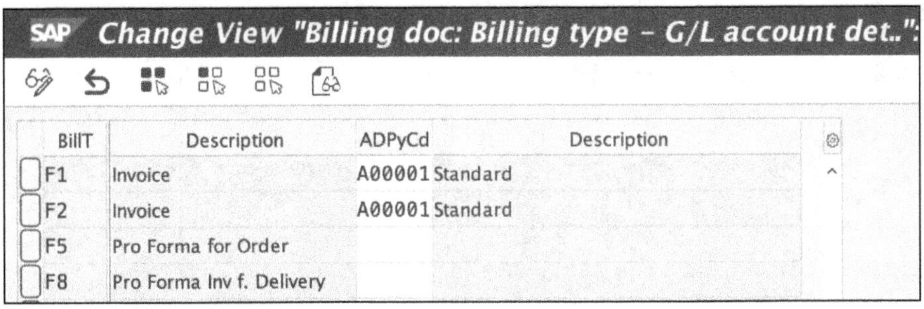

Change View "Billing doc: Billing type – G/L account det..":

| BillT | Description | ADPyCd | Description |
|---|---|---|---|
| F1 | Invoice | A00001 | Standard |
| F2 | Invoice | A00001 | Standard |
| F5 | Pro Forma for Order | | |
| F8 | Pro Forma Inv f. Delivery | | |

*Figure 6.35: Assign Procedures to Billing Document Types*

**Assigning General Ledger Accounts to Payment Card Types**

The final account determination configuration step assigns specific general ledger accounts to combinations of payment card types and organizational elements. This configuration uses the condition technique access sequences defined in the account determination procedure to find the appropriate account for each payment transaction.

Navigate to **Sales and Distribution • Billing • Payment Cards • Authorization and Settlement • Maintain Clearing House • Account Determination • Assign G/L Accounts**. The configuration presents multiple assignment options organized by condition table. The commonly used assignment table is **006, SlsOrg/Card Cat**, which determines accounts based on sales organization and card category combination (see Figure 6.36).

Assign G/L Accounts

Assign G/L Accounts

| Tab | Description |
|---|---|
| 004 | General |
| 006 | SlsOrg/Card cat. |

*Figure 6.36: Assign General Ledger Accounts*

Execute the configuration activity for table 006 and create assignment entries for each sales organization that processes payment cards. The assignment table includes columns for **App** (always VD for sales), condition type (**CndTy.**; always CC01), chart of accounts (**Chrt/Accts**), sales organization (**SOrg.**), card **Type**, general ledger account (**G/L Account**), and accruals account (**Accruals Acc.**; see Figure 6.37).

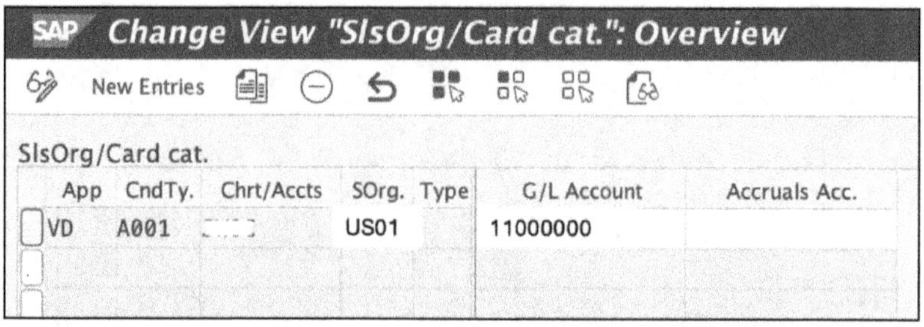

Change View "SlsOrg/Card cat.": Overview

New Entries

SlsOrg/Card cat.

| App | CndTy. | Chrt/Accts | SOrg. | Type | G/L Account | Accruals Acc. |
|---|---|---|---|---|---|---|
| VD | A001 | [illegible] | US01 | | 11000000 | |

*Figure 6.37: General Ledger Account Entry for Example Sales Org US01*

For implementations in which all payment types within a sales organization use the same receivables account, create a single entry per sales organization without specifying individual card types. This approach leverages the card category assignment so that the entry applies to all payment types in category 03.

The general ledger account is the open-item managed account in which authorized payment amounts are posted when billing documents are released to financial accounting. The account must be configured as open-item managed with line-item display enabled to support settlement processing and reconciliation activities.

If you require different receivables accounts for different payment types or payment method categories, you can create more granular assignments by specifying card types in individual table entries. The access sequence evaluates assignment entries in priority order, first checking for specific card type matches before falling back to category-level defaults. This behavior enables organizations to define card-type-specific accounts where needed while maintaining a simpler, category-based default for other payment types.

### Classification of Direct Capture Payment Types

External payment methods that use one-step direct capture processing require special configuration to inform SAP S/4HANA that no authorization step precedes settlement. This classification affects settlement program logic, refund processing workflows, and financial posting behavior. Configure direct capture classification through **Integration with Other SAP Components • Integration with the SAP Digital Payments Add-On • Classify Payment Card Type as Direct Capture**.

The configuration screen presents a table in which direct capture payment types are explicitly listed. Add entries for payment types that use one-step direct capture processing. The direct capture classification affects how the settlement program selects open items for processing. Two-step payment authorizations are selected based on the presence of valid authorization data in the billing document, while direct capture payments are selected based on the presence of payment transaction identifiers received during the original payment processing. The classification also influences refund-processing logic as direct capture refunds reference the original capture transaction identifier rather than an authorization ID.

For SAP S/4HANA 2023 and later, an additional configuration activity, **Assign Payment Card Type to Digital Payment Type for Manual Refunds**, enables mapping direct capture payment types to refund payment type categories used in the Create Refunds for Digital Payments app. This mapping simplifies manual refund processing by grouping payment types into logical categories for user selection.

### Business Partner Payment Card Mapping

SAP S/4HANA uses business partner master data as the central repository for customer information, while legacy customer master records exist for backward compatibility. Payment cards registered through the digital payments add-on are stored in business partner tables and must be mapped to customer master equivalents to ensure visibility across all sales processes. Configure this mapping through the **Cross-Application Components • Master Data Synchronization • Customer/Vendor Integration • Business Partner Settings • Settings for Customer Integration • Field Assignment for Customer/Vendor Integration • Assign Attributes • Assign Payment Cards** activity.

The configuration table requires creating entries that map business partner payment card types to customer/vendor integration (CVI) payment card types. In most cases, the codes are identical between business partner and CVI systems, resulting in a one-to-one mapping in which the **Card Type (BP)** value equals the **Card Type (CVI)** value.

Create assignment entries for all payment types that will be used in your implementation. This mapping configuration ensures that payment cards visible in the business partner master record are also accessible when processing sales documents using customer master functions and that payment transactions post correctly to customer accounts in financial accounting. Without proper mapping (see Figure 6.38), payment cards might be successfully registered in business partner records but fail to appear in sales order payment card selection lists or cause errors during billing document creation.

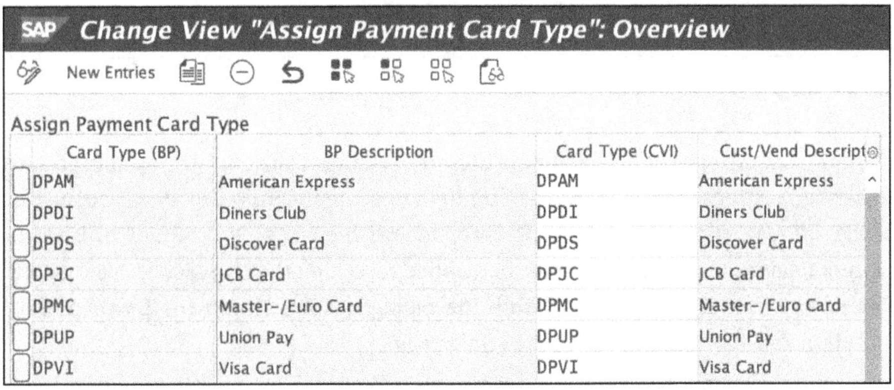

SAP Change View "Assign Payment Card Type": Overview

New Entries

Assign Payment Card Type

| Card Type (BP) | BP Description | Card Type (CVI) | Cust/Vend Descript |
|---|---|---|---|
| DPAM | American Express | DPAM | American Express |
| DPDI | Diners Club | DPDI | Diners Club |
| DPDS | Discover Card | DPDS | Discover Card |
| DPJC | JCB Card | DPJC | JCB Card |
| DPMC | Master-/Euro Card | DPMC | Master-/Euro Card |
| DPUP | Union Pay | DPUP | Union Pay |
| DPVI | Visa Card | DPVI | Visa Card |

*Figure 6.38: Payment Card Type Mapping*

### 6.2.3 Finance Central Settings

Financial accounting configuration establishes systemwide settings for payment card processing that affect document posting, clearing behavior, and settlement execution. These settings are maintained in SAP S/4HANA Finance and apply across all company codes unless company-specific overrides are configured.

#### Central Payment Card Settings

Access the central settings through **Financial Accounting • Accounts Receivable and Accounts Payable • Business Transactions • Payments with Payment Cards • Make Central Settings for Payment Cards**. The configuration screen contains several indicators and field assignments that control payment document creation and clearing behavior.

The **Retain Cust.Item** indicator (see Figure 6.39) determines whether customer open items remain posted when billing documents with payment cards are released to financial accounting. Select this indicator to maintain the customer account item in addition to the payment card receivables account posting. This setting enables reconciliation of customer payment card transactions through standard customer account reports and maintains historical transaction visibility at the customer level.

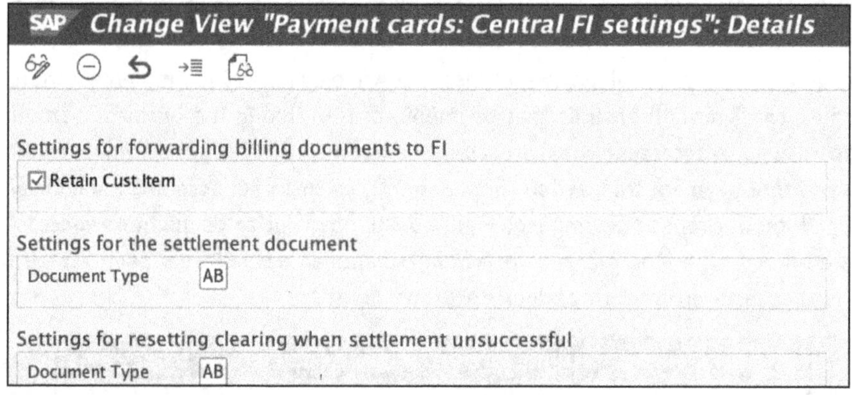


*Figure 6.39: Central Financial Settings Details*

When the retain customer item (**Retain Cust.Item**) indicator is selected, the system creates two offsetting postings during billing: a debit to the payment card receivables account for the billing amount, and a credit clearing entry against the customer account. This maintains a zero balance on the customer account while recording the transaction history. The clearing entry includes references to both the billing document and the payment card authorization, facilitating comprehensive audit trails.

The **Document Type** field under **Settings for the settlement document** specifies which document type the settlement program uses when posting captured payments from the payment card receivables account to bank clearing accounts. Enter a document type that will control the number range assignment, field status, and posting period validation for settlement documents.

The **Document Type** field under **Settings for resetting clearing when settlement unsuccessful** defines the document type used when reversing erroneous settlement postings. This situation occurs when settlements are rejected by the PSP after initial confirmation or when posting errors are discovered during reconciliation. Enter "AB" here, or a custom document type appropriate for reversal postings. The clearing reset function recreates open items on the payment card receivables account and reverses the clearing account postings, allowing corrected settlement processing. These settings apply globally across all company codes and payment types unless overridden by more specific configuration.

#### Clearing Account Assignment to Payment Card Receivables

The system must determine which bank clearing account to use when settlement programs post captured payments. This assignment links the payment card receivables account to the corresponding PSP clearing account, enabling automatic account determination during settlement processing. Access the configuration through **Financial Accounting • Accounts Receivable and Accounts Payable • Business Transactions • Payments with Payment Cards • Assign G/L Account to Cash Clearing Account**.

Create assignment entries in the configuration table (see Figure 6.40), linking receivables accounts to clearing accounts. In this example, **COA** is the example chart of accounts, **11000000** is the general ledger account for receivables, and **12000000** is the clearing general ledger account.

Change View "Clearing account/external functions": Overview

New Entries

| Char | Receivable | Short Text | Clearing | Short Text |
|---|---|---|---|---|
| COA | 11000000 | DPVI Visa card | 12000000 | A/R-CreditCard Sales |

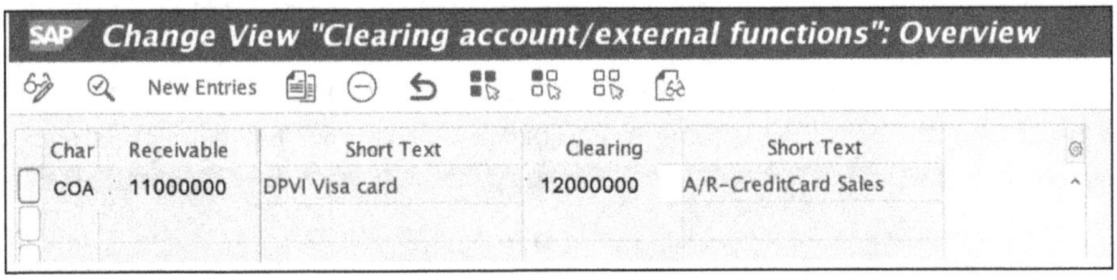

*Figure 6.40: Assignment Entries*

The receivable account values correspond to the payment card receivables, and the clearing account values are the PSP settlement clearing accounts. This configuration enables the settlement program to automatically post from receivables accounts to clearing accounts without requiring manual account entry during each settlement run execution.

### 6.2.4 Customer Master Payment Guarantee Procedure

This step involves assigning the customer payment guarantee flag in customer master data. Access customer master records through Transaction XD02 or the Maintain Business Partner app. Navigate to the **Sales Area Data** view and locate the **Billing Document** tab.

In the **Payment Guarantee Procedure** field, enter "0002" to indicate that the customer may use payment card payment methods. This field assignment applies at the sales area level, meaning that customers can have different payment guarantee procedures for different sales organizations, distribution channels, or divisions if business processes require such differentiation.

You can assign payment guarantee procedure 0002 to all customers who may potentially use payment cards (see Figure 6.41) as the procedure only activates when payment cards are assigned in sales documents. Customers without payment cards registered in their business partner records or sales documents without payment card assignments are not affected by the payment guarantee procedure setting.

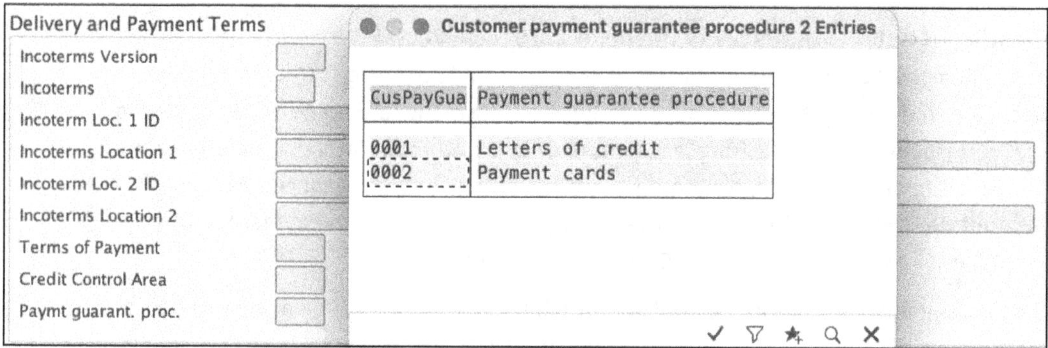


*Figure 6.41: Payment Guarantee Procedure Setting*

Alternatively, the customer flag assignment can be managed selectively, assigning the flag only to customers who routinely pay with cards or who have explicitly provided card-on-file authorization. This approach reduces the number of customers flagged for payment guarantee processing but requires ongoing maintenance as customer payment preferences change.

### 6.2.5 Electronic Bank Statement Configuration for Payment Advice

Payment advice processing uses electronic bank statement functionality to import and post settlement confirmations, PSP fees, refunds, and other payment-related transactions received from the digital payments add-on. This configuration establishes account symbols, posting rules, transaction type mappings, and bank account assignments that enable automated reconciliation of payment transactions.

#### Account Symbol Definition

Account symbols provide flexible account determination for electronic bank statement postings, allowing the same posting rule to determine different accounts based on company code or other criteria. Navigate to **Financial Accounting • Bank Accounting • Business Transactions • Payment Transactions • Electronic Bank Statement • Make Global Settings for Electronic Bank Statement** to access account symbol configuration.

Execute the **Define Account Symbols** activity to create symbol definitions for digital payments add-on payment advice processing. Table 6.5 lists the required account symbols for credit card and external payment advice processing.

| Account Symbol | Description | Purpose in Payment Processing |
|---|---|---|
| DPSETTLEMENT | Digital Payments Settlement | Settlement clearing account for payment card settlement program |
| DPCLEAR | Digital Payments Clearing | PSP clearing account for confirmed settlements from advice |
| DPTRANSFER | Digital Payments Transfer | PSP transfer account for funds pending bank disbursement |
| DPFEE | Digital Payments Fee | Expense account for PSP transaction fees |
| DPEXTCLEAR | DP Ext. Pymts Clear | External payment clearing account for alternative payments |
| DPEXTTRANSFER | DP Ext. Pymts Trans | External payment transfer account for non-card methods |
| DPEXTCROSS | DP Ext.P Cross Clear | Cross-clearing account for external payment reconciliation |
| DPEXTREFCLEAR | DP Ext.Ref.Bank Clea | Bank account clearing for external payment refunds |

*Table 6.5: Account Symbols for Credit Card and External Payment Advice Processing*

Create the DPSETTLEMENT account symbol (see Figure 6.42) for the clearing accounts for the settlement. Selecting the **SIP Relev.** (self-initiated payment relevant) checkbox allows the symbol to be used in the automatic payment program; it does not need to be checked for SAP digital payments add-on settlement processes as the add-on-related settlements are triggered automatically by the add-on.

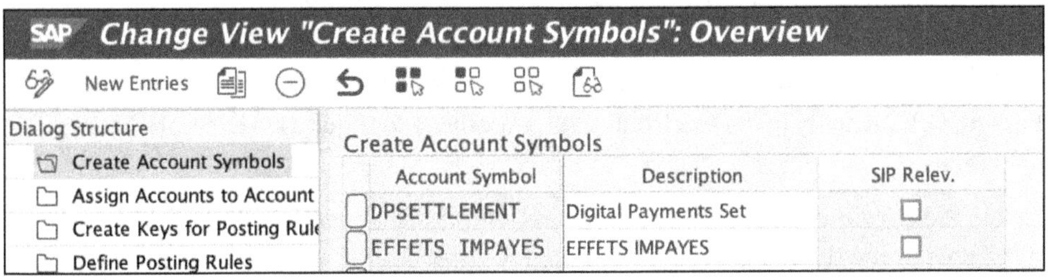

*Figure 6.42: DPSETTLEMENT Account Symbol*

Create account symbols for the clearing accounts for payment advice processing (see Figure 6.43) if you plan to use it in your implementation.

SAP New Entries: Overview of Added Entries

Dialog Structure
- Create Account Symbols
- Assign Accounts to Account
- Create Keys for Posting Rul
- Define Posting Rules
- Create Transaction Type
  - Assign External Transacti
- Assign Bank Accounts to Tr

Create Account Symbols

| Account Symbol | Description | SIP Relev. |
| --- | --- | --- |
| DPCLEAR | Digital Payments Cle | ☐ |
| DPTRANSFER | DP transfer | ☐ |
| DPFEE | DP fee | ☐ |
| DPEXTCLEAR | Ext payments clear | ☐ |
| DPEXTTRANSFER | Ext payment transfer | ☐ |

*Figure 6.43: Account Symbols for Payment Advice Processing*

After defining account symbols, assign actual general ledger accounts to each symbol through the **Assign Accounts to Account Symbol** activity. Create assignment entries specifying the chart of accounts, account modifier pattern, currency qualifier, and target general ledger account number for each symbol. The **Account Modifier** and **Currency** fields use plus signs as wildcards to create generic assignments applicable across all modifiers and currencies unless specific entries exist. Refer to SAP Notes 2524512 and 2730920 for the latest instructions to complete this configuration as per your implementation scope.

Account symbols enable consistent account determination across payment advice posting rules while accommodating organizational variations in account structures. The same posting rule can post to different accounts in different company codes by referencing an account symbol that resolves to company-specific accounts through the assignment configuration.

### Posting Rule Key and Posting Rule Definition

Define posting rule keys through the **Create Keys for Posting Rules** activity in the electronic bank statement configuration area. Posting rule keys provide mnemonic identifiers for posting rules and improve configuration readability compared to numeric rule identifiers. Create all posting rule keys by referring to SAP Notes 2524512 and 2730920.

Posting rules specify the detailed accounting logic, including that for the debit account, credit account, posting keys, document type, and posting area. Navigate to **the Define Posting Rules** activity in the electronic bank statement configuration to create posting rule definitions.

Each posting rule requires careful configuration to ensure that payment advice transactions post correctly and maintain balanced journal entries. The posting rule definition includes the posting rule key, posting area indicator, debit posting key, debit account or account symbol, credit posting key, credit account or account symbol, document type, and posting type classification.

**Virtual Bank Account Assignment to Transaction Type**

The final electronic bank statement configuration step assigns virtual bank accounts to transaction type CAMT053, establishing the link between payment advice import jobs and the virtual bank master data created as part of the prerequisite work. Access the configuration through the **Assign Bank Accounts to Transaction Types** activity in the electronic bank statement settings. Create assignment entries for each virtual bank. The table requires entries in the **Bank Key, Bank Account**, and **Transaction Type** columns. Enter the virtual bank identifier in the **Bank Key** column, the house bank account ID (matching the merchant alias) in the **Bank Account** column, and "CAMT053" in the **Transaction Type** column.

The virtual bank account assignments enable payment advice import jobs to correctly identify which merchant accounts the advice data relates to, ensuring that postings are made to the appropriate clearing accounts and that reconciliation processes can match payments against the correct receivables.

### 6.2.6 Tax Configuration for PSP Fees

PSPs charge transaction fees that may be subject to value-added tax or goods and services tax depending on jurisdiction and PSP location. When PSPs provide tax amounts in payment advice data, SAP S/4HANA must post these taxes correctly with appropriate tax codes. Configure tax codes for PSP fee postings through **Integration with Other SAP Components • Integration with the SAP Digital Payments Add-On • Accounts Receivable Settings • Maintain Tax Codes for Posting Charges.**

Figure 6.44 shows an example entry in which tax code 00 is configured, which means there are no PSP fees.

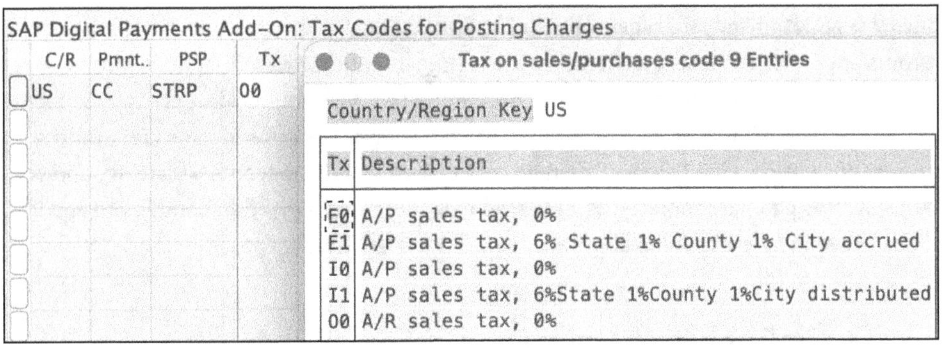

SAP Digital Payments Add-On: Tax Codes for Posting Charges

| C/R | Pmnt.. | PSP | Tx |
|---|---|---|---|
| US | CC | STRP | 00 |

Tax on sales/purchases code 9 Entries

Country/Region Key US

| Tx | Description |
|---|---|
| E0 | A/P sales tax, 0% |
| E1 | A/P sales tax, 6% State 1% County 1% City accrued |
| I0 | A/P sales tax, 0% |
| I1 | A/P sales tax, 6%State 1%County 1%City distributed |
| 00 | A/R sales tax, 0% |

*Figure 6.44: US Tax Code Setup*

The tax configuration supports both gross and net fee reporting from PSPs. Figure 6.45 shows an example configuration for Great Britain (GB) in which the standard VAT fee tax code V1 is being selected.

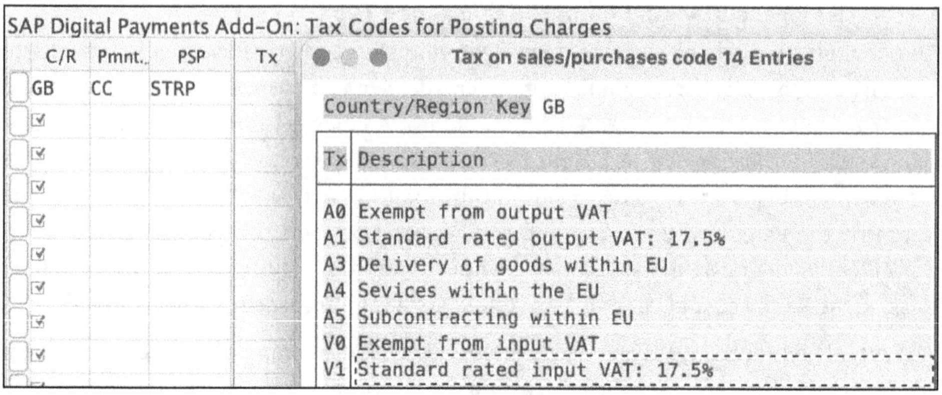

SAP Digital Payments Add-On: Tax Codes for Posting Charges

| C/R | Pmnt.. | PSP | Tx |
|---|---|---|---|
| GB | CC | STRP | |

Tax on sales/purchases code 14 Entries

Country/Region Key GB

| Tx | Description |
|---|---|
| A0 | Exempt from output VAT |
| A1 | Standard rated output VAT: 17.5% |
| A3 | Delivery of goods within EU |
| A4 | Sevices within the EU |
| A5 | Subcontracting within EU |
| V0 | Exempt from input VAT |
| V1 | Standard rated input VAT: 17.5% |

*Figure 6.45: GB Tax Code Setup*

## 6.3 SAP Fiori Apps and User Interface

SAP S/4HANA's digital payments add-on integration includes comprehensive SAP Fiori app coverage for all payment-processing activities, from master data management through operational transaction processing to exception handling and monitoring. SAP Fiori apps replace traditional SAP GUI transactions for most payment operations, though classic transactions remain available for specific technical activities or if users prefer them.

Each application is delivered with predefined business catalogs that bundle related apps for easy role assignment through business role templates. Understanding the SAP Fiori app landscape is essential for training planning, authorization design, and operational procedure development. We begin with master data management apps, which establish the secure foundation for card registration and tokenization. We then explore the Manage Sales Orders app, which utilizes registered cards to trigger real-time authorizations during document creation. From there, we shift to automated fulfillment through the Schedule Accounts Receivable Jobs app, where authorized transactions are batched for settlement and reconciliation. Finally, the remaining sections provide the necessary visibility and control, covering payment monitoring, settlement analysis, and exception handling to ensure a complete, end-to-end audit trail within the SAP Fiori landscape.

### 6.3.1 Master Data Management Apps

Payment card master data management occurs primarily through business partner applications enhanced with digital payments add-on integration. These applications provide secure interfaces for card registration, display of card portfolios, and card lifecycle management, including expiration updates and deletion processing. In the following sections, we'll specifically discuss the Manage Customer Master Data app and the Maintain Business Partner app.

### Manage Customer Master Data App

The Manage Customer Master Data app serves as the primary interface for creating and maintaining customer master data centrally for departments involved with sales. You can create, change, search, display, and copy customer master data. For digital payments, the application provides embedded payment card management capabilities integrated with the add-on's secure card entry services.

Access the application from the **Master Data Maintenance** tile on the SAP Fiori launchpad. The application opens to a search screen via which you can locate existing customers by number, name, or address components (see Figure 6.46).

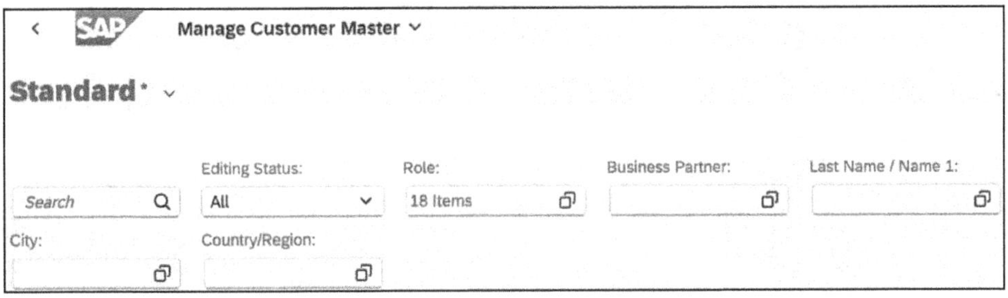

*Figure 6.46: Customer Master: Search Options*

Search results display as a list of entries showing the business partner identification, name, address, and role indicators. Select a business partner in the customer role and click the **Edit** button to open the record in change mode. Locate the **Payment Cards** section (see Figure 6.47), which shows all registered cards for the selected business partner.

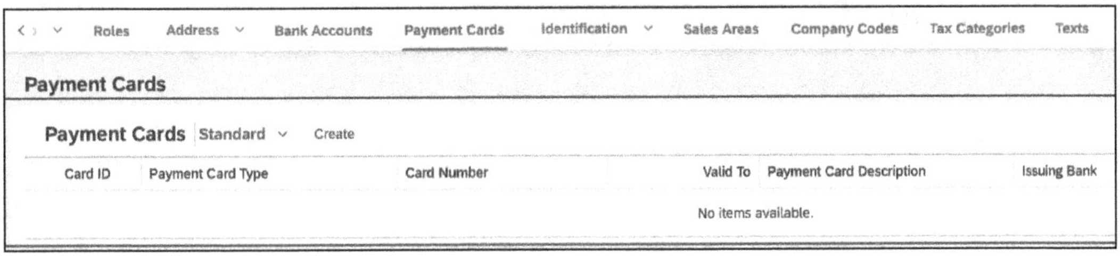

*Figure 6.47: Payment Cards Section*

The payment cards section presents stored cards (if present) in tabular format with columns for **Card ID, Card Type, Masked Card Number, Card Holder Name, Expiration Date,** and **Status**. Each row represents one registered card. Click the **Create** button to initiate new card registration (see Figure 6.48).

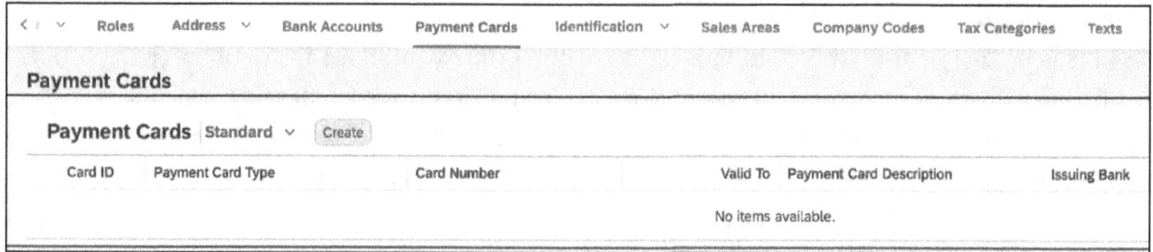


*Figure 6.48: Create Button: Payment Cards Section*

A window titled **Register New Payment Card** will appear (see Figure 6.49).

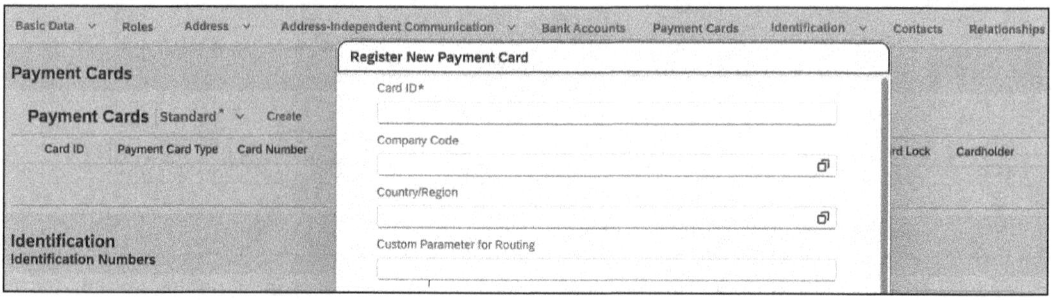

*Figure 6.49: Register New Payment Card*

Enter the **Card ID** (see Figure 6.50), **Company Code**, **Country/Region**, and, if applicable, any custom parameter to ensure the request is directed to the appropriate PSP.

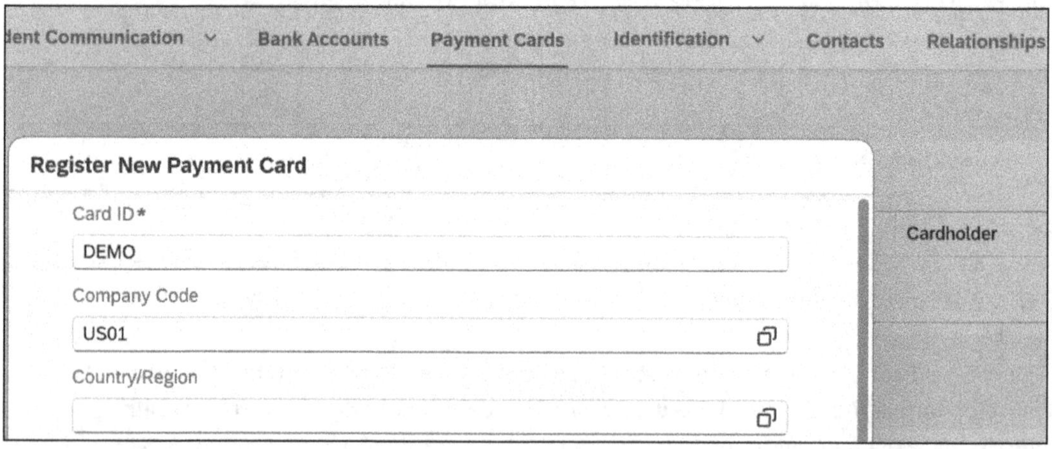

*Figure 6.50: Enter Card Details*

The application initiates an asynchronous request to the digital payments add-on to obtain a secure registration URL. Upon receipt of the registration URL from the add-on, the application displays the PSP's card entry interface within an iFrame (see Figure 6.51).

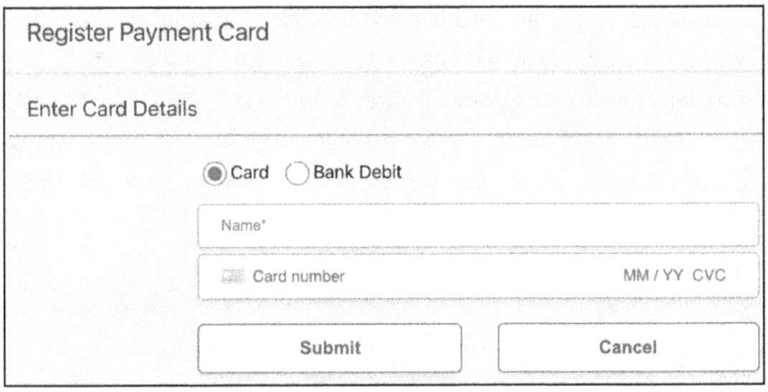

Register Payment Card
Enter Card Details
Card
Bank Debit
Name*
Card number
MM / YY CVC
Submit
Cancel

*Figure 6.51: Card Entry Interface*

The embedded card entry form requests standard payment card information, including the cardholder name, card number, expiration month and year (MM/YY), and card verification value (CVC). After entering all required information, click the **Submit** button (see Figure 6.52).

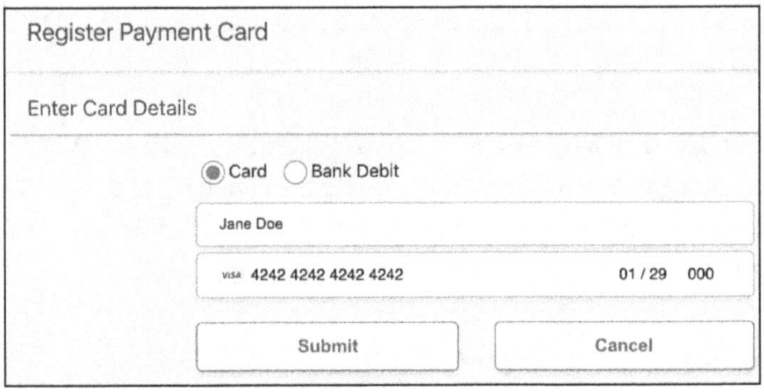

Register Payment Card
Enter Card Details
Card
Bank Debit
Jane Doe
VISA 4242 4242 4242 4242
01 / 29 000
Submit
Cancel

*Figure 6.52: Card Details for Registration*

A **Success** message appears to confirm the request was sent to the PSP (see Figure 6.53).

Card
Bank Debit
Jane Doe
VISA 4242 4242 4242 4242
01 / 29 000
Submit
Cancel
Success
Your request has been sent to the service provider.
Ok

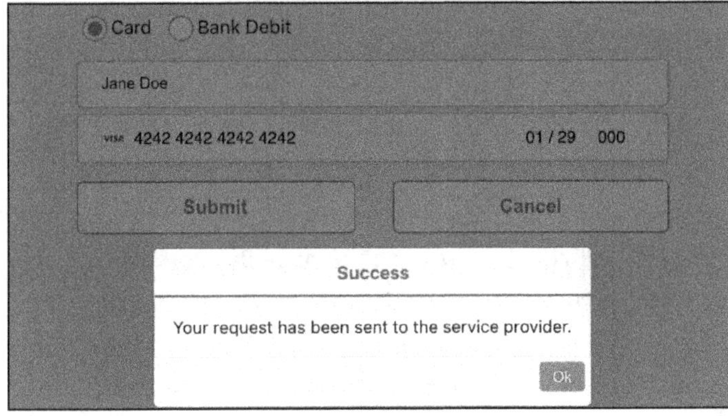

*Figure 6.53: Successful Card Registration*

The PSP registers the card and sends tokenized card metadata back to the digital payments add-on. The add-on then creates its own internal token mapping and forwards the card metadata—including the masked number, expiration date, and token—to SAP S/4HANA. After receiving the registration response, the application adds a new payment card entry to the business partner record, showing the newly registered card in the payment cards table (see Figure 6.54).

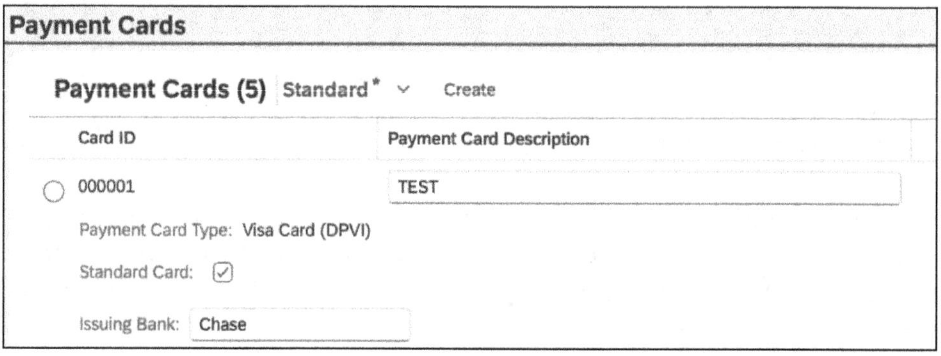

Payment Cards
Payment Cards (5) Standard* Create
Card ID | Payment Card Description
000001 | TEST
Payment Card Type: Visa Card (DPVI)
Standard Card:
Issuing Bank: Chase

*Figure 6.54: Newly Registered Card*

Selecting a card and clicking **Delete** (see Figure 6.55) sends a request through the add-on to the PSP, immediately invalidating the card token and removing the card from both the PSP vault and SAP S/4HANA storage. Deletion is instant and irreversible; to use a deleted card again, customers must re-register it.

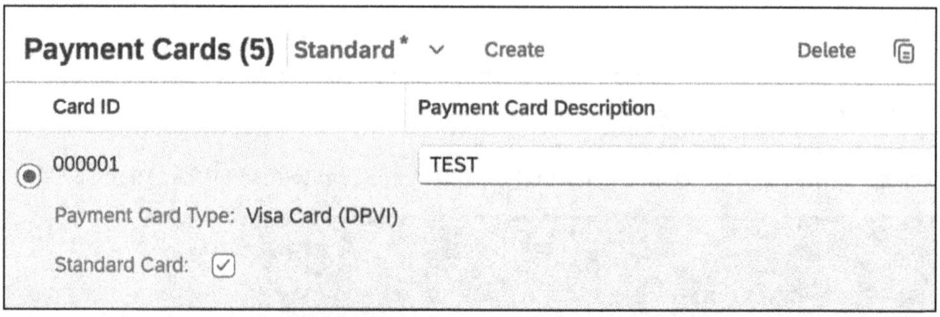

Payment Cards (5) Standard* Create Delete
Card ID | Payment Card Description
000001 | TEST
Payment Card Type: Visa Card (DPVI)
Standard Card:

*Figure 6.55: Payment Card: Delete Option*

#### Maintain Business Partner App

Transaction BP serves as the SAP GUI counterpart to the Manage Business Partner app, delivering a few payment card management features within the classic SAP screen format.

The initial screen prompts users to enter a business partner number or utilize the search functionality to locate records by name or address. After entering the business partner number, select either the **Display** or **Change** option as appropriate for your required mode. Access the **Payment Transactions** view (see Figure 6.56).

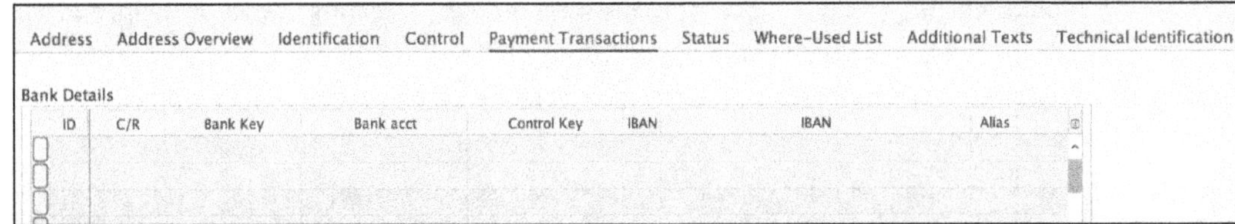


*Figure 6.56: Payment Transactions Tab*

The **Payment Cards** section in this tab shows registered cards listed in an ALV grid (see Figure 6.57). This grid shows the columns available in the SAP Fiori app, along with extra technical details helpful for troubleshooting when you select the **Card Details** button—such as card validity dates, issuing banks, the payment card lock indicator, and so on.

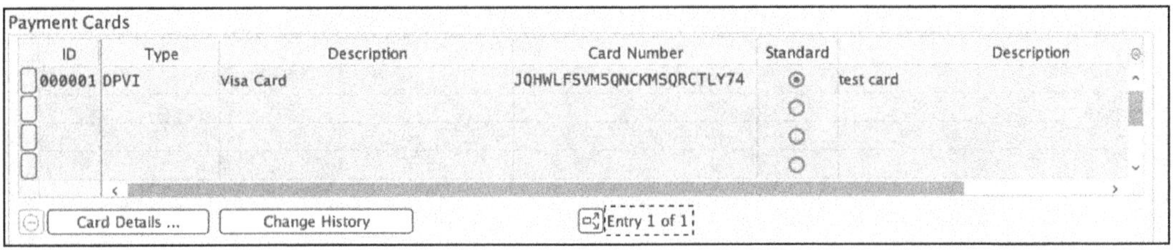

*Figure 6.57: Registered Cards List*

## 6.3.2 Manage Sales Orders App

Sales document applications enable payment card assignment during order entry, authorization status monitoring, and payment-specific order modifications. These applications integrate payment processing workflows into standard sales processes, making payment operations a natural part of order management rather than separate specialized activities.

The Manage Sales Orders app in particular provides comprehensive sales document management, including creation, modification, and display functions. You can create a new sales order by clicking the **Create** hyperlink (see Figure 6.58).

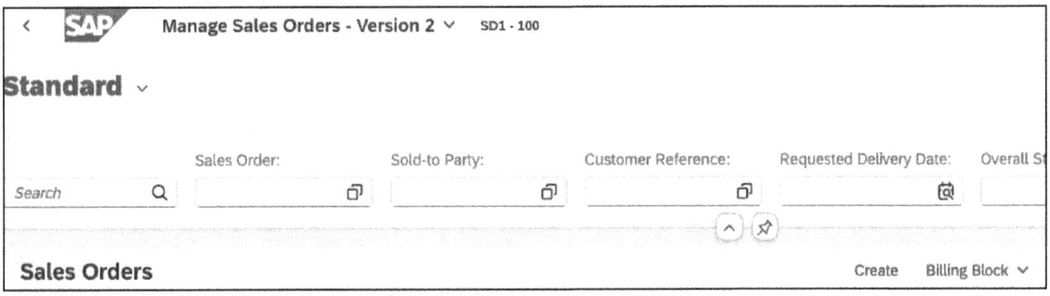

*Figure 6.58: Create New Sales Order*

For digital payments, the application includes the **Electronic Payments** tab in the document header (see Figure 6.59), where payment cards are assigned and authorization status is monitored. Assign payment cards to the sales order by selecting the search help, and the application displays a value help dialog listing all active, nonblocked payment cards from the sold-to party's business partner master data. Select the appropriate card and confirm the selection.

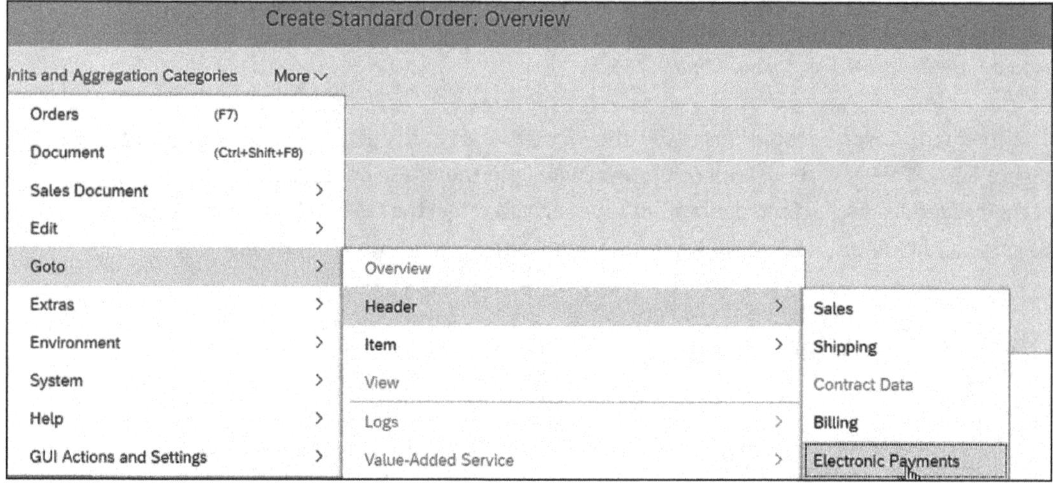


*Figure 6.59: Electronic Payments Tab in Order Header*

The system creates a payment card entry in the order header with card metadata populated from business partner master data. Save the sales order by clicking the **Save** button. The save operation triggers authorization evaluation based on payment guarantee schema determination, checking group requirements, and authorization horizon calculations. If conditions are met for authorization, then the system sends the request through the digital payments add-on to the PSP. Authorization responses update the **Electronic Payments** tab with result data.

### 6.3.3 Schedule Accounts Receivable Jobs App

Background job scheduling for settlement, payment advice processing, and authorization refresh operations occurs through the Schedule Accounts Receivable Jobs app. This application provides unified access to all payment card–related batch processes with guided parameter entry, job history management, and execution monitoring integrated into a single interface.

The Schedule Accounts Receivable Jobs app supports creating, scheduling, monitoring, and analyzing recurring and one-time background jobs for accounts receivable processes. For digital payments, the application provides specific job templates optimized for payment card settlement, payment advice processing, and authorization refresh operations. Access the application under the **Periodic Activities** section of the SAP Fiori launchpad.

The application opens to a job list view showing previously created and executed jobs with status indicators, execution timestamps, and result summaries. Filter the job list by job template type, execution status, creation date range, or responsible user to focus on relevant jobs. Click the **Create** button (see Figure 6.60) to define a new job.

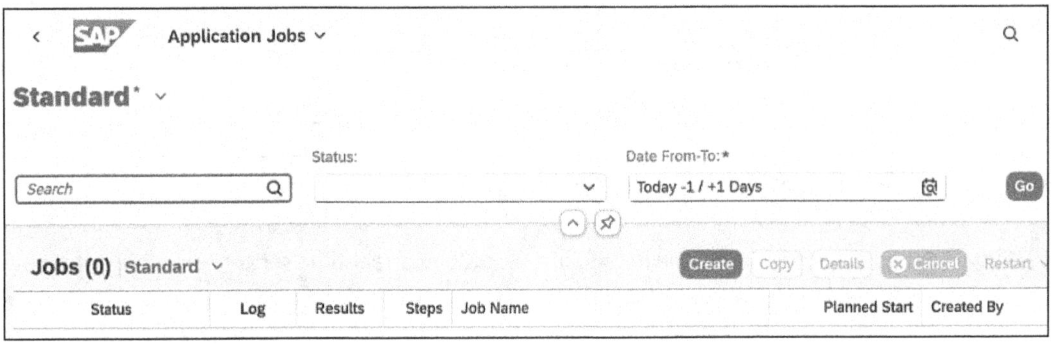

*Figure 6.60: Create Jobs Option*

The system presents a template selection dialog listing all available job templates for accounts receivable. Digital payment–specific templates include **Payment Card Settlement** for capture processing, **SAP Digital Payments: Advice Processing** for advice import, **Repeat Payment Card Settlement** for retry of failed settlements, and **Digital Payments Payment Card Authorizations for Customer Line Items** for authorization refresh.

### Payment Card Settlement

Select the **Payment Card Settlement** template (see Figure 6.61) to configure a settlement job.

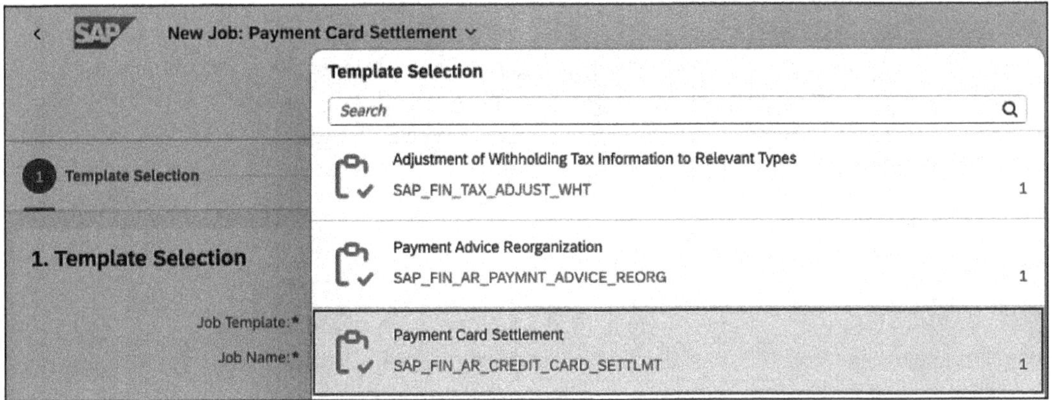


*Figure 6.61: Payment Card Settlement Job Template Selection*

The application navigates to a guided parameter-entry wizard with sections for naming the job (see Figure 6.62), scheduling options, run settings, and selection criteria.

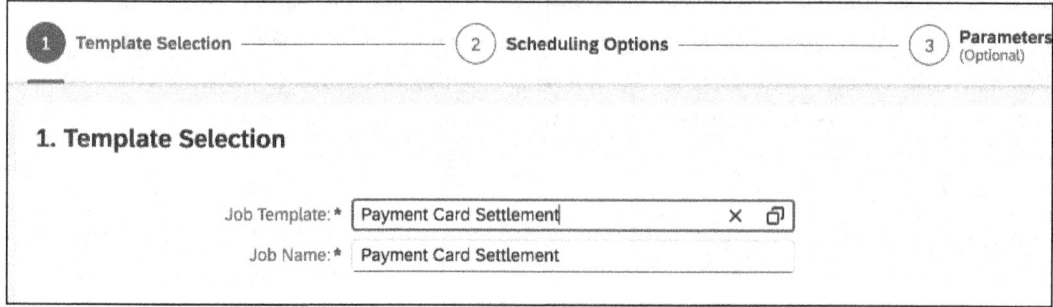


*Figure 6.62: Settlement Job Name Definition*

Scheduling options include immediate execution for real-time settlement testing or scheduled execution with date, time, and recurrence pattern specification (see Figure 6.63). Configure daily recurrence for production settlement operations to automate regular payment capture without manual intervention.

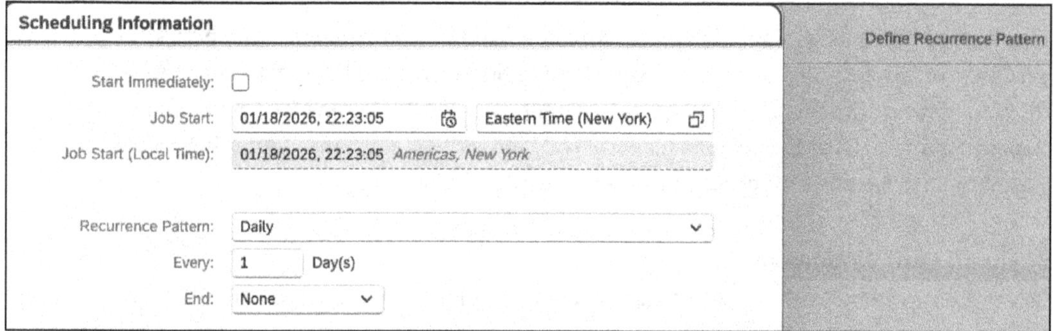


*Figure 6.63: Scheduling Options*

The **Company Code** parameter restricts settlement to specific organizational units. The other mandatory selection criterion is **G/L Account** (see Figure 6.64).

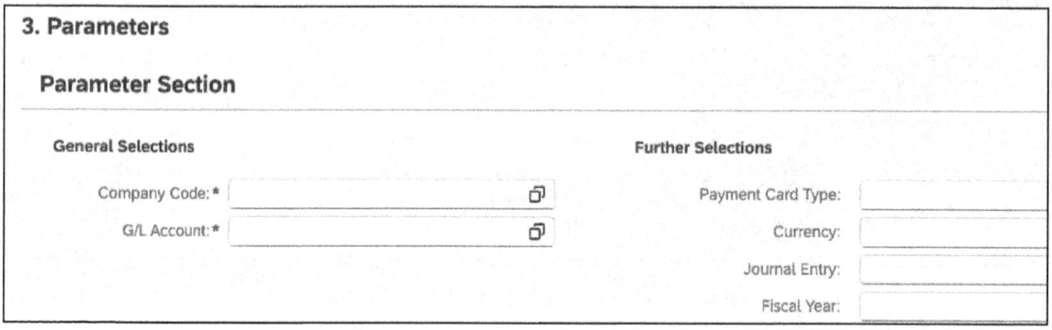

3. Parameters

Parameter Section

General Selections

Company Code: *

G/L Account: *

Further Selections

Payment Card Type:

Currency:

Journal Entry:

Fiscal Year:

*Figure 6.64: Parameter Section*

Click the **Schedule** button to create the job definition and initiate execution based on configured timing. For immediate execution, the application displays real-time progress indicators showing settlement request transmission and PSP response processing. Settlement execution time varies from seconds for small transaction volumes to several minutes for high-volume batch processing. The application remains responsive during execution, allowing users to navigate to other tasks while settlement processes in the background.

Review job results through the application log accessible from the job list view (see Figure 6.65). Click a completed job entry to display execution details including start and end timestamps, transaction counts, successful settlement count, failed settlement count, and a detailed message log. The message log lists each processed billing document with the settlement outcome, PSP response codes, and error descriptions for failed transactions.

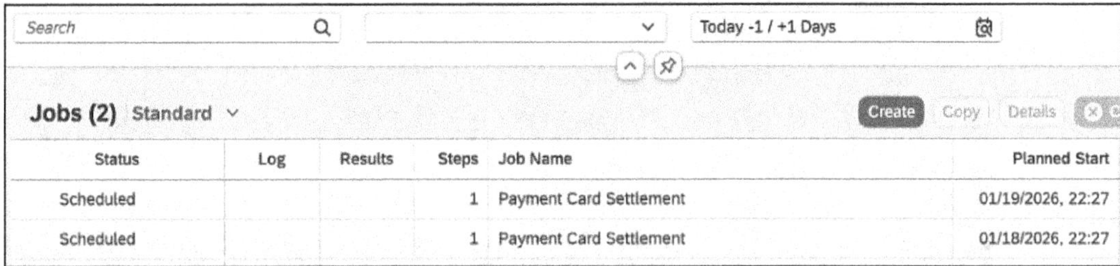

Search | Today -1 / +1 Days

Jobs (2) Standard | Create | Copy | Details

| Status | Log | Results | Steps | Job Name | Planned Start |
|---|---|---|---|---|---|
| Scheduled | | | 1 | Payment Card Settlement | 01/19/2026, 22:27 |
| Scheduled | | | 1 | Payment Card Settlement | 01/18/2026, 22:27 |

*Figure 6.65: Job List View*

### Repeat Payment Card Settlement

You can also select the **Repeat Payment Card Settlement** template (see Figure 6.66) to configure a new job.

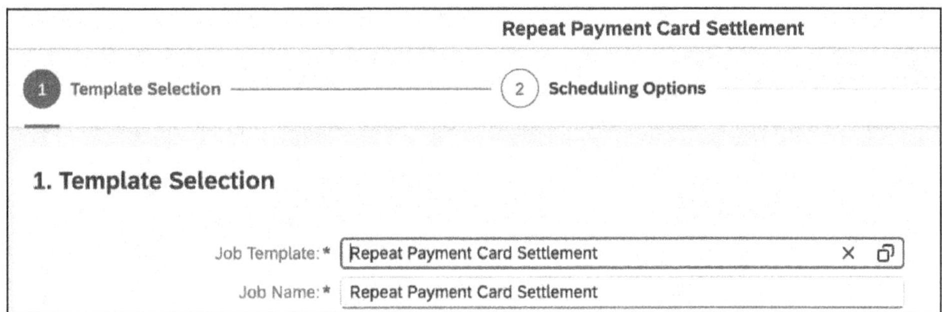


*Figure 6.66: Repeat Payment Card Settlement Template*

With this background job, you can repeat a settlement run that has failed. Specify the **Settlement run number** as the selection criterion (see Figure 6.67), then click **Schedule**.

Parameter Section

General selections

Settlement run number: *

Date:

Short Text:

Long Text:

Processing parameters

Send L2/L3 Data:

*Figure 6.67: Selection Criteria*

### SAP Digital Payments: Advice Processing

The **SAP Digital Payments: Advice Processing** job template enables automated retrieval and import of payment advice data from PSPs through the digital payments add-on. Select the **SAP Digital Payments: Advice Processing** template (see Figure 6.68) from the job template list.

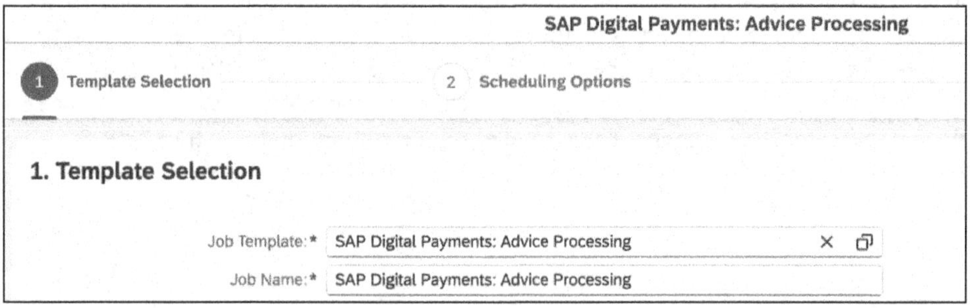

*Figure 6.68: Advice Processing Template*

The application presents specialized parameter sections for general selections, further selections, and PSP selections tailored to payment advice scenarios. **General Selections** (see Figure 6.69) include a virtual bank identifier that links to the bank master record representing the PSP relationship, which enables advice retrieval for specific PSP merchant accounts. The **Digital Payment Method** parameter distinguishes between credit card (CC) advice and external payment (EP) advice, allowing separate job configurations if different processing frequencies or selection criteria apply.

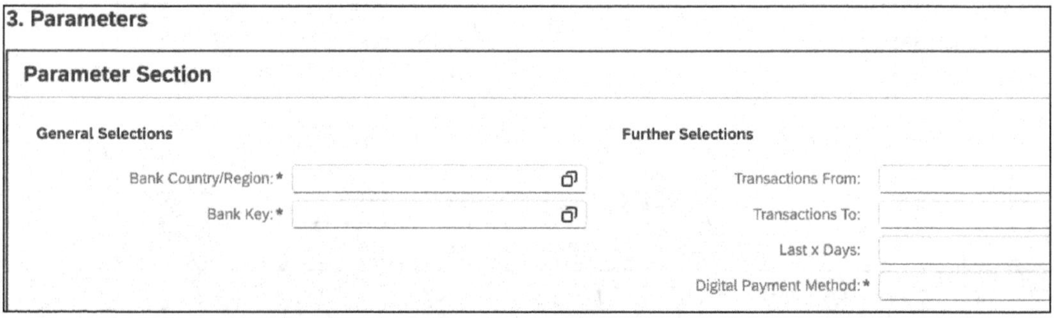

*Figure 6.69: General Selection Options*

The **Payment Service Provider** (see Figure 6.70) and **Merchant Alias** parameters enable filtering when multiple PSPs or merchant accounts exist. Leave these blank to retrieve advice for all configured PSPs and merchants, maximizing automation. Specify values when testing PSP-specific configurations or investigating merchant-specific reconciliation issues. The **Reconciliation** checkbox controls whether previously imported advice should be reimported, which is useful for reprocessing historical periods after configuration corrections.

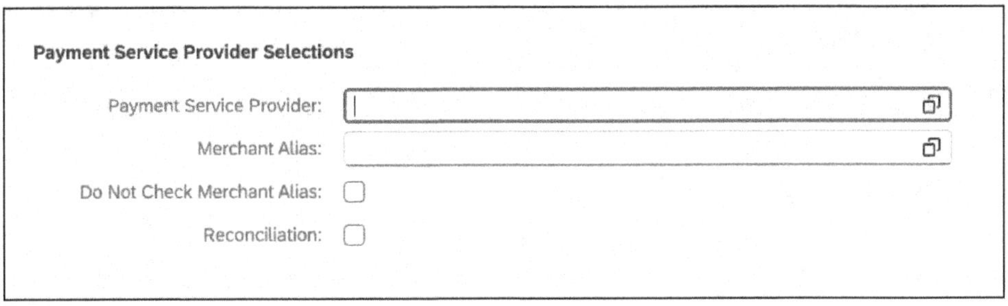

Payment Service Provider Selections

Payment Service Provider:

Merchant Alias:

Do Not Check Merchant Alias:

Reconciliation:

*Figure 6.70: Payment Service Provider Selection Options*

Schedule advice processing jobs to align with PSP advice generation timing. Many PSPs generate advice data hourly or at multiple times throughout the business day, enabling near-real-time reconciliation when import jobs execute frequently. Daily execution provides adequate reconciliation for most business scenarios while minimizing system load compared to hourly processing. Execute the job and review the application log to see the advice item counts retrieved from the add-on, items successfully matched to SAP S/4HANA billing documents, items matched to authorization data, and items parked for later matching.

### Payment Card Authorizations for Customer Line Items with Standard Card

Select the **Payment Card Authorizations for Customer Line Items with Standard Card** job template (see Figure 6.71) to create a new job.

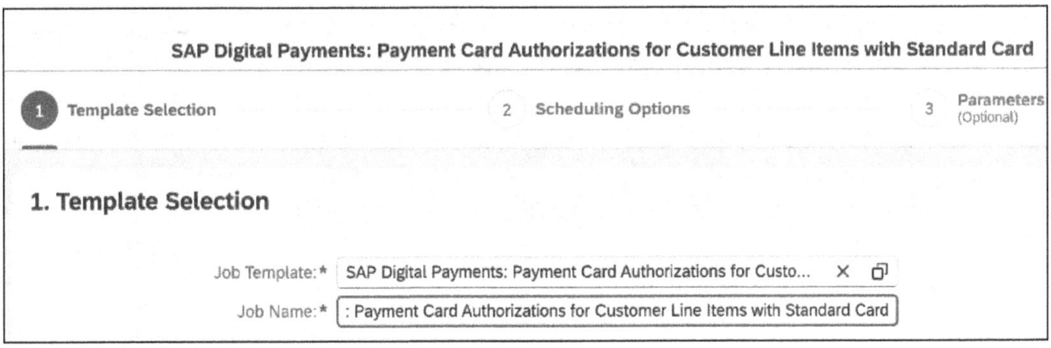

SAP Digital Payments: Payment Card Authorizations for Customer Line Items with Standard Card

1 Template Selection 2 Scheduling Options 3 Parameters (Optional)

1. Template Selection

Job Template:* SAP Digital Payments: Payment Card Authorizations for Custo...

Job Name:* : Payment Card Authorizations for Customer Line Items with Standard Card

*Figure 6.71: Payment Card Authorizations Job Template*

This background job authorizes payment cards for customer line items. You can select open items or down payment requests with various search criteria. Use this template only if the payment card in the customer master data is marked as a *standard card*. **Company Code** and **Customer** are mandatory selection parameters to schedule this job (see Figure 6.72).

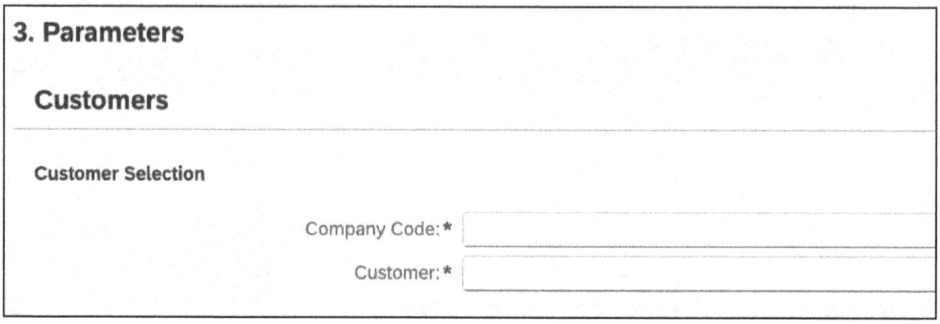

3. Parameters

Customers

Customer Selection

Company Code:*

Customer:*

*Figure 6.72: Customer Selection Parameters*

#### Payment Card Authorizations for Customer Line Items with Collection Authorization

Select the **Payment Card Authorizations for Customer Line Items with Collection Authorization** job template (see Figure 6.73) to create a new job.

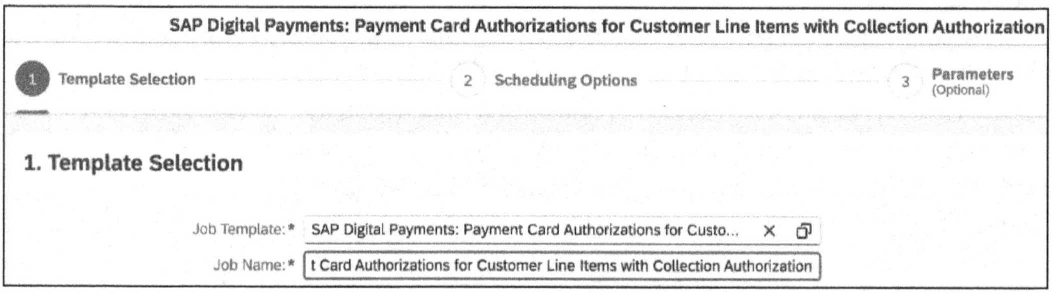

SAP Digital Payments: Payment Card Authorizations for Customer Line Items with Collection Authorization

1 Template Selection 2 Scheduling Options 3 Parameters (Optional)

1. Template Selection

Job Template:* SAP Digital Payments: Payment Card Authorizations for Custo...

Job Name:* t Card Authorizations for Customer Line Items with Collection Authorization

*Figure 6.73: Collection Authorization Job Configuration*

This background job enables payment card authorizations for customer line items. You can select open items or down payment requests with various search criteria. This template is applicable for payment card authorizations when the payment card specified in the relevant customer master data is identified as a *collection authorization payment card*. **Company Code** and **Customer** are mandatory selection parameters to schedule this job (see Figure 6.74).

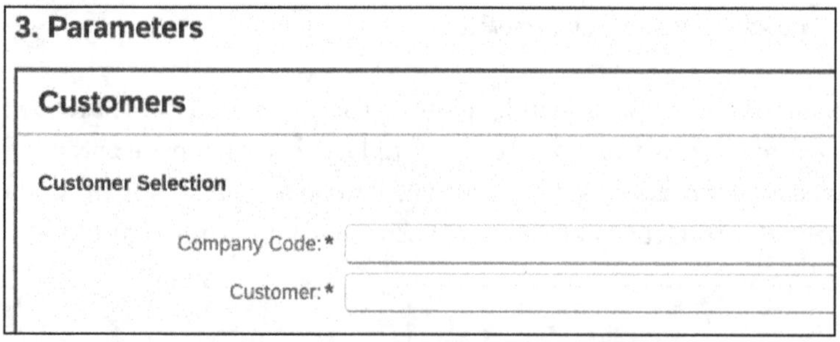

3. Parameters

Customers

Customer Selection

Company Code:*

Customer:*

*Figure 6.74: Customer Selection: Mandatory Parameters*

### 6.3.4 Display Payment Card Data App

Monitoring applications provide visibility into the payment processing status across all transaction lifecycle stages. These applications support operational oversight, exception identification, and analysis of payment trends or issues that require management attention.

The Display Payment Card Data app provides comprehensive inquiry functionality for payment card transactions, authorization statuses, settlement execution, and advice processing results. The application aggregates data from sales documents, billing documents, accounting postings, settlement runs, and payment advice tables into unified views organized by business process perspective. Launch the application from the **Accounts Receivable** tab within the **Payments and Clearing** section.

This app allows you to locate specific payments and their associated information through search filters like **Authorization Date, Card Number, Company Code,** and more (see Figure 6.75). The results show clearly presented details about **Entry, Authorization, Settlement,** and the **Settlement Run**, together with information regarding each payment and its payment card.

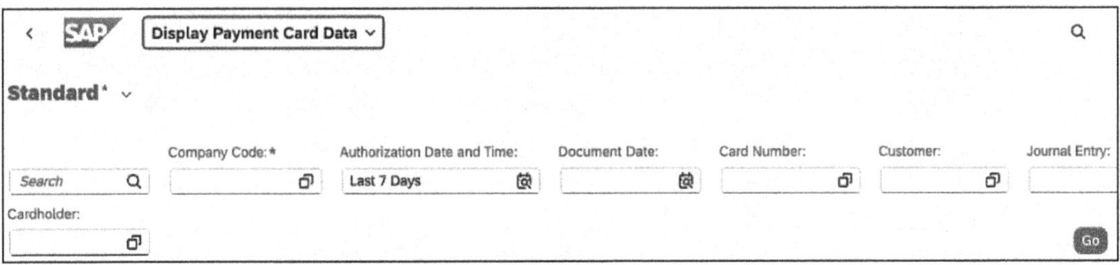


*Figure 6.75: Display Payment Card Data: Search Filters*

### 6.3.5 Exception Handling Apps

Exception-handling applications enable identification and resolution of payment processing problems including authorization declines, settlement failures, and advice reconciliation gaps. These applications provide workflow support for finance and customer service teams managing payment exceptions.

#### Resolve Payment Card Issues – Schedule Job

The Resolve Payment Card Issues – Schedule Job app lets you schedule jobs to process sales documents with payment cards needing authorization. Documents are selected if they are still open or in process and if their next electronic payment date (shown, for example, in a sales order header) falls within the authorization horizon. When this date arrives, the document requires reauthorization or, if only preauthorized earlier, full authorization.

You can choose to start the job right away or set it to run in the background. Running jobs when system usage is low can help decrease system load and speed up processing times.

### Resolve Payment Card Issues – Reauthorizations

The Resolve Payment Card Issues – Reauthorizations app (see Figure 6.76) enables you to search for sales orders and outbound deliveries that need payment card reauthorization and to trigger reauthorization as needed. In our example, **0** indicates that no payment card reauthorizations are currently needed.

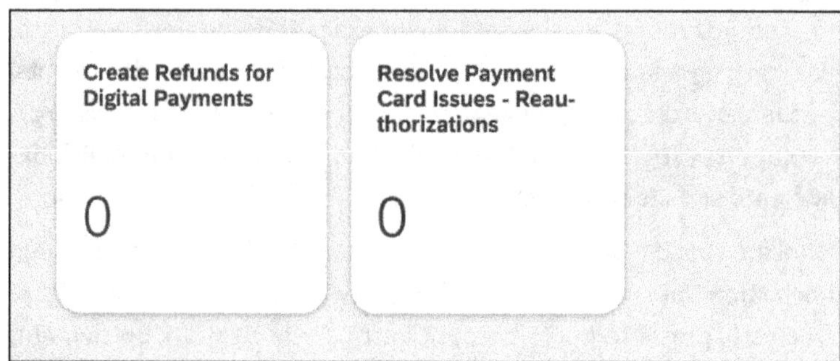


*Figure 6.76: Resolve Payment Card Issues – Reauthorizations App*

With this app, you can use or customize views and page variants, filter and search sales orders (see Figure 6.77), view payment details for each order, reauthorize payment cards individually or in bulk, access related sales order and delivery apps, navigate to customer pages and overviews, and export sales order lists to XLSX files.

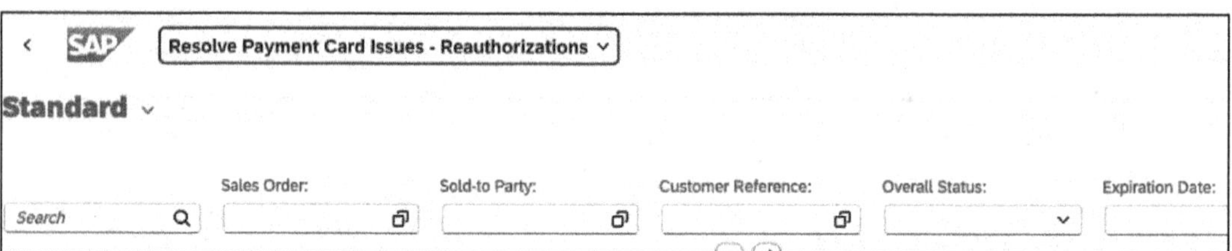


*Figure 6.77: Search Options*

## 6.3.6 Refund and Reversal Processing Apps

Refund processing applications enable the creation of payment returns to customers for canceled orders, returned merchandise, or billing corrections. These applications integrate with digital payments add-on refund APIs to process returns through PSPs back to the original payment methods.

### Create Refunds for Digital Payments

The Create Refunds for Digital Payments app lets you create refunds for payment card transactions made through the **Payment Card Authorizations for Customer Line Items** job in the Schedule Accounts Receivable Jobs app. You can check the refund process flow and see how refund-related documents such as customer invoices, payment records, and the refund statement are connected.

Use this app to search for refundable payments using different criteria (see Figure 6.78); create partial or full refunds (if the previous refund for a payment is completed); view the refund process flow by selecting a payment and navigating to its details page; and, for payments clearing multiple invoices, display one related invoice in the process flow.

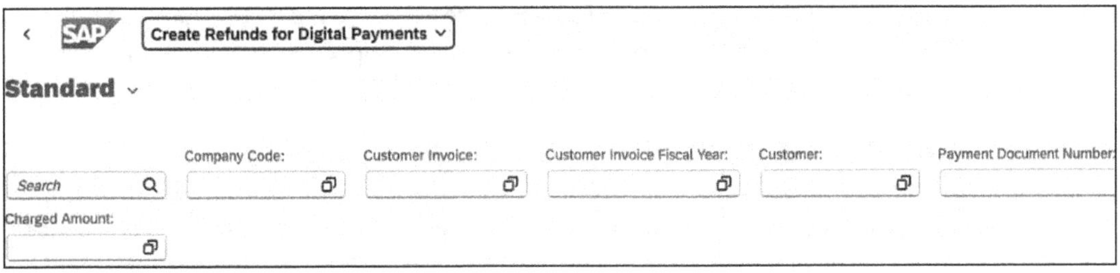

*Figure 6.78: Search Criteria in Create Refunds App*

**Reset Cleared Items – Payment Cards**

The Reset Cleared Items – Payment Cards app enables you to reverse settled payment card transactions that cannot be processed by the provider or are returned by the cardholder.

Use this app to search for items (see Figure 6.79) to reverse payment card postings, reset clearing of payments in accounts receivable, and clear payments with reversal entries. It supports both captures and refunds.

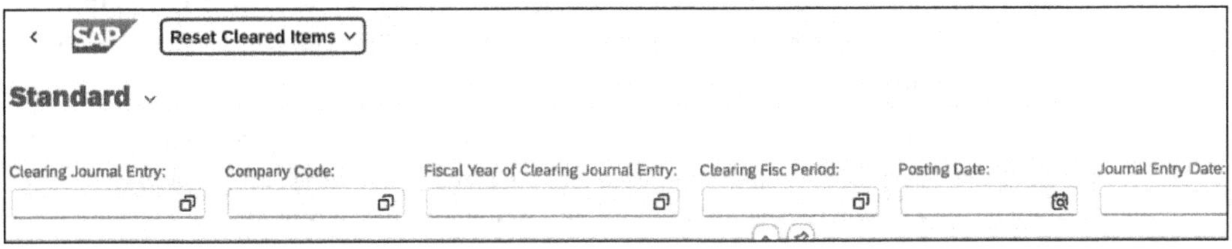

*Figure 6.79: Search Options: Reset Cleared Items App*

## 6.4 Order-to-Cash Process Integration

With the technical connectivity established and payment methods configured, the digital payments add-on integration becomes operational across the complete order-to-cash business process. This section demonstrates the end-to-end payment card processing workflow from initial card registration through final bank reconciliation, highlighting how SAP S/4HANA's native integration simplifies each process step compared to custom implementations. The process encompasses the end-to-end integration of digital payments within the order-to-cash cycle, illustrating a logical progression from initial setup to final reconciliation. We begin with master data preparation and sales document configuration, which establish the foundation for real-time authorization and automated billing. We then move into financial fulfillment, covering settlement execution, payment advice processing, and exception handling to ensure a seamless flow from the point of sale to the general ledger.

### 6.4.1 Payment Card Registration

Payment card registration in customer master data is the entry point for all subsequent payment processing. Registration occurs through secure interfaces provided by the digital payments add-on, ensuring that sensitive card data never enters SAP S/4HANA while maintaining tokenized references for transaction processing. The registration process validates card details with the PSP, creates encrypted token mappings, and stores nonsensitive metadata in business partner records.

To register a new payment card, follow the step-by-step instructions as illustrated in Section 6.3.1.

Payment card registration can fail for various reasons related to card validity, PSP connectivity, or data quality issues. Technical errors such as timeout exceptions indicate network connectivity problems between SAP S/4HANA and the add-on or between the add-on and PSP services. These errors typically resolve through retry attempts after verifying that all systems are operational. Configuration errors suggesting an invalid merchant ID or missing PSP activation indicate incomplete add-on tenant configuration that must be addressed through the add-on's PSP status and PSP determination configuration UIs.

Registered payment cards remain in business partner master data until explicitly deleted. Users can view existing cards by accessing the **Payment Cards** section in business partner management, providing customer service teams with reference information during customer interactions. To delete a payment card, select the card entry in the **Payment Cards** table and click the **Delete** button. Detailed instructions are provided in Section 6.3.1 in the Manage Customer Master Data App section.

Blocking payment cards provides an alternative to deletion when temporary payment method restrictions are needed. Choose a blocking reason from the configured values, such as **Card Lost** or **Other Reason**. Blocked cards remain visible in business partner master data but cannot be selected during sales order creation or settlement processing. Remove blocks by deselecting the block indicator when the underlying issue is resolved.

### 6.4.2 Sales Order Creation with Payment Authorization

Sales order processing represents the first transactional use of registered payment cards. Authorization requests validate customer payment capacity and establish payment commitments for order fulfillment.

Create a sales order using the Manage Sales Orders app or Transaction VA01 for SAP GUI–based processing. Enter standard sales order data including sold-to party, ship-to party, requested delivery date, material line items with quantities, and pricing information. The customer sold-to party must have payment guarantee procedure 0002 assigned in customer master data for payment card processing to activate.

Navigate to the sales document header by clicking the header navigation link or using the goto menu. Locate the **Electronic Payments** tab, which becomes visible when the sales document type has payment card plan type 03 assigned. The **Electronic Payments** tab displays payment card and authorization management functions organized into several sections that detail different aspects of payment-processing visibility.

The **Authorization Amount** section (see Figure 6.80) shows calculated values, including the total sales order value requiring authorization, any previously authorized amounts, amounts already settled through billing, and the remaining amount requiring authorization for the next delivery or service. These calculated fields update dynamically as order values change or authorizations are created, providing real-time visibility into payment coverage status.

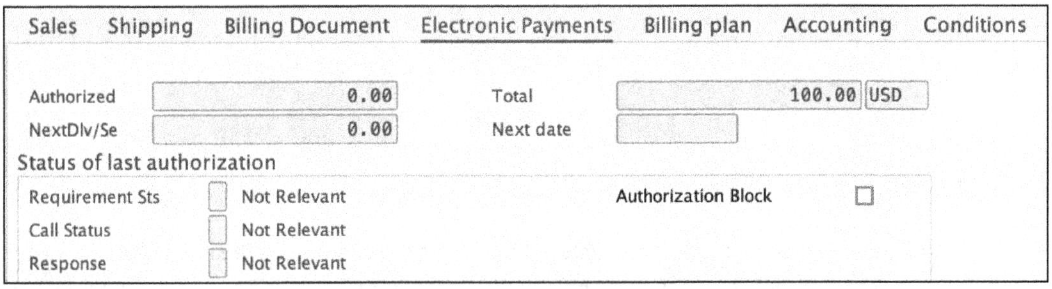

*Figure 6.80: Authorization Amount Section*

Click the **Assign Payment Card** button to process the sales order using a one-time payment card, rather than selecting a card from the customer master record. SAP launches a new window to securely enter the one-time payment card details (see Figure 6.81). This payment card will be used only for this sales document and will not be stored in the customer master record for future use. You should train your customer service representatives to have a clear understanding of when to use this feature compared to registering the card in the customer master record and using it in the sales order. This scenario is relevant if the customer specifically instructs you to not store a card in their profile. Another possible use is if you are creating a credit memo request in reference to an invoice, and the customer wants the refund to be issued against this new card rather than the original card that was used earlier during sales order creation.

*Figure 6.81: One-Time Payment Card Registration*

After you submit the card details, the window will close. Remember to click the **Confirm** button (see Figure 6.82) to assign the one-time payment card to the sales order.

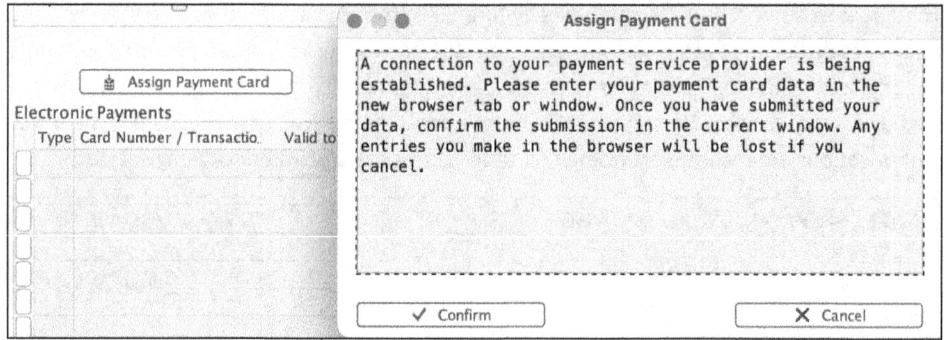


*Figure 6.82: Confirm Submission*

The other option is to select a registered card from the customer's business partner master data. The system presents a value help dialog (see Figure 6.83) listing all active, nonblocked payment cards registered for the sold-to party. The list displays masked card numbers, card types, expiration dates, and cardholder names, enabling order entry personnel to select the appropriate card based on a customer's verbal or written confirmation.

Select the desired payment card from the list and confirm the selection. The system creates a payment card entry in the sales order header, displaying the selected card's metadata including card type, masked number, expiration date, and initial authorization status indicators. At this point, the card is assigned to the order, but no authorization request has been sent to the PSP. Authorization processing occurs when the sales order is saved, provided authorization requirements are met.

Payment cards: Card number 4 Entries

Customer 6
Payment Card Type DPVI

| Card number | to | Def. | Block | Name of cardholder | Description |
|---|---|---|---|---|---|
| 5W6IQL3HFEY2WWPEFBR4R5C | 12/31/9999 | ☐ | | Test | |
| 5W6IQL3HFEY2WWPEFBR4R5R | 12/31/9999 | ☐ | | | |
| GRN5PKPCTTXXW3SFXRLDTBDM | 12/31/9999 | ☑ | | TEST | |

*Figure 6.83: Payment Card Dialog Window*

The **Electronic Payments** tab also includes an **Authorization Log** section that records all authorization attempts, responses, and status changes for the sales order. The log is initially empty for new orders but will populate with detailed entries as authorization processing occurs. Each log entry includes timestamps, authorization amounts, PSP response codes, and authorization status indicators, creating an audit trail of all payment validation activities.

Complete any remaining sales order data entry, including pricing, availability confirmation, and any special processing instructions. Save the sales order to trigger authorization processing. The system evaluates the payment guarantee schema, checking group requirements, and authorization horizon settings to determine whether immediate authorization should occur.

### 6.4.3 Delivery Processing with Authorization Validation

Delivery document creation represents a critical control point in the payment secured order-to-cash process. The system validates that sufficient payment authorization exists before allowing goods issue posting, preventing product shipment for orders that lack payment coverage. This validation protects the organization from fulfillment of potentially unpaid orders while maintaining efficient logistics operations for properly authorized transactions.

Delivery creation typically occurs through automated workflows such as collective delivery processing or through manual creation using Transaction VL01N or the Create Outbound Deliveries app. The delivery document inherits references to payment card and authorization data from the source sales order, though payment details are not duplicated in delivery tables. Instead, the delivery maintains pointers to sales order payment data for validation purposes.

When a delivery document is created with reference to a sales order containing payment card assignments, the system immediately validates the authorization status during the delivery save operation. The validation check retrieves the current authorization data from the sales order, calculates the delivery value that will be fulfilled, and compares the available authorization amount against delivery value requirements.

Authorization validation examines the authorization validity period to confirm the authorization has not expired. The system compares the authorization timestamp plus the validity period in days against the current date. If the authorization is still within the validity window, validation succeeds and delivery creation proceeds. If the authorization has expired, the system generates a warning message indicating insufficient authorization for the sales document.

For deliveries created with expired or insufficient authorizations, the system permits delivery document creation but blocks goods issue posting. Users can save and view the delivery but cannot execute picking confirmation or post goods movement. The delivery remains in **Created** status with blocking indicators until the sales order authorization is refreshed to cover the delivery value with a valid authorization.

When authorization validation warnings occur, warehouse or logistics personnel cannot proceed with goods issue and must coordinate with sales or customer service teams to resolve authorization issues. The Resolve Payment Card Issues app provides centralized visibility into all sales orders and deliveries with authorization problems, allowing you to systematically work through authorization refresh requirements.

### 6.4.4 Billing Document Creation and Financial Posting

Billing document creation converts delivered goods or completed services into financial receivables, transferring payment card and authorization data from sales and logistics documents into accounting postings. The billing process validates authorization currency and inheritance, determines payment-specific general ledger accounts, and creates financial documents that initiate settlement workflows.

Billing documents are typically created through collective billing programs such as Transaction VF04 or through individual billing using Transaction VF01 and the Create Billing Documents app. When creating billing documents for deliveries or sales orders containing payment card data, the system automatically includes payment card and authorization information in the billing document without requiring manual entry.

The billing document inherits the payment card token, masked card number, card type, authorization code, authorization amount, and authorization validity date from the source sales order. This inheritance ensures billing documents maintain the complete payment context necessary for settlement processing and financial posting. The billing document header includes an **Electronic Payments** section like that for sales orders, displaying inherited payment information in read-only format.

During billing document creation, the system performs final authorization validation to confirm that the authorization remains valid and covers the billing amount. If the authorization has expired between delivery processing and billing creation, the system generates error messages that prevent the billing document's release to financial accounting.

After completing billing document data entry and performing final validations, release the billing document to financial accounting by clicking the **Post** button or saving the document if autorelease is configured for the billing type. The release operation generates accounting documents that contain customer receivable postings, revenue recognition entries, tax postings, and payment card–specific postings to designated accounts.

When billing documents with payment card data are released to financial accounting, the system applies account determination procedures to identify the correct general ledger accounts for payment receivable postings. Instead of posting receivables to standard customer accounts, the account determination procedure assigns postings to the payment card receivables account.

The accounting document includes payment card supplement data stored in table BSEGC, which links the financial document to the original sales order authorization. The supplement contains the payment card token, authorization code, PSP merchant alias, authorization validity period, and settlement status indicators. This supplement data enables settlement programs to select appropriate receivables for capture processing and facilitates authorization-to-settlement tracking.

### 6.4.5 Settlement Processing and Capture

Settlement is the payment capture stage in which authorized amounts are collected from cardholders and receivables are transferred from the payment card receivables account to PSP clearing accounts. Settlement processing in SAP S/4HANA uses background jobs configured through the Schedule Accounts Receivable Jobs app, enabling automated batch processing of eligible payment transactions.

Access the Schedule Accounts Receivable Jobs application from the SAP Fiori launchpad. Click the **Create** button to define a new job execution. The system presents a job template selection dialog; choose the **Payment Card Settlement** template from the available options. This template provides preconfigured selection and processing parameters optimized for digital payments add-on settlement processing. The **Company Code** parameter restricts settlement to specific organizational units, enabling company-specific settlement timing if required. For step-by-step instructions and screenshots, refer to Section 6.3.3.

The settlement job selection logic identifies billing documents that have been released to financial accounting, contain payment card supplement data with valid authorization codes, have not been previously settled or are marked for resettlement after settlement failures, and match the job's selection parameters. The system excludes billing documents with canceled authorizations, expired payment cards, or blocked settlement indicators.

Review the settlement job results through the application log accessible from the Schedule Accounts Receivable Jobs app. The log displays transaction counts including successfully sent settlements, failed settlements, and skipped transactions (those that do not meet the selection criteria). Detailed entries show individual billing document processing results, PSP response messages, and error descriptions for failed settlements.

For failed settlements, analyze error messages to determine the appropriate corrective action. Temporary PSP communication failures can be retried by reexecuting the settlement job with parameter selections that target previously failed transactions. For step-by-step instructions, refer to Section 6.3.3. Permanent failures such as authorization declines or card closures require manual intervention to arrange alternative payment collection methods with customers.

### 6.4.6 Payment Advice Processing and Reconciliation

Payment advice data provides PSP confirmation of successfully processed settlements. Importing and processing payment advice enables automated reconciliation of PSP clearing accounts and final confirmation of received payments, completing the accounts receivable cycle for payment card transactions.

Schedule payment advice import through the Schedule Accounts Receivable Jobs app using the **SAP Digital Payments: Advice Processing** job template. This job connects to the digital payments add-on API to retrieve payment advice data generated by PSPs for specified date ranges and merchant accounts.

The job log displays import statistics including total advice items retrieved, items successfully matched to SAP S/4HANA billing documents, items matched to authorizations, and items parked for later matching when corresponding SAP S/4HANA documents are not yet available. Review the log to verify the successful import and identify any matching warnings that require your attention.

### 6.4.7 Exception Handling and Problem Resolution

Payment card processing exceptions occur throughout the order-to-cash cycle, requiring systematic monitoring and resolution procedures to maintain operational efficiency. SAP S/4HANA provides SAP Fiori applications and background jobs for exception identification, analysis, and correction, enabling finance and customer service teams to manage payment issues proactively.

Authorization decline exceptions occur when PSPs reject payment authorization requests during sales order processing. Declined authorizations prevent order fulfillment and may require coordination with customers to resolve underlying payment issues:

- Access declined authorization exceptions through the Resolve Payment Card Issues – Reauthorizations app. You can search for and filter sales orders and their subsequent outbound deliveries with payment cards that need to be authorized again and trigger the reauthorizations. When you reauthorize a payment card, the system makes a dummy change in the sales order and saves it to trigger another authorization of the payment card.
- Review decline reasons to determine appropriate resolution actions. Generic declines often indicate insufficient funds or exceeded credit limits, requiring customers to provide alternative payment cards or arrange increased credit limits with card issuers. Specific decline reasons such as an invalid card number, expired card, or card reported lost indicate data quality issues that can be corrected by updating business partner payment card information.
- For customers willing to provide alternative payment methods, access the sales order in change mode, navigate to the **Electronic Payments** tab, delete the declined authorization entry and associated payment card entry, assign a different payment card from the customer's registered cards or register a new card, and save the order to trigger authorization on the new payment card. Successful authorization on the alternative card removes the credit block and enables delivery processing.

Settlement failures occur when PSPs decline capture requests despite valid authorizations, network communication disrupts settlement transmission, or add-on service issues prevent settlement completion. Settlement failures create PSP clearing account open items without corresponding fund collection, requiring exception management to maintain accurate receivables reporting:

- Access settlement information through the Display Payment Card Data app, filtering by the **Settlement Failed** settlement status. The application lists billing documents with failed settlement attempts, displaying PSP error codes, decline reasons, settlement request timestamps, and authorization details.
- Temporary technical failures such as timeout errors, network connection resets, or PSP service availability issues can be resolved through resettlement processing. Execute the **Payment Card Settlement** job with selection parameters configured to **Items with Errors** rather than standard item selection. This targets only previously failed settlements for retry processing, avoiding duplicate settlement of successful transactions.
- The Repeat Payment Card Settlement job template provides enhanced retry logic with configurable retry counts, exponential backoff timing, and error threshold management. Use this template for systematic retry of settlement failures rather than manual repeated execution of standard settlement jobs. The repeat settlement job tracks retry attempts and permanently marks transactions as failed after exhausting retry limits, enabling escalation to manual intervention.

### 6.4.8 Bank Statement Reconciliation

The final step in payment card processing reconciles PSP fund disbursements to bank accounts through standard bank statement processing. PSPs transfer collected funds to merchant bank accounts after deducting fees, chargebacks, and other adjustments, requiring bank statement import and clearing procedures to complete the receivables-to-cash cycle.

When PSP disbursements are received in your company bank accounts, the amounts appear on bank statements as incoming transfers with reference information provided by PSPs. The reference data quality varies between PSPs, ranging from comprehensive transaction detail enabling automated matching to generic references requiring manual reconciliation. You can negotiate reference data standards with PSPs during merchant agreement discussions to maximize reconciliation automation.

Import bank statements through standard SAP bank statement processing using formats such as MT940, BAI2, or CAMT.053 configured for the receiving bank. Use the Manage Bank Statements app to manage manual and electronic bank statements. The app provides an overview of all bank statements for all house bank accounts. You can also view detailed information for each bank statement. The bank statement import creates bank statement items representing the PSP disbursement transactions. These items require clearing against the PSP transfer clearing account (account symbol DPTRANSFER) open items created during payment advice processing.

Some PSPs provide separate bank statement detail files or reconciliation reports showing transaction-level breakdowns of disbursement amounts. These reports can be manually correlated with SAP S/4HANA payment advice processing results to verify that disbursement amounts match expected values.

An alternative method utilizes electronic bank statements, which are integrated into the system through one of three approaches:

- The system may automatically import statements via SAP Multi-Bank Connectivity.
- If bank statements are received in formats such as spreadsheets or MT940 files, they can be uploaded using the Manage Incoming Payment Files app.
- Alternatively, statements can be imported through the Bank Statement - Post API.

Access the Clear GL Accounts – Manual Clearing app to match bank statement disbursement items against PSP transfer account open items. Enter your transfer clearing account, set the company code, and specify the posting date range, encompassing the disbursement transaction date.

The application displays all open items on the PSP transfer account, including individual payment advice postings from DPIN processing, fee postings from DPFI processing, and refund postings from DPOU processing. Review the open item list to identify items that should sum to the bank statement disbursement amount. Select items for clearing by checking the selection boxes, verify that selected item total matches the bank statement amount, and post the clearing document to finalize the reconciliation.

Manual clearing is required when PSP reference data on bank statements does not enable automatic matching through standard bank statement clearing algorithms. Some PSPs provide enhanced reference data in bank statements that can be mapped to SAP S/4HANA clearing fields through bank statement Customizing. Configure posting rules in electronic bank statement settings to automatically clear PSP transfer accounts when reference data matching is reliable.

The complete order-to-cash process with payment card integration concludes when PSP disbursements are cleared against transfer account open items. At this point, the organization has received funds in bank accounts, all receivable accounts are cleared, revenues are recognized, and the payment card transaction cycle is complete. The integrated process from card registration through bank reconciliation demonstrates SAP S/4HANA's comprehensive support for digital payment workflows with minimal custom development or manual intervention.

## 6.5 Cloud Versus On-Premise Considerations

The digital payments add-on integration with SAP S/4HANA functions across all deployment models, including SAP S/4HANA Cloud Public Edition, SAP S/4HANA Cloud Private Edition, and on-premise systems. The distinction between deployment models centers on the management boundary separating customer responsibilities from SAP responsibilities. SAP S/4HANA Cloud Public Edition operates via a pure software-as-a-service (SaaS) model, with SAP managing the infrastructure, security patches, quarterly feature releases, and system availability. On-premise deployments place complete responsibility on customer IT organizations for server hardware, database management, security maintenance, and software upgrades. SAP S/4HANA Cloud Private Edition occupies a middle ground, with SAP or hyperscaler partners managing the infrastructure while customers retain greater configuration flexibility than the public cloud permits.

The following sections explore the critical distinctions between deployment models, focusing on how configuration strategies and feature rollouts vary across different SAP environments. Understanding these nuances, from SaaS-based activations to traditional Transaction SPRO Customizing, is essential for aligning payment-processing capabilities with specific organizational needs and technical landscapes.

### 6.5.1 Configuration Approach Differences

Add-on configuration varies between SAP S/4HANA Cloud Public Edition and on-premise or SAP S/4HANA Cloud Private Edition deployments; the key difference is between SaaS-based scope item activation and traditional Transaction SPRO Customizing approaches.

SAP S/4HANA Cloud Public Edition configurations are managed through scope item activation, custom business configuration applications, and communication arrangements. Scope item 1S2 activates the digital payments add-on integration, including business partner payment card management, sales document payment processing, billing settlement, and accounts receivable reconciliation.

Configuration changes in SAP S/4HANA Cloud Public Edition are constrained to SAP-supported modification paths including parameter value changes, optional feature activation, and extension fields within allowed Customizing areas. You cannot modify delivered ABAP code, create custom function modules for payment processing, or alter core payment-authorization logic. This restriction ensures clean core principles to support seamless quarterly upgrades without regression-testing custom modifications.

On-premise and SAP S/4HANA Cloud Private Edition systems use traditional Transaction SPRO Customizing for digital payments add-on configuration, providing granular control over every configuration setting at the cost of increased configuration complexity. You must execute individual configuration activities for payment card type definition, card category creation, payment card plan type assignment, authorization horizon specification, checking group definition, account determination procedure maintenance, and electronic bank statement setup.

Custom development capabilities in on-premise systems enable implementation of customer-specific payment processing requirements through BAdI implementations, user exits, and custom programs as needed. The on-premise configuration model requires deeper SAP technical expertise including Transaction SPRO navigation knowledge, SAP Note application skills, RFC destination understanding, and payment card processing technical architecture comprehension.

### 6.5.2 Feature Availability and Release Timing

Payment card processing capabilities evolve continuously through SAP development, delivering new PSP adapters, enhanced authorization workflows, expanded payment methods, and improved reconciliation processing. The timing and mechanisms for accessing these innovations differ substantially between SAP S/4HANA Cloud Public Edition and on-premise deployments.

SAP S/4HANA Cloud Public Edition customers receive quarterly feature releases automatically delivered by SAP without customer action required. Each quarterly release includes functional enhancements, PSP adapter updates, security patches, and bug fixes activated through scope item updates and application version upgrades.

The continuous innovation model ensures that SAP S/4HANA Cloud Public Edition systems remain current with payment industry evolution, PSP capability enhancements, and regulatory requirement changes. You avoid the technical debt accumulation that affects on-premise systems operating on older software versions, though sacrifice control over upgrade timing and feature adoption pace.

SAP has moved to a biennial major release strategy since SAP S/4HANA 2023. Innovations are delivered via Feature Pack Stacks (FPS) approximately every six months during the first two years of a major release. You have control over when to apply major releases and FPS innovations, with the flexibility to defer upgrades while remaining within SAP's seven-year mainstream maintenance window. This control enables coordination of SAP S/4HANA upgrades with other major IT initiatives, business cycle timing, and organizational change capacity.

### 6.5.3 Migration Scenarios and Upgrade Paths

Migration approaches vary depending on source system version, business process complexity, data volume, and deployment target selection.

SAP ERP systems with existing payment card processing through custom development or third-party payment gateway integration must evaluate whether to migrate existing payment logic or adopt the digital payments add-on as part of an SAP S/4HANA transformation:

- Greenfield migration approaches that implement SAP S/4HANA with a fresh configuration provide opportunities to adopt the digital payments add-on from project inception, avoiding migration of legacy payment customizations. You can decommission SAP ERP payment card–processing custom code, terminate legacy PSP connections, and implement clean payment processing through the add-on without inheriting technical debt from prior implementations.
- Brownfield migration that preserves existing configurations and custom code can incorporate the digital payments add-on through selective adoption strategies. Organizations might continue legacy payment processing for existing customers while adopting the add-on for new customers or new sales channels, gradually transitioning the payment

volume to the modern integration as business rules allow. This phased approach reduces migration risk while enabling eventual full add-on adoption.

Payment card data migration from SAP ERP to SAP S/4HANA requires careful handling of token references, authorization data, and historical transaction records. Tokenized payment cards registered through legacy PSP integrations cannot be directly migrated to the digital payments add-on as the token formats and encryption keys differ between systems. You must reregister payment cards during or after migration, potentially impacting the customer experience and requiring communication strategies that explain the need to reregister.

### 6.5.4 Hybrid Deployment Strategies

Some organizations implement hybrid strategies, running both SAP S/4HANA Cloud and on-premise SAP S/4HANA systems concurrently, using each deployment model for appropriate business unit or geographic requirements. Hybrid strategies enable customization for complex headquarters operations while providing standard cloud solutions for subsidiaries or maintaining on-premise systems for core legacy processes while adopting cloud systems for new business models.

The digital payments add-on integrates with both cloud and on-premise systems through standardized APIs, supporting hybrid scenarios in which some organizational units use SAP S/4HANA Cloud Public Edition and others use on-premise systems, all connecting to the same or separate add-on tenants. Payment processing remains consistent across deployment models from customer and PSP perspectives despite backend infrastructure differences.

## 6.6 Summary

This chapter provided comprehensive guidance for implementing the SAP digital payments add-on in SAP S/4HANA environments, demonstrating how SAP S/4HANA's native integration capabilities significantly reduce implementation complexity compared to SAP ERP while delivering enhanced functionality and a superior user experience. The out-of-the-box integration approach eliminates extensive custom development requirements, enabling faster deployment and lower total cost of ownership for organizations that are modernizing their payment-processing infrastructure.

Native integration in SAP S/4HANA eliminates the custom function modules for authorization and settlement that SAP ERP implementations must develop and maintain. SAP Fiori application coverage provides modern user interfaces for all payment processing activities, replacing multiple SAP GUI transactions and custom screens. The role-based applications improve user productivity through guided workflows, embedded help, and responsive designs.

Standard SAP Fiori applications for job scheduling, payment monitoring, and exception handling replace the custom programs and transaction codes that SAP ERP implementations typically developed for operational support. The delivered applications provide capabilities that would require significant custom development investment in SAP ERP, accelerating time to value and reducing total implementation cost. Business partner integration in SAP S/4HANA provides unified payment card storage with automatic synchronization to customer master records.

The cumulative advantages of SAP S/4HANA's native integration suggest that organizations operating SAP ERP should prioritize SAP S/4HANA migration when planning payment processing modernization. Implementing custom payment processing in aging SAP ERP systems creates technical debt that will require replacement within mainstream maintenance support timelines. Accelerating an SAP S/4HANA migration enables payment modernization through delivered capabilities rather than custom development.

With SAP S/4HANA and the digital payments add-on integration activated, configured, and validated, you are positioned to begin production payment processing. The remaining implementation work focuses on security hardening, comprehensive testing, go-live preparation, and operational enablement, which are covered in subsequent chapters.

# Chapter 7
# Security and Compliance

*Protecting sensitive payment data and maintaining regulatory compliance are foundational requirements for any digital payment implementation, and the SAP digital payments add-on is specifically designed to address these concerns through built-in security controls. This chapter examines the practical aspects of PCI DSS compliance, tokenization, encryption, audit logging, data privacy, and security testing that organizations must configure and manage throughout the lifecycle of their deployment.*

Organizations managing payment processing in their landscape have two fundamental responsibilities: protecting sensitive cardholder data and maintaining compliance with industry regulations. These responsibilities directly impact system architecture, operational procedures, and business continuity. Security failures can result in data breaches, financial penalties, reputational damage, and loss of payment processing privileges. Compliance violations trigger mandatory audits, remediation costs, and potential service interruptions.

The SAP digital payments add-on provides built-in security controls specifically designed to reduce the Payment Card Industry Data Security Standard (PCI DSS) compliance scope and simplify regulatory obligations. Through tokenization, the add-on is designed to prevent storage of sensitive card data within SAP systems. Through encryption and secure key management, it protects data in transit and at rest. Through comprehensive audit logging, it enables compliance monitoring and forensic investigation. Understanding how to properly configure and maintain these security controls determines whether an implementation succeeds or creates ongoing operational and compliance burdens.

In this chapter, we will discuss the practical implementation of security and compliance requirements, including the following:

- How the add-on reduces the PCI DSS scope and provides a compliance strategy framework
- Tokenization implementation, including token lifecycle management and integration patterns
- Encryption architecture, key management practices, and certificate handling
- Audit-logging configuration, monitoring approaches, and retention policies
- General Data Protection Regulation (GDPR) requirements, including data subject rights and privacy controls
- Security testing methodologies and validation procedures

## 7.1 PCI DSS Compliance Strategy

PCI DSS establishes mandatory security requirements for organizations that store, process, or transmit cardholder data. Achieving and maintaining compliance requires significant investment in security controls, annual assessments, quarterly vulnerability scans, and ongoing remediation activities. The digital payments add-on fundamentally changes the compliance scope by removing cardholder data from SAP systems through tokenization, thereby reducing the number of systems, networks, and applications subject to PCI DSS requirements. Systems that can influence the security of the tokenization process may still be considered in scope under PCI DSS v4.0.

We begin by examining how the add-on reduces PCI DSS scope and then look at how compliance levels and self-assessment questionnaire types determine your organization's specific validation requirements. From there, we present a compliance strategy framework covering the key technical and operational controls, followed by a responsibility matrix that clarifies obligations across your organization, SAP, and your payment service provider (PSP). Finally, we walk through annual assessment preparation and the ongoing quarterly activities needed to maintain compliance.

### 7.1.1 Understanding the PCI DSS Scope

The PCI DSS compliance scope encompasses all system components that store, process, or transmit cardholder data, plus any systems connected to these components. Without tokenization, this scope typically includes the following:

- SAP ERP or SAP S/4HANA application servers
- SAP database servers containing cardholder data
- All networks connecting these systems
- Workstations used to access cardholder data
- Backup systems containing cardholder data
- Development and test systems if they contain production data
- Identity management systems with access to cardholder data
- Integration middleware touching payment data

Table 7.1 compares the PCI DSS scope before and after implementing the digital payments add-on.

| Scope Area | Traditional Implementation | With Digital Payments Add-On |
|---|---|---|
| Cardholder data storage | SAP database tables store full card numbers | No cardholder data in SAP; only tokens stored |

*Table 7.1: PCI DSS Scope Comparison*

| Scope Area | Traditional Implementation | With Digital Payments Add-On |
|---|---|---|
| In-scope SAP systems | All SAP servers processing payments | Only SAP components directly initiating or handling tokenization API calls, subject to Qualified Security Assessor interpretation |
| Network segmentation | Complex isolation of payment-processing networks | Simplified; SAP systems outside primary cardholder data environment |
| SAP customizations | All ABAP code accessing card data in scope | Minimal; token-handling code has reduced requirements |
| Database backups | Backups contain sensitive data; require encryption/protection | Backups contain only tokens; significantly reduced risk |
| Development/test systems | Require data masking or production-equivalent controls | Can use production tokens safely (no cardholder data exposure) |
| Annual assessment effort | Extensive documentation of all components | Focused on add-on configuration and integration points |
| Quarterly vulnerability scanning | All SAP systems require scanning | Reduced scan scope; SAP systems outside primary focus |

*Table 7.1: PCI DSS Scope Comparison (Cont.)*

### 7.1.2 PCI DSS Compliance Levels

The PCI Security Standards Council assigns compliance levels based on annual transaction volume. Your compliance level determines your assessment requirements and validation frequency. Table 7.2 summarizes PCI DSS compliance levels and their requirements.

| Level | Annual Transaction Volume | Assessment Requirements |
|---|---|---|
| Level 1 | Visa/Mastercard/Discover: Over 6 million transactions<br>American Express (Amex): Over 2.5 million transactions | ■ Annual Report on Compliance (ROC) by Qualified Security Assessor<br>■ Quarterly network scans by Approved Scanning Vendor (ASV)<br>■ Attestation of Compliance (AOC) |

*Table 7.2: PCI DSS Compliance Levels*

| Level | Annual Transaction Volume | Assessment Requirements |
|---|---|---|
| Level 2 | Visa/MC/Discover: 1 to 6 million transactions<br>Amex: 50K to 2.5 million transactions | ▪ Annual Self-Assessment Questionnaire (SAQ)<br>▪ Quarterly network scans by ASV<br>▪ AOC<br>▪ Some card brands may require Qualified Security Assessor assessment |
| Level 3 | Visa/MC: 20,000 to 1 million e-commerce transactions<br>Amex: Less than 50K | ▪ Annual SAQ<br>▪ Quarterly network scans by ASV<br>▪ AOC |
| Level 4 | Fewer than 20,000 e-commerce transactions or up to 1 million other transactions (face to face) | ▪ Annual SAQ (requirements vary by card brand)<br>▪ Quarterly network scans (may be recommended rather than required) |

*Table 7.2: PCI DSS Compliance Levels (Cont.)*

Most enterprise SAP implementations processing card payments operate at Level 1 or Level 2, requiring formal assessments and extensive documentation. The digital payments add-on architecture specifically addresses Level 1 and Level 2 compliance requirements by implementing controls that satisfy PCI DSS requirements while minimizing scope. Card brands may override level classification based on breach history or risk profile. For the latest updates, refer to the following:

- *https://www.americanexpress.com/content/dam/amex/us/merchant/new-data-security/DSOP_United_States_EN.pdf* for American Express
- *https://corporate.visa.com/en/resources/security-compliance.html#6c3dbcefc9* for Visa
- *https://www.mastercard.com/us/en/business/cybersecurity-fraud-prevention/site-data-protection-pci.html* for Mastercard
- *https://www.discoverglobalnetwork.com/solutions/pci-compliance/pci-overview/* for Discover

### 7.1.3 SAP Self-Assessment Questionnaire Type

The digital payments add-on enables most organizations to complete Self-Assessment Questionnaire A (SAQ A) or Self-Assessment Questionnaire A for E-commerce Payment (SAQ A-EP) rather than the full Self-Assessment Questionnaire D (SAQ D), which applies to traditional implementations. Understanding which SAQ applies to your implementation depends on how cardholder data flows through your environment. Table 7.3 compares the applicable SAQ types as per PCI DSS v4.0.

| SAQ Type | Applicable Scenario | Number of Requirements (v4.0) | Key Characteristics |
|---|---|---|---|
| SAQ A | E-commerce with outsourced payment page (redirect or iFrame) | Approx. 31 requirements | ■ Customer redirected to PSP payment page<br>■ No cardholder data touches merchant systems<br>■ Merchant website does not directly receive payment data |
| SAQ A-EP | E-commerce with API integration using tokenization | Approx. 191 requirements | ■ Payment data posted to merchant web server<br>■ Immediately tokenized via API<br>■ No local storage of cardholder data<br>■ Covers digital payments add-on with web/mobile channels |
| SAQ D | Full cardholder data environment | Full standard (350+ requirements) | ■ Traditional SAP implementation storing card data<br>■ Card data processed, transmitted, or stored<br>■ Most comprehensive assessment<br>■ Required without proper tokenization |

*Table 7.3: PCI DSS Self-Assessment Questionnaire Types*

After implementing the digital payments add-on for e-commerce with proper API tokenization, you can qualify for SAQ A-EP. Under PCI DSS v4.0, this reduces the scope to approximately 180 to 200 applicable requirements, compared to the full standard's 350+ requirements.

If you use a redirect-based integration (iFrame or full redirect), you qualify for SAQ A. However, under PCI DSS v4.0, SAQ A is no longer just a documentation exercise. It now includes over 30 requirements, mandating active technical controls such as maintaining an inventory of scripts loaded in the consumer's browser (Requirement 6.4.3) and deploying tamper-detection mechanisms on payment pages (Requirement 11.6.1) to prevent e-skimming attacks. For the latest information, refer the SAQ instructions and guidelines document at *https://www.pcisecuritystandards.org/document_library/*.

### 7.1.4 Compliance Strategy Framework

Successful PCI DSS compliance requires a structured approach addressing technical controls, operational procedures, and ongoing validation. The following framework provides a practical implementation strategy for organizations deploying the digital payments add-on:

- **Scope definition and network segmentation**
  - Begin by clearly defining which systems handle cardholder data before and after add-on implementation.
  - Create network diagrams showing data flows and trust boundaries.
  - Document which SAP systems will integrate with the add-on and how tokenization removes them from the primary PCI scope.

  Your compliance strategy can start with network segmentation analysis. Before add-on implementation, document every system touching payment data in the legacy environment: SAP ERP application servers, database servers, the payment gateway VM (virtual machine), web servers processing orders, and all network segments connecting these systems. After add-on implementation, you will see that only the SAP application servers making API calls to the digital payments add-on remain candidates for scope evaluation, significantly simplifying quarterly vulnerability scanning and annual assessments.
- **Tokenization implementation validation**
  Verify that no cardholder data persists in SAP systems after tokenization. Review all database tables, custom code, interface files, logs, and temporary storage locations. Confirm that only tokens appear in SAP data structures. Test that tokens cannot be reversed to reveal original card numbers without proper authorization. Key validation checkpoints include the following:
  - SAP database tables contain only tokens (no primary account numbers [PANs]).
  - Custom ABAP code does not log or store cardholder data.
  - Interface files and IDocs contain tokens instead of card numbers.
  - Application logs contain tokens only (not full PANs).
  - Backup and archive data contain tokens only.
  - Development and test systems can use production tokens safely.
- **Encryption and key management**
  - Implement strong encryption for all data in transit between SAP systems and the digital payments add-on. Configure TLS 1.2 or higher with strong cipher suites.
  - Establish proper certificate validation and renewal processes.
  - For any residual data requiring encryption at rest (such as encryption keys themselves), implement proper cryptographic key management following PCI DSS Requirement 3.

- **Access control implementation**
  - Configure role-based access controls that limit who can initiate payment transactions, view payment data, or access add-on configuration settings.
  - Implement multifactor authentication for administrative access.
  - Establish procedures for adding, modifying, and removing user accounts.
  - Document all roles and their associated privileges.
- **Audit logging and monitoring**
  - Enable comprehensive logging in the digital payments add-on, covering all payment transactions, administrative actions, and security events.
  - Configure SAP audit logging to track access to tokenized payment data.
  - Establish log-review procedures and retention policies that meet PCI DSS Requirement 10.
- **Vulnerability management**
  - Establish procedures for applying SAP security patches, add-on updates, and operating system patches.
  - Configure quarterly vulnerability scanning by an approved scanning vendor.
  - Implement a process for tracking and remediating identified vulnerabilities within defined timeframes.
- **Security testing program**
  - Conduct annual penetration testing covering the digital payments add-on integration, SAP payment transactions, and network segmentation.
  - Perform application security testing on custom code that handles tokens.
  - Test disaster recovery and business continuity procedures for payment-processing capabilities.
- **Documentation and evidence collection**
  - Maintain documentation supporting compliance, including network diagrams, system configurations, security policies, access control matrices, change management records, and testing results.
  - Organize evidence to facilitate annual assessments and respond efficiently to auditor requests.

### 7.1.5 Compliance Responsibility Matrix

Understanding which party is responsible for specific PCI DSS requirements prevents gaps in compliance coverage. When using the digital payments add-on, responsibilities are shared between your organization, SAP (as the add-on provider), and your PSP. Table 7.4 outlines the compliance responsibility matrix.

| PCI DSS Requirement Area | Your Organization | SAP (Add-On Provider) | Payment Service Provider |
|---|---|---|---|
| Secure network architecture | Configure SAP system security, network segmentation | Provide secure multitenant architecture | Maintain secure PSP infrastructure |
| Cardholder data protection | Ensure tokens are used exclusively; validate no data leakage | Implement tokenization engine; manage token vault | Protect cardholder data in vault |
| Vulnerability management | Patch SAP systems; maintain security updates | Maintain add-on security patches | Secure PSP platform |
| Access control | Manage SAP user access; implement authentication | Provide role-based access control (RBAC) capabilities in add-on | Control PSP account access |
| Network monitoring | Monitor SAP system logs; review payment events | Provide audit logging capabilities | Monitor PSP transaction activity |
| Security testing | Test SAP integration; conduct penetration testing | Security testing of add-on platform | Test PSP payment processing |
| Compliance documentation | Maintain SAP security documentation | Provide PCI attestation for add-on | Maintain PSP PCI certification |

*Table 7.4: PCI DSS Responsibility Matrix*

Note that the responsibility allocation provided here must be validated against your contractual agreements and SAP Trust Center documentation.

### 7.1.6 Annual Assessment Preparation

Organizations at Level 1 compliance undergo annual on-site assessments by a Qualified Security Assessor. Preparation should begin 90 days before the scheduled assessment to ensure all documentation, evidence, and testing is complete. Consult the following checklist for the key preparation activities:

- [ ] Update network diagrams showing current environment and data flows
- [ ] Document all system changes since previous assessment
- [ ] Review and update security policies and procedures
- [ ] Complete internal vulnerability scans; remediate findings
- [ ] Review all user access; remove inactive accounts

- [ ] Verify logging is enabled and logs are being retained properly
- [ ] Confirm encryption is functioning for all data in transit
- [ ] Test incident response procedures
- [ ] Organize evidence folders by PCI DSS requirement number
- [ ] Conduct preassessment gap analysis
- [ ] Schedule assessor site visit and participant interviews
- [ ] Prepare executive briefing on compliance status

During the assessment, assessors will request evidence demonstrating compliance with applicable PCI DSS requirements. Evidence typically includes configuration screenshots, log samples, policy documents, test results, and user access reports. Organizing evidence in advance significantly reduces the assessment duration and prevents delays caused by missing documentation. PCI DSS compliance is not an annual event but an ongoing operational responsibility. Quarterly activities maintain compliance between annual assessments and ensure continuous security monitoring. Table 7.5 outlines quarterly compliance activities.

| Activity | Responsible Role | Key Tasks |
|---|---|---|
| ASV vulnerability scanning | Security team | ■ Schedule scans with ASV and review scan results<br>■ Remediate critical/high vulnerabilities<br>■ Document exception justifications<br>■ Obtain passing scan report |
| Access review | SAP security admin | ■ Review all user accounts with payment access<br>■ Verify multifactor authentication (MFA) is functioning<br>■ Remove inactive accounts<br>■ Document access changes |
| Log review | Security operations | ■ Sample transaction logs for anomalies<br>■ Review failed authentication attempts<br>■ Investigate suspicious activities<br>■ Document review findings |
| Security awareness training | HR/compliance | ■ Deliver training to staff with payment data access<br>■ Document training completion and update training materials as needed |
| Policy review | Compliance team | ■ Review security policies for necessary updates and communicate policy changes<br>■ Obtain management approval for updates |

*Table 7.5: Quarterly Compliance Activities*

## 7.2 Tokenization Implementation

Tokenization is the foundational security control in the SAP digital payments add-on architecture. It replaces sensitive cardholder data with nonsensitive substitutes called *tokens*, eliminating the need to store actual card numbers in SAP systems. The add-on manages the entire tokenization lifecycle—from initial token creation through usage in transactions to eventual expiration or deletion. Understanding how tokenization works, how tokens flow through SAP business processes, and how to properly manage token lifecycles is essential for both security and operational success.

We start by examining the tokenization architecture and how cardholder data flows through the add-on to the PSP's vault without ever persisting in SAP systems. We then explore the different token types and formats available, followed by the lifecycle states that govern how tokens are created, used, and eventually retired. Finally, we look at how tokens are stored within SAP database tables across different transaction types and system landscapes.

### 7.2.1 Architecture

The digital payments add-on implements a centralized tokenization service running on SAP BTP. When cardholder data enters the system through a web storefront, call center, mobile app, or API, it is immediately sent to the add-on for tokenization. The add-on forwards the cardholder data to the connected PSP, which stores it securely in its PCI-compliant vault and returns a unique token. This token is then passed back to the requesting SAP system for storage and future use.

The critical architectural principle is that actual cardholder data never persists in SAP systems. Card numbers may transiently exist in application memory during API invocation, but they are never written to database tables, logs, or temporary files. Only tokens are stored, and these tokens have no mathematical relationship to the original card numbers, making them useless if intercepted or stolen. Figure 7.1 illustrates the token creation flow.

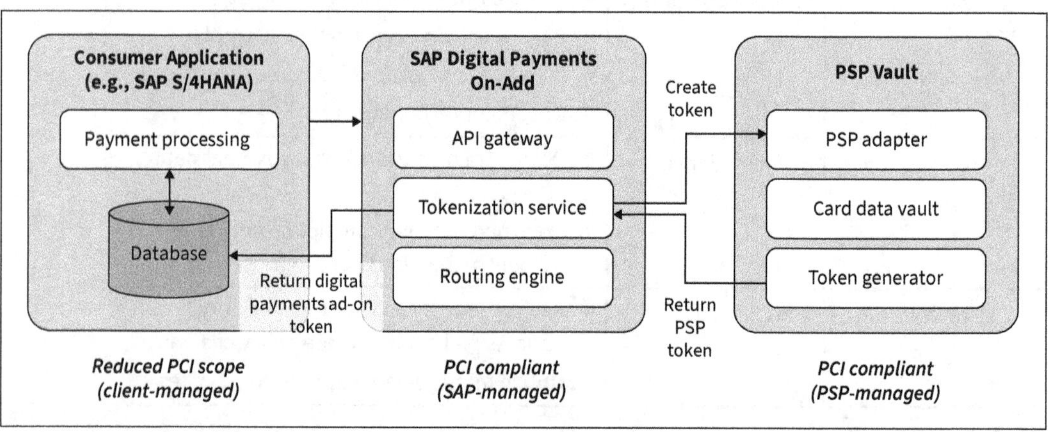


*Figure 7.1: Token Creation Flow*

### 7.2.2 Types and Format

The digital payments add-on supports multiple token types depending on your PSP's capabilities and your business requirements. Understanding which token type applies to your implementation affects how you handle tokens in SAP customizations and integrations. Table 7.6 compares common token types.

| Token Type | Format | Example | Use Case | Key Characteristics |
|---|---|---|---|---|
| Opaque token | Random alphanumeric string | tok_1JX3kL2eZ-vKYlo2C8H9Q3fGd | Most secure; no relationship to PAN | ■ No visual similarity to card number<br>■ Cannot be reversed without vault access<br>■ Recommended for most implementations |
| Format-preserving token | Matches PAN format with BIN (bank identification number) preservation | 4242-4242-XXXX-1234 | Legacy systems requiring a PAN-like format | ■ May preserve BIN and last four digits, depending on PSP configuration<br>■ Maintains Luhn checksum<br>■ May be required for older integrations |
| Network token | Issued by card networks (Visa, Mastercard) | DPAN with cryptogram | EMV-compliant mobile payments | ■ Enhanced security for digital wallets<br>■ Supports Apple Pay, Google Pay<br>■ Requires network token service integration |

*Table 7.6: Token Types and Characteristics*

Most implementations use *opaque tokens*, which provide the strongest security by eliminating any visual or mathematical relationship to the original card number. *Format-preserving tokens* may be necessary when integrating with legacy systems that perform PAN validation or routing based on card number patterns, but they should be avoided when possible as they reveal partial cardholder data (first six and last four digits).

### 7.2.3 Lifecycle States

Tokens progress through distinct lifecycle states from creation to deletion. The digital payments add-on tracks the token state and enforces rules about which operations are permitted in each state. Understanding these states is critical for troubleshooting payment failures and managing token expiration. Table 7.7 outlines token lifecycle states.

| State | Description | Permitted Operations | Typical Duration |
|---|---|---|---|
| Created | Token has been generated and stored in vault | Authorization, capture, charge | Until first transaction or expiration |
| Active | Token has been successfully used in at least one transaction | Authorization, reauthorization, capture, refund | Until expiration or deletion |
| Expired | Token has exceeded its validity period | None (token cannot be used) | Permanent state; requires new tokenization |
| Failed Verification | Token failed card verification check (card verification value [CVV], address verification service [AVS]) | None (token flagged as invalid) | Typically treated as permanent unless reverification is supported by the PSP |
| Deleted | Token has been explicitly removed from system | None | Permanent state; cannot be recovered |
| Suspended | Token temporarily disabled due to fraud concern | None (pending investigation) | Until manual reactivation or deletion |

*Table 7.7: Token Lifecycle States*

Tokens typically remain in the Created or Active state for the duration of their usefulness. They transition to Expired when the underlying card reaches its expiration date or when the PSP enforces token lifetime limits. They move to Deleted when explicitly removed through API calls, usually triggered by customer account closure or payment method updates.

### 7.2.4 Storage in SAP

Tokens are stored in SAP database tables alongside other payment-related data. The specific tables depend on whether you are using SAP ERP or SAP S/4HANA, and whether you are processing sales orders, subscriptions, or other transaction types. However, the principle remains consistent: Tokens are stored as strings in standard SAP payment fields that previously held card numbers.

In SAP ERP implementations, custom fields are often added to standard tables to store tokens as older SAP releases did not include dedicated token fields. Certain financial contract accounting (FI-CA) tables and payment framework extensions support token storage fields; actual field availability depends on release and configuration. Regardless of table structure, tokens should always be stored in fields of adequate length (minimum 50 characters recommended) to accommodate different token formats.

### 7.2.5 API Workflow

The digital payments add-on exposes REST APIs for token creation, retrieval, and deletion. SAP systems call these APIs during payment processing workflows, typically triggered by sales order creation, subscription billing, or manual payment entry by customer service representatives. The basic tokenization workflow follows these steps:

1. Data collection: Customer provides payment details through web form, mobile app, or phone to CSR
2. API request: SAP system constructs API request containing cardholder data (PAN, expiration, CVV)
3. Token creation: Add-on forwards data to PSP; PSP stores data in vault and returns token
4. Token storage: SAP system receives token and stores it in appropriate database table
5. Data disposal: Temporary variables containing cardholder data are cleared from memory
6. Confirmation: Customer receives confirmation that payment method has been saved

The most critical step is ensuring that cardholder data is never logged or persisted during this workflow. Custom ABAP code must be carefully reviewed to prevent accidental logging of card numbers to application logs, system traces, or debugging outputs.

### 7.2.6 Validation and Verification

When tokens are created, the PSP typically performs card verification to ensure the card is valid and active. This verification may include the following:

- Luhn check: Mathematical validation of the card number format
- CVV: Verification that the CVV matches card issuer records
- AVS: Confirmation that the billing address matches card issuer records
- Zero-dollar authorization: Temporary authorization to verify the card is active and has available credit

Not all verification methods are mandatory, and configuration options in the digital payments add-on control which checks are enforced. However, stricter verification reduces the risk of storing tokens for invalid or fraudulent cards, which later fail when attempting actual transactions. Table 7.8 compares verification methods and their implications.

| Verification Method | What It Validates | When to Use | Impact of Failure |
|---|---|---|---|
| Luhn check | Card number format is mathematically valid | Always (automatic) | Token creation fails immediately |
| CVV verification | Security code matches card issuer records | During initial tokenization | Token may be created but flagged as unverified |
| AVS check | Billing address matches card records | During initial tokenization | Token created; warnings logged for address mismatch |
| Zero-dollar authorization | Card is active and can be authorized | High-risk or subscription scenarios | Token creation fails if card cannot be authorized |
| 3D Secure | Customer identity verified through issuer | E-commerce transactions | Authentication may occur before authorization and token usage, depending on integration pattern |

*Table 7.8: Token Verification Methods*

### 7.2.7 Reuse and Multiuse Tokens

Tokens can be single-use (consumed during one transaction) or multiuse (reusable for multiple transactions). The token type depends on business requirements and PSP capabilities (see Table 7.9).

| Single-Use Tokens | Multiuse Tokens |
|---|---|
| ▪ Created during checkout for immediate payment<br>▪ Consumed during authorization or charge<br>▪ Cannot be reused for subsequent transactions<br>▪ Lower security risk; shorter validity period<br>▪ Suitable for guest checkout scenarios | ▪ Created when customer saves payment method<br>▪ Stored in customer master data or subscription records<br>▪ Reused for recurring billing, subscription renewals, or future orders<br>▪ Higher value target; require stronger access controls<br>▪ Suitable for registered customer accounts |

*Table 7.9: Token Types in Payment Processing*

Most enterprise implementations require multiuse tokens to support subscription billing, recurring orders, and customer self-service scenarios in which customers manage saved payment methods. Multiuse tokens are typically associated with customer account numbers in SAP, creating a permanent link between the customer and their stored payment method. Table 7.10 compares single-use and multiuse token characteristics.

| Characteristic | Single-Use Tokens | Multiuse Tokens |
|---|---|---|
| Validity period | Minutes to hours | Years (until card expiration) |
| Storage location | Temporary session data | Customer master data tables |
| Typical use cases | Guest checkout, one-time payments | Subscriptions, registered accounts, autorenewals |
| Security controls | Minimal (short-lived) | Strong access controls, encryption at rest |
| Customer visibility | Not visible after transaction | Visible in customer account (last four digits) |
| Deletion trigger | Automatic after use | Customer request or account closure |
| PCI scope impact | Minimal | Requires secure storage practices |

*Table 7.10: Single-Use Versus Multiuse Tokens*

### 7.2.8 Expiration Handling

All payment tokens eventually expire when the underlying credit card reaches its expiration date. The digital payments add-on does not automatically update expired tokens; instead, customers must provide updated payment information when their cards expire. This creates operational considerations for subscription billing and recurring payment scenarios. Handling token expiration requires the following:

- Expiration monitoring: Regularly check stored tokens for upcoming expiration dates
- Customer notifications: Alert customers 30–60 days before card expiration
- Dunning management: Handle failed payments due to expired cards gracefully
- Self-service updates: Provide customers with easy methods to update payment information
- Grace periods: Configure subscription systems to allow short grace periods before cancellation

### 7.2.9 Deletion and Data Retention

Tokens should be deleted when they are no longer needed, both to reduce the data retention liability and to comply with privacy regulations. Common deletion triggers include a customer closing their account, removing a saved payment method, canceling a subscription, card expiration, and so on.

The digital payments add-on provides API methods for token deletion. When a token is deleted, it is removed from the PSP vault and can no longer be used for transactions. The deletion is permanent and cannot be reversed. SAP systems should remove corresponding token records from database tables when deletion is confirmed.

Establish clear data retention policies that define how long payment tokens are retained after associated accounts or transactions are closed. For regulatory compliance, some tokens may need to be retained for dispute resolution (typically 120–180 days after the last transaction) or tax-audit purposes (varies by jurisdiction). However, tokens should not be retained indefinitely beyond legitimate business or legal requirements.

### 7.2.10 Integration Patterns

Different SAP business processes require different tokenization integration patterns. The pattern you implement depends on whether customers provide payment details during order entry, account registration, subscription signup, or other scenarios. Table 7.11 outlines common tokenization integration patterns.

| Pattern | Implementation Approach | Key Considerations |
|---|---|---|
| Checkout tokenization (customer provides payment during order) | Web/mobile frontend calls add-on API before submitting order | ▪ Minimize PCI scope by tokenizing in browser<br>▪ SAP receives only the token, never cardholder data<br>▪ Requires JavaScript software development kit (SDK) integration |
| Account registration tokenization (customer saves payment method to account) | Tokenization triggered during account setup | ▪ Store token in customer master (table KNA1/KNB1)<br>▪ Associate token with customer number<br>▪ Enable self-service management in portal |
| Subscription tokenization (payment method stored for recurring billing) | Initial tokenization during subscription signup | ▪ Store token in subscription master data<br>▪ Handle expiration proactively<br>▪ Support dunning for failed renewals |

*Table 7.11: Tokenization Integration Patterns*

| Pattern | Implementation Approach | Key Considerations |
|---|---|---|
| Call center tokenization (CSR enters payment details over phone) | CSR application calls add-on API in real-time | ■ Implement timeout handling for slow API responses<br>■ Display clear error messages for invalid cards<br>■ Avoid logging cardholder data in CSR screens |
| Batch tokenization (legacy data migration to tokenized format) | Bulk API calls for historical payment data | ■ Process in small batches to avoid rate limits<br>■ Implement retry logic for failures<br>■ Securely delete source data after successful tokenization |

*Table 7.11: Tokenization Integration Patterns (Cont.)*

The checkout tokenization pattern is the most common in e-commerce implementations. It uses client-side JavaScript to capture payment details in the browser, tokenize them by calling the add-on API directly, and pass only the resulting token to the SAP backend. This ensures cardholder data never touches SAP servers, minimizing the PCI scope.

## 7.3 Encryption and Key Management

Encryption protects sensitive data both in transit between systems and at rest in storage. The SAP digital payments add-on implements multiple layers of encryption to protect payment data, API credentials, and cryptographic keys themselves. Proper encryption configuration is mandatory for PCI DSS compliance and represents a critical defense against data breaches. However, encryption is only as strong as the key management practices supporting it. Weak key management—storing keys alongside encrypted data, using default passwords, failing to rotate keys regularly—undermines even the strongest encryption algorithms. This section covers encryption protocols for data in transit, encryption mechanisms for data at rest, cryptographic key management practices, and certificate lifecycle management for secure communications.

### 7.3.1 Encryption in Transit

All communication between SAP systems and the digital payments add-on must use Transport Layer Security (TLS) version 1.2 or higher. TLS encrypts data as it travels across networks, preventing interception or tampering by unauthorized parties. Older protocols like SSL 3.0 and TLS 1.0 contain known vulnerabilities and are explicitly prohibited by PCI DSS. Table 7.12 lists required TLS configuration parameters.

| Configuration Parameter | Required Value | Purpose | Verification Method |
|---|---|---|---|
| Minimum TLS version | TLS 1.2 | Prevents use of vulnerable protocols | Test connection with TLS 1.0/1.1; should fail |
| Cipher suites | Strong encryption only (AES-GCM, ChaCha20) | Ensures strong encryption algorithms | Review SAP system cipher suite configuration |
| Certificate validation | Enabled | Prevents man-in-the-middle attacks | Test with invalid certificate; should fail |
| Certificate chain validation | Enabled | Validates entire certificate trust chain | Test with incomplete chain; should fail |
| Hostname verification | Enabled | Confirms certificate matches target host | Test with certificate for different hostname; should fail |
| Forward secrecy | Enabled (elliptic curve Diffie-Hellman ephemeral [ECDHE] key exchange) | Protects past communications if keys compromised | Review configured cipher suites for ECDHE support |

*Table 7.12: TLS Configuration Requirements*

SAP systems use the CommonCryptoLib (SAPCRYPTOLIB) library for TLS functionality. The library version must support TLS 1.2 and modern cipher suites. Table 7.13 identifies the minimum library versions required for proper TLS support.

| Component | Minimum Requirement | Recommended |
|---|---|---|
| CommonCryptoLib | 8.5.48 | 8.5.50 or higher (older versions lack full support for modern GCM cipher suites required by major payment gateways) |
| SAP kernel | 7.53 (NetWeaver 7.5), 7.77 (S/4HANA 1809/1909) | 7.89 or higher (SAP S/4HANA 2022+) |
| TLS protocol | TLS 1.2 | TLS 1.3 where supported by the SAP kernel and cryptographic library |
| Cipher suites | TLS_ECDHE_RSA_WITH_AES_256_GCM_SHA384 | TLS_AES_256_GCM_SHA384 (TLS 1.3) |

*Table 7.13: Required Cryptographic Library Versions*

Organizations running older SAP releases must verify their cryptographic library versions and apply necessary kernel patches before implementing the digital payments add-on. Testing TLS connectivity should be part of prerequisite validation during project planning phases.

### 7.3.2 Cipher Suite Selection

Cipher suites define the specific algorithms used for encryption, message authentication, and key exchange during TLS sessions. Not all cipher suites provide equivalent security. Some use weak encryption, vulnerable key exchange, or broken hash functions. PCI DSS requires strong cipher suites and prohibits known weak algorithms.

SAP systems should be configured to prefer GCM cipher suites and disable weak or broken suites entirely. This configuration is managed through the `ssl/ciphersuites` SAP profile parameter and the Trust Manager (Transaction STRUST).

### 7.3.3 Certificate Management

Certificate management for SAP digital payments integrations follows standard Trust Manager (Transaction STRUST) practices. You must ensure valid trust chains, timely certificate renewal, and appropriate use of client certificates when mutual TLS is required. Your Basis teams should follow SAP security hardening guides and certification authority (CA) policies to maintain uninterrupted payment connectivity.

### 7.3.4 Encryption at Rest

Although the digital payments add-on stores cardholder data in PSP vaults rather than SAP systems, certain sensitive data remains in SAP, requiring encryption at rest:

- API credentials (client IDs, client secrets, API keys)
- Client certificates and private keys
- Configuration data containing PSP account information
- Tokens stored in SAP database tables

SAP provides multiple mechanisms for encrypting data at rest. Table 7.14 compares encryption-at-rest options.

| Method | Scope | Performance Impact | Implementation Complexity | Best Used For |
|---|---|---|---|---|
| Transparent data encryption (TDE) | Entire database tablespaces | Moderate (5–10% overhead) | Low (add-on configuration) | Whole database protection; compliance requirement |

*Table 7.14: Encryption-at-Rest Options*

| Method | Scope | Performance Impact | Implementation Complexity | Best Used For |
|---|---|---|---|---|
| Secure storage | Individual fields | Minimal (encryption on demand) | Moderate (application code changes) | Specific sensitive fields; API credentials |
| File system encryption | File system directories | Low (hardware acceleration) | Low (OS configuration) | Backup files; interface files; logs |
| Application-level encryption | Custom ABAP logic | Variable | High (development effort) | Specific business requirements; legacy compatibility |

*Table 7.14: Encryption-at-Rest Options (Cont.)*

### 7.3.5 Key Management for Encryption

Encryption keys are the secret values used by encryption algorithms to transform plaintext into ciphertext. The security of encrypted data depends entirely on protecting encryption keys. If keys are compromised, encrypted data can be decrypted by attackers. PCI DSS Requirement 3 specifies detailed key management practices that organizations must follow. Key management encompasses several operational practices:

- **Key generation**
  Encryption keys must be generated using cryptographically secure random number generators. Weak or predictable keys can be guessed by attackers through brute-force attacks. Keys should meet minimum length requirements: 128 bits for symmetric encryption (AES-128), 256 bits for stronger symmetric encryption (AES-256), and 2,048 bits for asymmetric encryption (RSA-2048).
- **Key storage**
  Encryption keys must be stored securely, separate from the data they protect. Storing encryption keys in the same database as encrypted data provides minimal security benefit. Keys should be stored in hardware security modules (HSMs), key management systems (KMS), or encrypted key stores with restricted access. SAP systems store cryptographic keys in personal security environments (PSEs), which are encrypted files protected by passwords or PINs. The PSE encryption password becomes the critical secret requiring protection. This password should be stored in a secure location, accessible only by authorized administrators, and changed periodically according to security policies.
- **Key access controls**
  Only authorized systems and users should have access to encryption keys. Access should be logged for audit purposes. SAP provides authorization objects (see Table 7.15) for controlling access to PSEs and cryptographic operations.

| Authorization Object | Controls Access To | Recommended Users |
|---|---|---|
| S_TCODE | Transaction STRUST | Basis administrators only |
| S_DATASET | File system access to PSE files | Operating system administrators only |
| S_CRYPTOKEY | Cryptographic key operations | Application users requiring encryption (limited) |
| S_SSF_KEY | Secure Store and Forward (SSF) key access | Integration users for digital signatures |

*Table 7.15: SAP Authorization Objects for Cryptographic Keys*

- **Key rotation**
  Encryption keys should be rotated (replaced) periodically to limit the impact of potential key compromise. PCI DSS does not mandate specific rotation intervals but recommends rotation at least annually or when personnel with key access leave the organization. Rotation procedures must ensure uninterrupted service; old keys must remain available to decrypt existing data while new keys encrypt new data.
- **Key backup and recovery**
  Encryption keys must be backed up to prevent data loss if primary keys are corrupted or unavailable. However, key backups are high-value targets and must be protected with equivalent security to production keys. Backups should be stored in physically separate locations with restricted access. Testing key recovery procedures during disaster recovery drills ensures that keys can be restored when needed.
- **Key destruction**
  When encryption keys are no longer needed (such as when encrypted data is deleted or migrated to new keys), keys should be securely destroyed to prevent future unauthorized decryption. Destruction methods include cryptographic erasure (overwriting keys with random data) or physical destruction of storage media containing keys.

Table 7.16 summarizes key management lifecycle activities.

| Activity | Frequency | Responsible Team | Documentation Required |
|---|---|---|---|
| Key generation | When new keys are needed | Security team | Key inventory record created |
| Key storage | Ongoing | Basis team | PSE backup procedures documented |

*Table 7.16: Key Management Lifecycle*

| Activity | Frequency | Responsible Team | Documentation Required |
|---|---|---|---|
| Access control review | Quarterly | Security team | Authorization audit log |
| Key rotation | Annually (minimum) | Security + Basis teams | Rotation schedule and completion records |
| Key backup | After any key change | Basis team | Backup verification testing |
| Key recovery testing | Semiannually | Basis + disaster recovery team | Recovery test results documented |
| Key destruction | When keys are retired | Security team | Destruction certificates created |

*Table 7.16: Key Management Lifecycle (Cont.)*

## 7.4 Audit Logging and Monitoring

Audit logging and monitoring are core requirements under PCI DSS Requirement 10 and play a critical role in demonstrating that payment-related activities within SAP systems are controlled, traceable, and reviewable. In the context of the SAP digital payments add-on, logging focuses on token usage, payment API interactions, authentication events, and configuration changes that could impact the security of payment processing.

PCI DSS Requirement 10 mandates logging and monitoring of all access to cardholder data and all actions taken by users with administrative privileges. Table 7.17 lists required log elements by event type.

| Event Type | Required Log Elements |
|---|---|
| Token creation | User ID, timestamp, source IP, token ID, success/failure |
| Payment authorization | User ID, timestamp, transaction ID, amount, currency, result |
| Administrative change | Admin ID, timestamp, change type, old value, new value |
| Authentication event | User ID, timestamp, source IP, success/failure, failure reason |
| Log access | User ID, timestamp, log file accessed, access type |

*Table 7.17: Required Audit Log Elements*

Each log entry must include sufficient detail to reconstruct what happened, who performed the action, when it occurred, where it originated, and whether it succeeded or failed. Timestamps must be synchronized across all systems using Network Time Protocol (NTP) to ensure accurate correlation of events across distributed systems.

The digital payments add-on provides configurable logging levels for controlling the verbosity of log output. Higher logging levels capture more detailed information, which is useful for troubleshooting, but also generate larger log volumes—consuming storage and affecting performance. Production systems typically use INFO or WARN levels, while development systems use DEBUG levels for detailed troubleshooting. Table 7.18 describes the available logging levels.

| Level | Events Logged | Typical Volume | Recommended Usage | Example Events |
|---|---|---|---|---|
| ERROR | Critical failures requiring immediate attention | Very low | Always enabled | Payment gateway unreachable, database connection failure |
| WARN | Concerning events that don't prevent operation | Low | Always enabled | Token validation warning, certificate expiring soon |
| INFO | Normal operational events | Medium | Production default | Successful authorization, token created, configuration loaded |
| DEBUG | Detailed diagnostic information | High | Development only | API request payload, PSP response details, routing decisions |
| TRACE | Very detailed execution flow | Very high | Troubleshooting only | Function entry/exit, variable values, loop iterations |

*Table 7.18: Logging Levels and Use Cases*

Logging configuration is managed through the digital payments add-on administration interface. Logging levels can be adjusted globally or per component (API Gateway, Token Vault, PSP Adapter, etc.). Changes take effect immediately without requiring a system restart.

SAP systems generate multiple log types relevant to payment operations. Table 7.19 summarizes SAP logging mechanisms for payment operations.

| Log Type | Transaction Code | What It Captures | Retention Period | Review Frequency |
|---|---|---|---|---|
| Security audit log | SM19 (config), SM20 (view) | Authentication, authorization, sensitive data access | 90 days minimum (PCI requirement) | Daily for errors, weekly for full review |
| Application log | SLG1 | Business events from custom code | 30–90 days (configurable) | Daily for errors, on-demand for troubleshooting |
| System log | SM21 | System errors, ABAP dumps, database issues | 14–30 days | Daily for critical errors |
| Job log | SM37 | Background job execution details | Until job is deleted | After each job execution |
| Change documents | SCDO/SCU3 | Configuration changes to payment-related settings | One year minimum | Monthly audit review |

*Table 7.19: SAP Logging Mechanisms*

Generating logs is necessary for security and compliance, but insufficient on its own. Logs must be actively reviewed to detect anomalies, security incidents, and operational issues. PCI DSS requires a daily log review for critical systems, though automated monitoring can supplement manual review. Table 7.20 outlines log monitoring responsibilities.

| Activity | Responsible Team | Frequency | Tools Used | Escalation Criteria |
|---|---|---|---|---|
| Real-time alert response | Security operations center | Continuous | Security Information and Event Management (SIEM), monitoring dashboards | Critical alerts escalate to senior security |
| Daily log review | Security analysts | Daily | SIEM queries, Transaction SM20 | Suspicious patterns escalate to security management |

*Table 7.20: Log Monitoring Responsibilities*

| Activity | Responsible Team | Frequency | Tools Used | Escalation Criteria |
|---|---|---|---|---|
| Payment operations review | Payment operations team | Daily | Digital payments add-on dashboard | Transaction failures escalate to SAP support |
| Administrative action audit | Compliance team | Weekly | Change log reports | Unauthorized changes escalate to management |
| Comprehensive security analysis | Security architects | Monthly | SIEM analytics, custom reports | Trends indicating systemic issues escalate to CISO (chief information security officer) |

*Table 7.20: Log Monitoring Responsibilities (Cont.)*

PCI DSS Requirement 10.7 mandates retaining the audit log history for at least one year, with at least three months immediately available for analysis. Table 7.21 outlines a log-retention strategy.

| Retention Period | Storage Tier | Accessibility | Storage Method | Use Case |
|---|---|---|---|---|
| 0–3 months | Hot (online) | Immediate search/analysis | SIEM database, SAP database | Active investigations, daily monitoring |
| 3–12 months | Warm (nearline) | Retrieval within hours | Compressed files, secondary storage | Historical analysis, compliance audits |
| 1–7 years | Cold (archive) | Retrieval within days | Tape, cloud archive storage | Regulatory compliance, legal holds |
| 7+ years | Disposal | N/A | Secure deletion | No longer required |

*Table 7.21: Log-Retention Strategy*

Log archival processes should run automatically, compressing and moving older logs to archive storage without manual intervention. Archived logs must remain tamper-evident; any modification or deletion of archived logs should be detectable. Technologies like write once, read many (WORM) storage or cryptographic checksums provide tamper evidence.

Many organizations integrate SAP audit logs with centralized Security Information and Event Management (SIEM) platforms to support PCI DSS monitoring and reporting requirements. SAP logs related to payment processing, authentication, and system errors can be forwarded to external monitoring solutions using standard SAP interfaces or agent-based approaches. Although SIEM integration is not mandatory, it significantly simplifies alerting, correlation, and audit evidence collection for SAP digital payments add-on environments.

## 7.5 GDPR and Data Privacy

The General Data Protection Regulation (GDPR) establishes comprehensive data protection requirements for organizations processing the personal data of European Union residents. GDPR applies broadly to all personal data, but payment implementations face unique challenges given that payment card data is both personal data under GDPR and regulated financial data under PCI DSS. Organizations must satisfy both regulatory frameworks simultaneously, and requirements occasionally conflict or overlap in complex ways.

GDPR grants individuals extensive rights over their personal data, including rights to access, correct, delete, and port their data. Payment data presents challenges for exercising these rights because financial regulations require retaining transaction records for tax, audit, and fraud-prevention purposes. Balancing individual privacy rights against regulatory retention requirements requires careful analysis and documented justification.

This section covers GDPR principles applicable to payment processing, implementing data subject rights within payment systems, lawful bases for processing payment data, data retention and deletion policies, and cross-border data transfer considerations.

### 7.5.1 GDPR Principles for Payment Processing

GDPR establishes six core principles that govern all personal data processing. Payment implementations must demonstrate compliance with each principle. Table 7.22 maps GDPR principles to payment implementation controls.

| GDPR Principle | Application to Payments | Implementation Controls | Verification Method |
|---|---|---|---|
| Lawfulness | Valid legal basis for processing payment data | Document lawful basis in privacy policy; obtain consent where required | Review privacy policy annually; audit consent records |
| Fairness | Process payment data in expected ways | Provide clear payment terms; no hidden fees; transparent pricing | Customer feedback analysis; complaint review |

*Table 7.22: GDPR Principles and Payment Controls*

| GDPR Principle | Application to Payments | Implementation Controls | Verification Method |
|---|---|---|---|
| Transparency | Clear notice of payment data processing | Privacy policy disclosure; payment screen notices | Privacy policy review; user interface audit |
| Purpose limitation | Use payment data only for transactions | Technical controls preventing repurposing; access restrictions | Access log review; data flow mapping |
| Data minimization | Collect only required payment information | Remove optional fields; tokenize card data immediately | Form review; database schema analysis |
| Accuracy | Keep payment records accurate | Customer self-service updates; regular data quality checks | Data quality metrics; customer update rates |
| Storage limitation | Delete payment data when no longer needed | Automated deletion after retention period; retention policy | Retention policy review; deletion job verification |
| Integrity/confidentiality | Protect payment data security | Encryption, access controls, audit logging, penetration testing | Security audit; PCI compliance assessment |

*Table 7.22: GDPR Principles and Payment Controls (Cont.)*

### 7.5.2 Data Subject Rights in Payment Systems

GDPR grants individuals eight fundamental rights regarding their personal data. Payment systems must support these rights while respecting legitimate business needs and other legal obligations. Table 7.23 summarizes data subject rights implementation.

| Right | Implementation Approach | Response Timeline | Exceptions for Payments |
|---|---|---|---|
| Right to be informed | Privacy policy, checkout notices, terms | Provided at data collection | None |
| Right of access | Self-service portal, manual export for GDPR requests | 30 days (one month) | Cannot disclose fraud-detection logic |

*Table 7.23: Data Subject Rights Implementation*

| Right | Implementation Approach | Response Timeline | Exceptions for Payments |
|---|---|---|---|
| Right to rectification | Self-service updates, customer service corrections | 30 days | Cannot alter historical transaction records |
| Right to erasure | Automated deletion after retention period, manual for GDPR requests | 30 days | Cannot delete data under legal retention |
| Right to data portability | Export to CSV/JSON format | 30 days | Limited to data provided by customer |
| Right to object | Opt out for marketing, profiling | Immediate | Cannot object to contract-necessary processing |
| Right to restrict processing | Temporary processing flag, manual resolution | Immediate | Limited during active contract |
| Rights regarding automated decisions | Human review option for declined payments | Immediate | Fraud prevention justified as necessary |

*Table 7.23: Data Subject Rights Implementation (Cont.)*

### 7.5.3 Lawful Bases for Processing Payment Data

GDPR requires a valid legal basis for processing personal data. Multiple lawful bases may apply to payment processing. Table 7.24 maps payment activities to lawful bases.

| Payment Activity | Primary Basis | Secondary Basis | Consent Required? |
|---|---|---|---|
| Transaction processing | Contract performance | N/A | No |
| Fraud detection | Legitimate interest | Contract performance | No |
| Transaction record retention | Legal obligation | Legitimate interest | No |
| Payment method storage | Contract performance | Consent (for future use) | Depends on context |

*Table 7.24: Lawful Bases for Payment Activities*

| Payment Activity | Primary Basis | Secondary Basis | Consent Required? |
|---|---|---|---|
| Payment history analysis for risk | Legitimate interest | Contract performance | No |
| Marketing based on purchase history | Consent | Legitimate interest | Yes |
| Sharing data with affiliates | Consent | Legitimate interest | Yes |

*Table 7.24: Lawful Bases for Payment Activities (Cont.)*

### 7.5.4 Cross-Border Data Transfers

GDPR restricts transferring personal data outside the European Economic Area (EEA) unless destination countries provide adequate data protection. Payment data frequently crosses borders when using cloud-based payment services or international PSPs. Organizations can rely on several mechanisms for lawful cross-border transfers. Table 7.25 compares cross-border transfer mechanisms.

| Mechanism | When to Use | Implementation Effort | Flexibility | Regulatory Risk |
|---|---|---|---|---|
| Adequacy decision | Transfer to adequate countries | None (automatic) | Limited to approved countries | Very low |
| Standard contractual clauses | Transfer to non-adequate countries | Low (template contracts) | Broad applicability | Low (if properly implemented) |
| Binding corporate rules | Internal transfers within corporate group | High (approval process) | Flexible within group | Low (after approval) |
| Consent | Specific one-time transfers | Medium (valid consent required) | Case-by-case basis | Medium (consent validity questions) |
| Derogations (necessity) | Emergency or contract-necessary transfers | Low | Very limited circumstances | High (must truly be necessary) |

*Table 7.25: Cross-Border Data Transfer Mechanisms*

### 7.5.5 Privacy by Design and Default

GDPR requires implementing privacy protection throughout system design and configuration. For payment implementations, *privacy by design* principles include the following:

- Tokenizing card data immediately upon collection (minimization)
- Encrypting payment data in transit and at rest (confidentiality)
- Restricting access to payment data based on roles (need to know)
- Deleting payment data automatically after retention period (storage limitation)
- Providing self-service privacy controls to customers (transparency, empowerment)
- Conducting privacy-impact assessments before implementing new payment features (accountability)

Privacy by default means that systems should be configured with most privacy protective settings enabled automatically, without requiring customers to opt into privacy protections.

## 7.6 Security Testing and Validation

Security testing validates that implemented security controls function correctly and identifies vulnerabilities before attackers exploit them. PCI DSS mandates specific testing activities, including annual penetration testing, quarterly vulnerability scanning, and security testing of custom code. Organizations implementing the digital payments add-on must test both the add-on integration and SAP system security to ensure comprehensive protection.

Security testing should occur throughout the implementation lifecycle, not just before go-live. Testing during development identifies issues early when they are easier and cheaper to fix. Testing during user acceptance validation ensures security controls do not negatively impact business processes. Testing after go-live confirms production configurations match tested development environments. This section covers security-testing methodologies, vulnerability assessment requirements, penetration-testing procedures, code security review practices, and ongoing validation programs.

### 7.6.1 Security Testing Methodology

Effective security testing follows a structured methodology, progressing from automated scanning to manual expert assessment:

1. **Automated vulnerability scanning**
   Automated tools scan systems for known vulnerabilities, misconfigurations, and missing patches. These tools provide broad coverage with minimal effort, but they also generate false positives, requiring human validation.

2. **Manual configuration review**
   Security analysts manually review system configurations against security baselines and best practices. This catches misconfigured security controls that automated tools miss.
3. **Application security testing**
   Specialized tools and manual testing assess application layer vulnerabilities in custom code, APIs, and user interfaces. This includes testing for injection flaws, authentication bypasses, and business logic vulnerabilities.
4. **Penetration testing**
   Security experts simulate real-world attacks, attempting to exploit identified vulnerabilities and discover new attack paths. Penetration testing validates whether vulnerabilities are exploitable and assesses the impact if successful.
5. **Red team assessment**
   This refers to an advanced adversary simulation in which security teams attempt to achieve specific objectives (exfiltrate payment data, gain admin access) using any means necessary. This tests the overall security posture and incident response capabilities.

Most organizations conduct Level 1, Level 2, Level 3, and Level 4 testing annually, with Level 5 reserved for mature security programs or high-risk environments.

### 7.6.2 Vulnerability Scanning Requirements

PCI DSS Requirement 11.2 mandates quarterly vulnerability scans by an ASV. Organizations can also conduct internal vulnerability scans using their own tools, but ASV scans are required for compliance validation. ASV scans must cover all systems in the CDE and systems directly connected to the CDE.

PCI DSS requires remediating all critical and high-level vulnerabilities within defined timeframes. Organizations must rescan after remediation to confirm vulnerabilities are resolved. A passing grade ASV scan shows that no critical or high vulnerabilities exist. Table 7.26 outlines vulnerability remediation timelines.

| Severity | Common Vulnerability Scoring System Score Range | Remediation Deadline | Rescan Required | Example Vulnerabilities |
|---|---|---|---|---|
| Critical | 9.0–10.0 | Immediate (within 24 hours) | Yes | Remote code execution, unauthenticated database access |
| High | 7.0–8.9 | Within 30 days | Yes | SQL injection, authentication bypass, insecure deserialization |

*Table 7.26: Vulnerability Remediation Requirements*

| Severity | Common Vulnerability Scoring System Score Range | Remediation Deadline | Rescan Required | Example Vulnerabilities |
|---|---|---|---|---|
| Medium | 4.0–6.9 | Within 90 days | Recommended | Cross-site scripting (XSS), weak ciphers, missing patches |
| Low | 0.1–3.9 | No specific deadline | Not required | Version disclosure, verbose error messages |

*Table 7.26: Vulnerability Remediation Requirements (Cont.)*

### 7.6.3 Penetration Testing Procedures

PCI DSS Requirement 11.3 requires annual penetration testing of the CDE and systems connected to it. Penetration testing must be performed by qualified individuals, either internal resources with the necessary training and independence or external security firms. The penetration testing scope for digital payments add-on implementations includes network layer testing, application layer testing, SAP-specific testing, and social engineering (if in scope).

### 7.6.4 Code Security Review

Custom ABAP code interacting with payment data or the digital payments add-on API should undergo a security review to identify vulnerabilities before deployment. Such a code review combines automated static-analysis tools with manual expert assessment. Table 7.27 lists common code vulnerabilities in payment implementations.

| Vulnerability Type | Risk | Secure Alternative |
|---|---|---|
| SQL injection | Attacker can extract payment data | Use parameterized queries with bind variables |
| Hard-coded credentials | Credentials exposed in source code | Store credentials in secure configuration or vault |
| Logging sensitive data | Card numbers exposed in log files | `log.write('Processing card ' + tokenize(card_number))` |
| Weak error messages | Card numbers in error messages | `throw 'Invalid payment method provided'` |

*Table 7.27: Common Payment Code Vulnerabilities*

| Vulnerability Type | Risk | Secure Alternative |
|---|---|---|
| Race condition | Double spending or authorization bypass | Use database transactions with proper isolation |
| Mass assignment | Users can manipulate price or payment amount | Explicitly whitelist allowed fields |

*Table 7.27: Common Payment Code Vulnerabilities (Cont.)*

Static analysis tools like Code Inspector (Transaction SCI) in SAP can automatically identify some vulnerabilities. However, manual code review by experienced developers remains essential for identifying business logic flaws and contextual issues that automated tools miss.

### 7.6.5 Security Testing During Development

Security testing should not wait until production deployment. Integrating security testing into the development lifecycle (part of the DevSecOps approach) identifies issues early, when they are easier and cheaper to fix. Table 7.28 maps security testing to development phases.

| Development Phase | Security Testing Activity | Tools/Methods | Exit Criteria |
|---|---|---|---|
| Design | Threat modeling, security requirements definition | Whiteboard sessions, data flow diagrams | Security requirements documented and approved |
| Development | Unit testing of security controls, static code analysis | JUnit/ABAP Unit, SonarQube, Code Inspector | All unit tests pass, no critical findings |
| Integration | Integration security testing, API security testing | Postman, Burp Suite, custom test scripts | Authentication/authorization tests pass |
| User acceptance testing | Security regression testing, business process validation | Manual testing, test scripts | No security issues blocking business processes |
| Preproduction | Vulnerability scanning, configuration review | Nessus, Qualys, manual review | No critical or high vulnerabilities |

*Table 7.28: Security Testing in Development Lifecycle*

| Development Phase | Security Testing Activity | Tools/Methods | Exit Criteria |
|---|---|---|---|
| Production | Penetration testing, production validation | Compliance verification | PCI compliance validated, no critical issues |

*Table 7.28: Security Testing in Development Lifecycle (Cont.)*

### 7.6.6 Ongoing Validation and Continuous Monitoring

Security is not a one-time implementation but an ongoing operational responsibility. Continuous monitoring and periodic validation ensure security controls remain effective as systems evolve. Table 7.29 summarizes an ongoing security validation schedule.

| Activity | Frequency | Responsible Team | Documentation | Compliance Requirement |
|---|---|---|---|---|
| Vulnerability scanning (ASV) | Quarterly | External ASV | Passing scan report | PCI DSS 11.2.2 |
| Internal vulnerability scanning | Monthly | Internal security team | Scan results, remediation tracking | PCI DSS 11.2.1 |
| Penetration testing | Annually | External firm or internal team | Penetration test report | PCI DSS 11.3 |
| Code security review | Per release | Development + security teams | Code review checklist, findings log | PCI DSS 6.3.2 |
| Access review | Quarterly | SAP security team | User access reports, approval records | PCI DSS 7.1.2 |
| Configuration audit | Monthly | Basis team | Configuration baseline, change log | PCI DSS 2.2 |
| Security awareness training | Annually (minimum) | HR/compliance | Training-completion records | PCI DSS 12.6 |
| Log review | Daily (automated), weekly (manual) | Security operations | Review reports, investigation notes | PCI DSS 10.6 |

*Table 7.29: Security Validation Schedule*

## 7.7 Summary

This chapter highlighted the essential security and compliance requirements for successfully implementing the SAP digital payments add-on. It emphasized that payment security is a mandatory foundation, not an optional enhancement, for reducing risks and meeting regulatory obligations. Key strategies include minimizing PCI DSS compliance scope through cardholder data tokenization, enforcing robust encryption and key management for data protection, and maintaining comprehensive audit logging and monitoring to support security operations and compliance.

The chapter also addressed GDPR and data privacy concerns, ensuring that organizations balance individual rights with regulatory demands. Security validation is continuous, involving regular vulnerability scans, penetration tests, and code reviews to keep systems secure. Ultimately, organizations must establish ongoing processes for monitoring, compliance, and incident response to maintain a secure payment environment.

# Chapter 8
# Testing and Troubleshooting

*Before deploying the SAP digital payments add-on into production, thorough testing and a clear troubleshooting methodology are essential to validate that every component—from API connectivity to end-to-end transaction processing—functions as expected. This chapter provides a structured approach to testing strategies, test scenarios, performance validation, and diagnostic techniques that help ensure a smooth and confident go-live.*

In an SAP digital payments add-on implementation, testing and troubleshooting are critical phases in which configuration transforms into operational capability. A comprehensive testing strategy validates that payment processing functions correctly across all integration points, including SAP ERP, the digital payments add-on, payment service providers (PSPs), and external systems like e-commerce platforms. Troubleshooting skills developed during testing become essential operational competencies, enabling rapid resolution of payment failures that directly impact revenue and customer satisfaction.

Unlike traditional SAP implementations in which testing focuses primarily on internal system functionality, digital payments testing must account for external dependencies beyond your control. PSPs may experience downtime, network connectivity can fail intermittently, and cloud services have their own operational characteristics that differ from those of on-premise systems. Your testing strategy must validate not only that your configuration is correct but also that your implementation handles external system failures gracefully without data loss or customer impact.

This chapter provides comprehensive guidance for testing the digital payments add-on integration across multiple dimensions. You will learn how to establish test environments that mirror production configurations, develop test scenarios covering authorization through settlement and refund processing, execute performance testing to validate the transaction volume capacity, diagnose common integration issues using systematic troubleshooting approaches, and analyze application logs and debugging tools to isolate the root causes of payment failures.

## 8.1 Test Strategy and Planning

A structured test strategy defines what will be tested, when testing occurs, who performs testing activities, and how test results will be evaluated and documented. For digital payments add-on implementations, the test strategy must address the unique characteristics of payment processing, including real-time API dependencies on external cloud services, the financial impacts if test failures could result in actual payment processing errors that affect customers, security and compliance requirements that mandate careful test data management, and multicomponent integration spanning SAP ERP, SAP Business Technology Platform (SAP BTP) cloud services, PSP systems, and e-commerce platforms.

Digital payments testing follows a phased approach aligned with standard SAP implementation methodologies, progressing from isolated component testing through integrated end-to-end validation, as follows:

- **Unit testing (developer-led)**
  Unit testing validates individual components in isolation before integration. For digital payments implementations, unit testing typically focuses on custom developments and API testing using tools like Postman.
- **Functional testing**
  Functional testing validates that the system meets business requirements from the end user perspective, focusing on usability, business process compliance, and requirement fulfillment. Examples include manual order processing via Transaction VA01 with payment cards, e-commerce order integration with preauthorized payments, credit memo and refund processing, settlement processing via Transaction FCC1, payment reconciliation procedures, exception handling (declined cards, expired authorizations), and more.

  Users execute test scripts in a quality assurance (QA) system, following documented procedures. Tests validate that screens display correctly with payment card fields visible, users can select cards from the business partner master via the F4 help, authorization succeeds for valid cards and fails appropriately for declined cards, billing creates proper accounting documents, settlement processes correctly with proper general ledger posting, refunds process through successfully, and so on. Functional testing identifies usability issues, missing configuration, business process gaps, and training requirements before production deployment.
- **Integration testing**
  Integration testing validates that multiple components work together correctly, focusing on data flow between SAP ERP modules, digital payments add-on APIs, and PSP systems. Examples include SAP ERP Sales and Distribution to SAP ERP Financials integration (sales order to billing to accounting), SAP ERP to digital payments add-on API communication, digital payments add-on to PSP routing and processing, electronic bank statement processing with payment advice, business partner master data to sales order integration, authorization function integration with Transaction VA01 sales order saving, and settlement function integration with Transaction FCC1 processing.

  Integration testing happens in the QA system after transporting all configuration and code from development. Use realistic test data that are representative of production scenarios, including various card types, multiple currencies, different customer master data configurations, and sales documents with complex pricing. Best practice is to define the following:

- Entry criteria: All unit tests passed, configuration transported to QA, test data loaded, RFC destinations configured
- Exit criteria: All integration scenarios pass from end to end, no data corruption, proper error handling validated

- **User acceptance testing**
  User acceptance testing (UAT) represents the final validation by business stakeholders that the system is ready for production use. This formal testing phase requires sign-off before go-live approval. The testing scope involves end-to-end business scenarios with real business user participation, exception scenarios and error handling, report validation and reconciliation procedures, approval workflow testing, month-end close procedures with payment processing, and so on.

  Business stakeholders execute predefined test cases in a QA system that mirrors the production configuration exactly. Test cases document expected results, actual results, pass/fail status, and any issues requiring resolution. UAT typically runs from two to four weeks, with daily testing sessions and issue triage meetings.

  UAT success criteria include all critical business scenarios passing without errors, performance meeting business requirements, reports providing required business information, exception-handling procedures working correctly, and stakeholders providing formal sign-off for production deployment. Best practice is to define the following:
  - Entry criteria: Functional and integration testing complete, QA system matches production configuration, UAT test scripts prepared
  - Exit criteria: Business stakeholder sign-off obtained, all critical defects resolved, production cutover plan approved

Table 8.1 defines the testing scope across the integrated digital payments landscape, identifying components, interfaces, and validation points.

| Component | Testing Focus | Integration Points | Validation Requirements |
|---|---|---|---|
| SAP | ABAP objects, configuration, transactions | API calls to add-on, table updates | SAP payment processing transactions, jobs, reports, and financial postings function correctly |
| Digital payments add-on | API functionality, routing, PSP selection | SAP ERP inbound, PSP outbound | Correct routing, proper error handling |
| PSP | Authorization, settlement, refund processing | Add-on adapter interface | PSP responses handled correctly |

*Table 8.1: Test Scope Definition*

| Component | Testing Focus | Integration Points | Validation Requirements |
|---|---|---|---|
| E-commerce platform | Order posting, payment integration | SAP ERP order creation | Preauthorized orders process correctly |
| Electronic bank statements | Advice processing, reconciliation | Payment advice files, clearing | Automated clearing succeeds |

*Table 8.1: Test Scope Definition (Cont.)*

Payment testing requires carefully managed test data that simulates production scenarios without exposing real customer information or processing actual payments. Typical types of test data include the following:

- **Test card numbers**
  Use PSP-provided test cards that simulate different authorization scenarios. Stripe, a commonly used PSP, provides test card numbers for various outcomes, as documented at *https://stripe.com/docs/testing*.
- **Test customer master data**
  Create test business partners as follows:
  - Test customer numbers must be in appropriate number ranges (typically starting with 9) and have the sales role (`FLCU01`) and the finance role (`FLCU00`) assignments.
  - Provide valid addresses in different countries for tax jurisdiction testing, provide a credit control area assignment for credit management integration testing, and ensure the **Payment Cards** tab is populated with test card tokens.
- **Test sales orders**
  Create test sales documents using the following:
  - Test material master data with pricing and tax configuration, various order types (standard orders, returns, credit memos)
  - Multiple delivery scenarios (single delivery, split delivery, partial delivery) and different pricing conditions, including discounts and surcharges

Finally, it's important to establish defect tracking and resolution procedures before testing begins. Use defect management tools to log issues discovered during testing with standardized severity classifications.

## 8.2 Setting Up Test Environments

Test environments are isolated landscapes where testing occurs without risk to production systems or actual payment processing. Digital payments testing requires coordinated environment setup across multiple systems: an SAP ERP test system, a digital payments add-on test tenant on SAP BTP, PSP test accounts and merchant configurations, and e-commerce or portal test instances if applicable. This section first covers configuring Postman as a standalone API testing tool to verify add-on connectivity, then walks through setting up your SAP ERP quality system with the necessary transports and test-specific settings.

### 8.2.1 Postman Setup for API Testing

Postman provides a developer-friendly interface for testing digital payments add-on APIs independently from SAP ERP, thus enabling validation of API functionality, authentication, and responses before integrating with SAP. Download Postman from *https://www.postman.com* and install the application on your laptop. Generate an add-on test tenant service key to be used for Postman testing. Best practice is to have a separate key and not reuse the key that will be used in SAP. Download the add-on API specification in JSON format from *https://api.sap.com/api/DPCoreAPI/overview*. Follow these steps:

1. Select the **API Specification** tab and click the **Download** icon to the right of the **OpenAPI JSON** file (see Figure 8.1). You may be prompted to log in with your SAP Universal ID credentials.

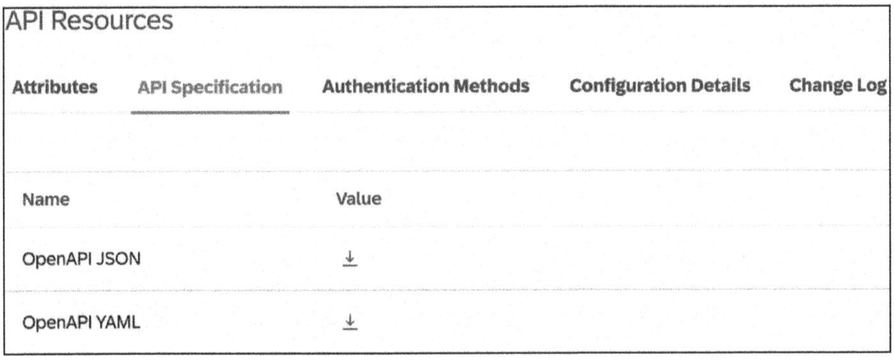

*Figure 8.1: Add-on API JSON Download*

2. Launch Postman and click the **Import** button in the top toolbar. Drag and drop the downloaded OpenAPI JSON file, then select **Postman Collection** as the import type (see Figure 8.2), and click **Import**.

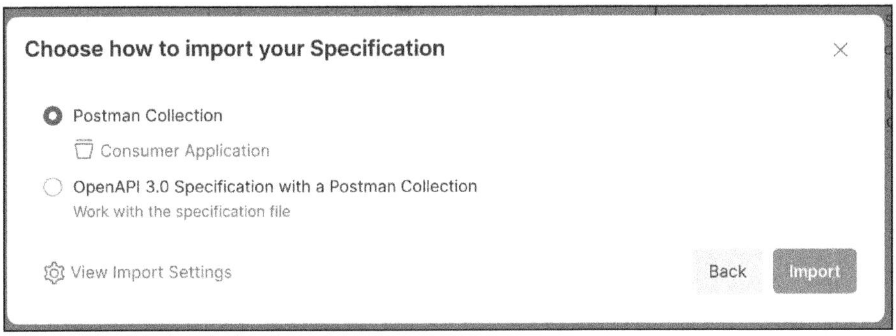

*Figure 8.2: Import Postman Collection*

3. Postman creates a collection named **Consumer Application** (see Figure 8.3) containing all digital payments add-on API endpoints organized by function (authorization, charges, refunds, cards, etc.).

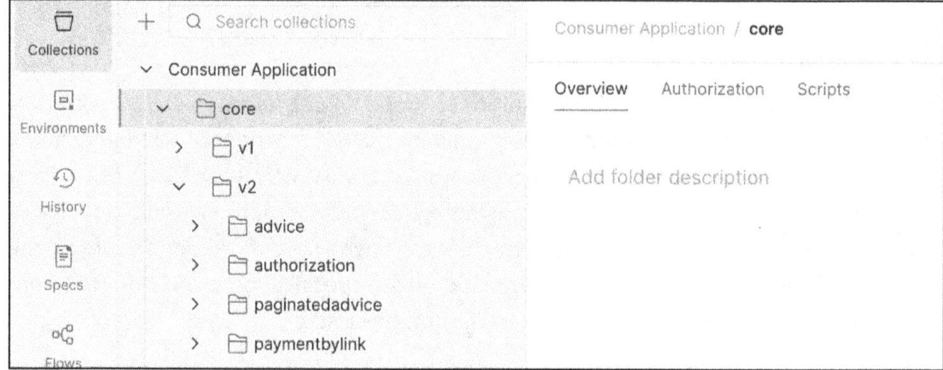


*Figure 8.3: Add-On API Collection Import Success*

4. Rename the collection to "SAP DP Add-On APIs" (see Figure 8.4) for clarity by right-clicking the collection and then choosing **Rename**.

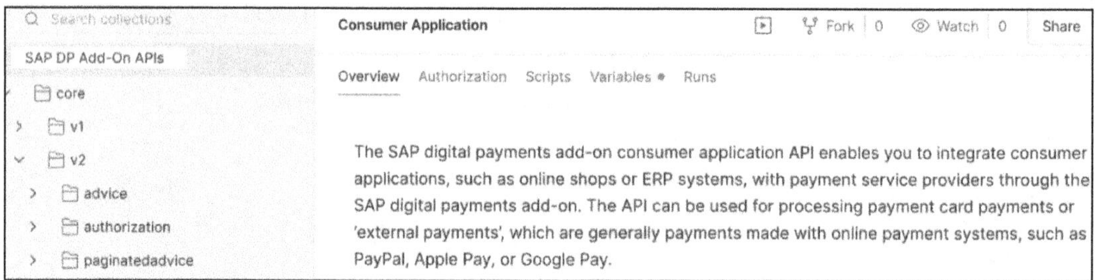


*Figure 8.4: Add-On API Collection Naming*

5. Select the collection, click the **Variables** tab, and locate the **baseUrl** variable showing the current value. The default URL points to the demo environment. Modify it for your test environment (see Figure 8.5). If your test tenant, for example, is in US10, change the URL to "https://digitalpayments-core.test-digitalpayments-sap.cfapps.us10.hana.ondemand.com".

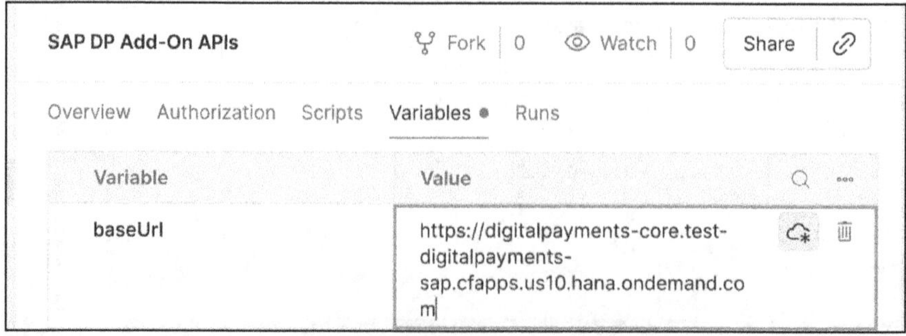


*Figure 8.5: Base URL Setup Example for US10*

6. Select the collection, click the **Authorization** tab (see Figure 8.6), set **Auth Type** dropdown to **OAuth 2.0**, and set the **Add auth data to** dropdown to **Request Headers**.

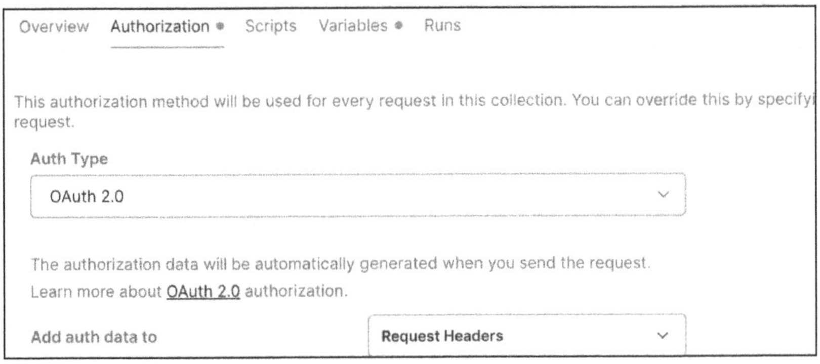

*Figure 8.6: OAuth 2.0 Authentication Setup in Postman Collection*

7. Scroll to the **Configure New Token** section (see Figure 8.7). Enter these values from your test add-on tenant service key and save:
   - **Token Name**: Enter "dp_token"
   - **Grant Type**: Select **Client Credentials**
   - **Access Token URL**: Combine the uaa.url from the service key with /oauth/token (e.g., https://your-tenant.authentication.us10.hana.ondemand.com/oauth/token)
   - **Client ID**: Enter the `clientid` value from the service key
   - **Client Secret**: Enter **clientsecret** value from service key
   - **Scope**: Leave blank
   - **Client Authentication**: **Send as Basic Auth header**

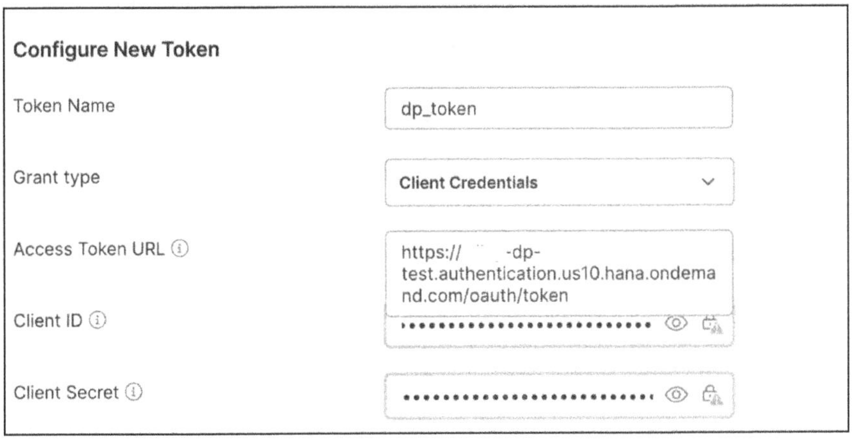

*Figure 8.7: Configure New Token in Postman*

8. Click the **Get New Access Token** button (see Figure 8.8). Postman sends an authentication request to the SAP BTP UAA service. If the configuration is correct, you will see a

green checkmark with the message **Authentication complete**, and the token details will be displayed showing the **access_token** value, **token_type** (Bearer), and **expires_in** (typically 1,800 seconds, which is 30 minutes). Click the **Use Token** button. The token is now active for API requests.

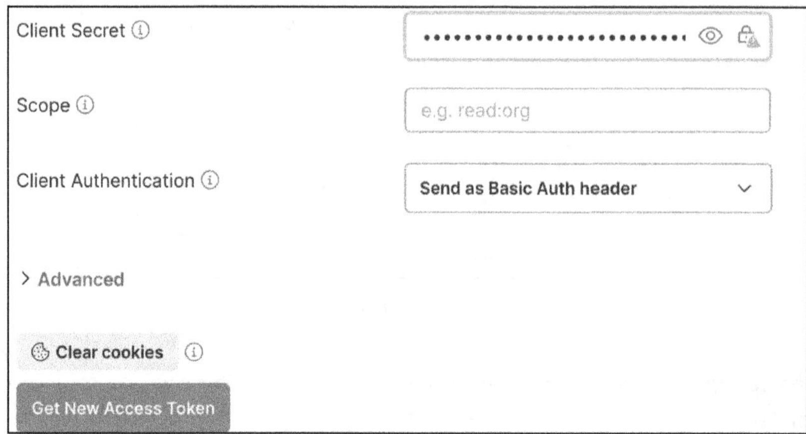


*Figure 8.8: Get New Access Token Button*

9. The digital payments OAuth tokens expire 30 minutes after creation. Before executing API calls, check the **Current Token** section (see Figure 8.9) and use the **Refresh** link to obtain a new token if the current one has expired. Expired tokens result in HTTP 401 Unauthorized errors. You can also enable the **Auto-refresh Token** option for autorefresh of expired tokens.

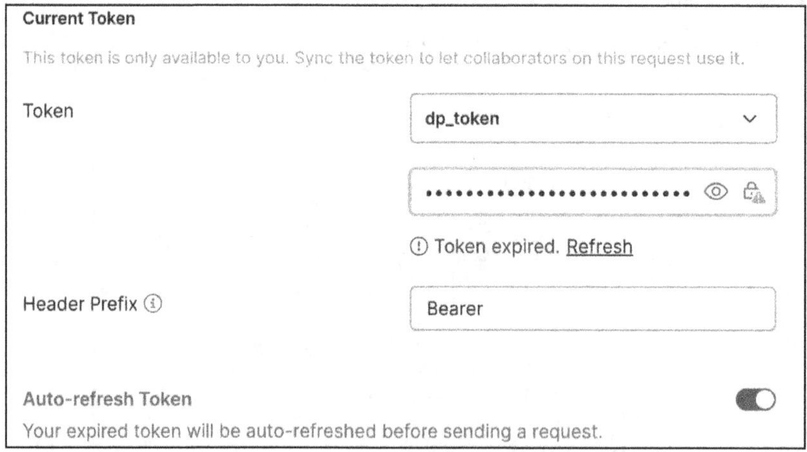


*Figure 8.9: Add-On Token Refresh and Autorefresh Settings*

10. Navigate to the **/core/v1/serviceconfiguration** API endpoint in the collection, verify that the **Authorization** tab shows **Inherit auth from parent** (uses collection-level OAuth config), and click the **Send** button (see Figure 8.10).

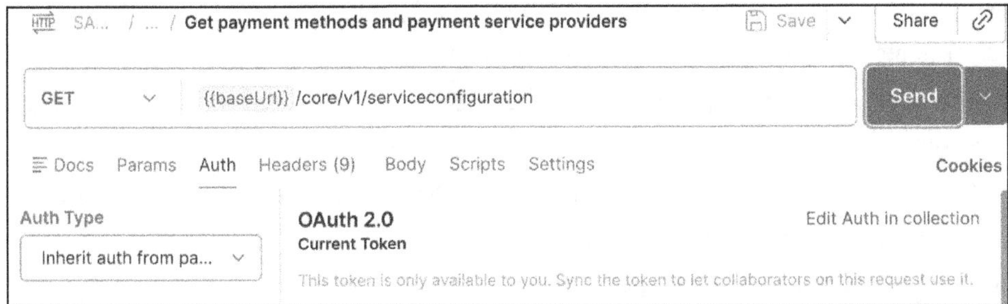


*Figure 8.10: Test API Execution*

11. If successful, you will see a response (200 OK) JSON containing a `PaymentServiceProviders` array with PSP codes and a `PaymentMethods` array (see Figure 8.11). This validates that OAuth authentication works correctly and your RFC destinations in SAP ERP can successfully call the add-on.

Body | 200 OK · 149 ms · 1.05 KB

{} JSON | Preview | Visualize

```
    "PaymentMethods": [
        {
            "Code": "EP",
            "Descriptions": [
                {
                    "Language": "en",
                    "Description": "External Payment"
                }
            ]
        },
        {
            "Code": "CC",
```

*Figure 8.11: Example Response for serviceconfiguration API*

The `PaymentServiceProviders` code value (e.g., `STRP`) is the four-character PSP code used in SAP ERP configuration and API calls.

### 8.2.2 SAP ERP Test System Configuration

Your SAP ERP QA system requires identical configuration to the production system to ensure realistic testing, with modifications for test-specific settings like RFC destinations pointing to the test add-on tenant.

Transport all digital payments configuration from development to QA using your implementation transport request. Verify that the transport includes all SAP Note objects (classes, function modules, tables, views), table configuration (`TCCAA`, `FAR_DP_T100_MAP`, etc.), Transaction SPRO configuration (payment card types, account symbols, posting rules), general ledger accounts, and virtual bank setup.

After transport import, manually configure test-specific settings that do not transport the following:

- RFC destinations (Transaction SM59); use test URLs and test service key credentials
- Test customer master data (Transaction BP)
- Test sales documents for regression testing
- User authorizations for testers

Finally, execute report FAR_DP_UPDATE_PSP_PAYMENT_METH via Transaction SE38. Successful execution displays a log showing that the PSP list was retrieved successfully with PSP codes and payment methods. This validates that RFC destinations, OAuth authentication, and basic API connectivity all function correctly.

## 8.3 Test Scenarios and Scripts

Test scenarios define specific business situations that must be validated during testing, with detailed steps, expected results, and validation criteria. For digital payments add-on implementations, test scenarios span the complete payment lifecycle: card registration, token management, and authorization through settlement to refunds, covering both successful processing and exception scenarios that test error-handling and recovery mechanisms.

### 8.3.1 Card Registration and Token Management

A card registration test validates that the add-on and consumer applications can successfully register payment cards and are able to manage subsequent payment processing in a PCI-compliant manner using tokens.

The `getregistrationurl` API endpoint provided by the add-on enables the consumer applications to get the URL of the payment card registration UI from a PSP (e.g., Stripe). Let's review the parameters that are available in the API endpoint to get the registration URL of the PSP. As you can see in Figure 8.12, there are multiple parameters and values that can be leveraged by consumer applications to route their requests to the correct PSP and the corresponding registration URL. This ensures that the cards are registered with the correct PSP and support payment processing in the intended regions and currencies as per your requirements.

Not all parameters are mandatory. The parameters to be provided depend on your PSP routing setup to get a valid registration URL.

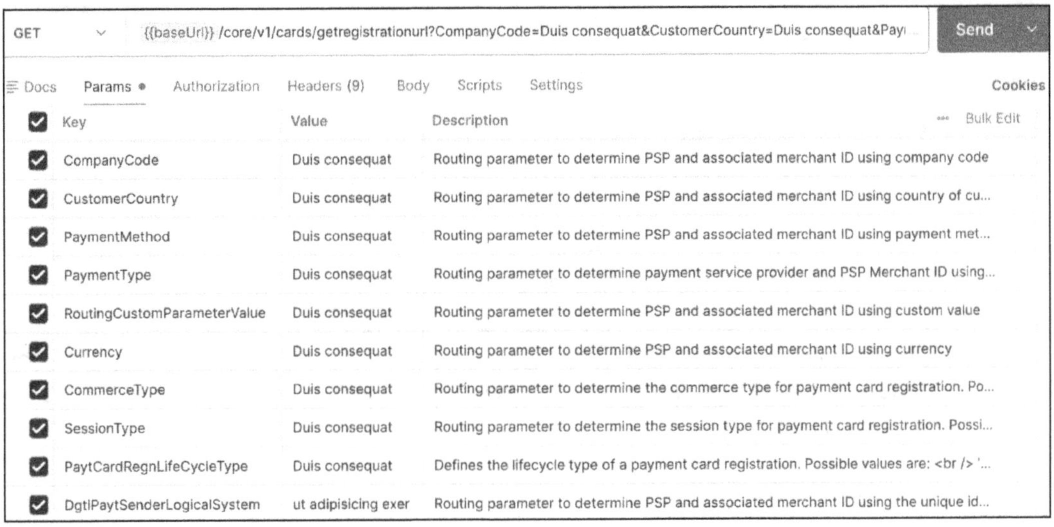

GET {{baseUrl}} /core/v1/cards/getregistrationurl?CompanyCode=Duis consequat&CustomerCountry=Duis consequat&Payi Send

Docs Params • Authorization Headers (9) Body Scripts Settings Cookies

| Key | Value | Description |
|---|---|---|
| CompanyCode | Duis consequat | Routing parameter to determine PSP and associated merchant ID using company code |
| CustomerCountry | Duis consequat | Routing parameter to determine PSP and associated merchant ID using country of cu... |
| PaymentMethod | Duis consequat | Routing parameter to determine PSP and associated merchant ID using payment met... |
| PaymentType | Duis consequat | Routing parameter to determine payment service provider and PSP Merchant ID using... |
| RoutingCustomParameterValue | Duis consequat | Routing parameter to determine PSP and associated merchant ID using custom value |
| Currency | Duis consequat | Routing parameter to determine PSP and associated merchant ID using currency |
| CommerceType | Duis consequat | Routing parameter to determine the commerce type for payment card registration. Po... |
| SessionType | Duis consequat | Routing parameter to determine the session type for payment card registration. Possi... |
| PaytCardRegnLifeCycleType | Duis consequat | Defines the lifecycle type of a payment card registration. Possible values are: <br /> '... |
| DgtlPaytSenderLogicalSystem | ut adipisicing exer | Routing parameter to determine PSP and associated merchant ID using the unique id... |

*Figure 8.12: getregistrationurl API Request Parameters*

Prepare your test data with different variations to test all possible PSPs, regions, and currencies. In this test scenario, we are simulating a request from a consumer application (e.g., SAP) in which the company code for the US region is configured as US01. It is requesting the registration URL for a customer residing in the US, trying to register a card that supports USD transactions. When you populate the required parameters, as shown in Figure 8.13, you will get the URL and the corresponding session ID.

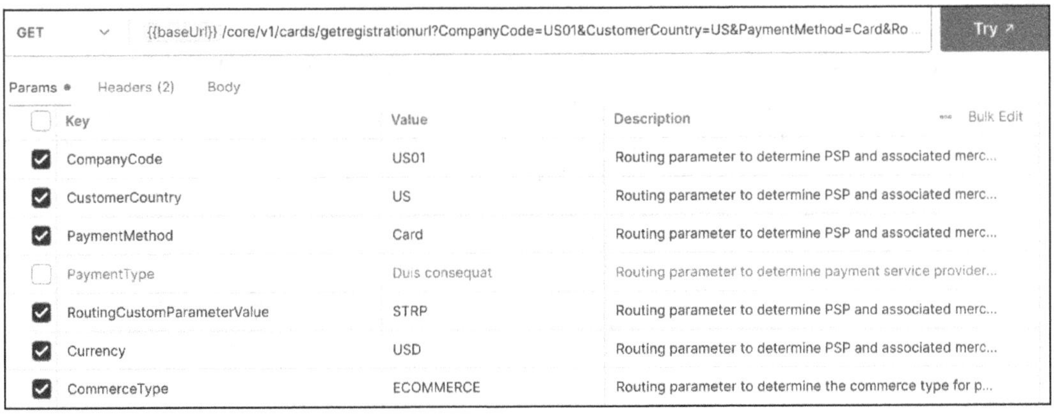

GET {{baseUrl}} /core/v1/cards/getregistrationurl?CompanyCode=US01&CustomerCountry=US&PaymentMethod=Card&Ro Try ↗

Params • Headers (2) Body

| Key | Value | Description |
|---|---|---|
| CompanyCode | US01 | Routing parameter to determine PSP and associated merc... |
| CustomerCountry | US | Routing parameter to determine PSP and associated merc... |
| PaymentMethod | Card | Routing parameter to determine PSP and associated merc... |
| PaymentType | Duis consequat | Routing parameter to determine payment service provider... |
| RoutingCustomParameterValue | STRP | Routing parameter to determine PSP and associated merc... |
| Currency | USD | Routing parameter to determine PSP and associated merc... |
| CommerceType | ECOMMERCE | Routing parameter to determine the commerce type for p... |

*Figure 8.13: Test Scenario for US Company Code*

Copy the registration URL from the Postman response (see Figure 8.14).

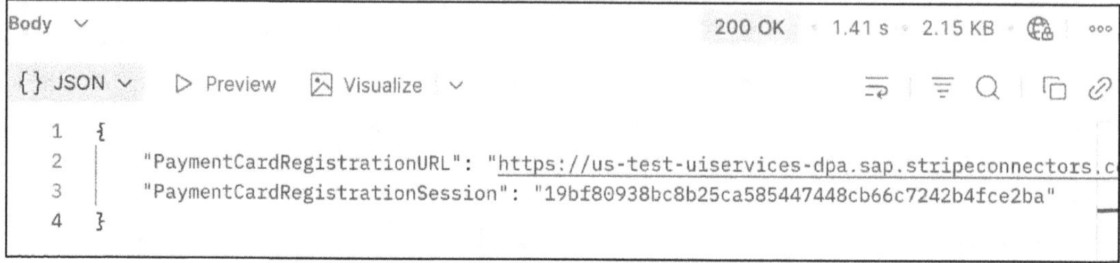


*Figure 8.14: Postman Response with Registration URL and Session ID*

Launch a browser window and paste the URL into the address bar. If the registration request was successful, then you should see the PSP-provided iFrame (see Figure 8.15) to enter the card details.

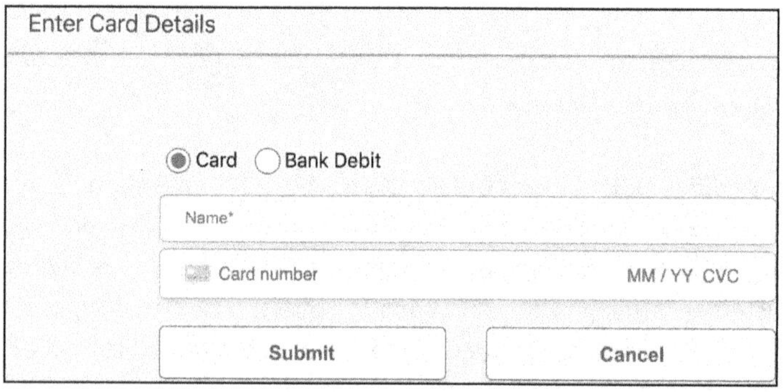

Enter Card Details

Card Bank Debit

Name*

Card number MM / YY CVC

Submit Cancel

*Figure 8.15: PSP-Provided Secure iFrame*

Once you enter the card details (e.g., a PSP-provided test card) and submit the card for registration, you should see a successful registration message in the browser. To check if the registration request is successful, use the `poll` API endpoint (see Figure 8.16) and provide the session ID that was received in the previous API response.

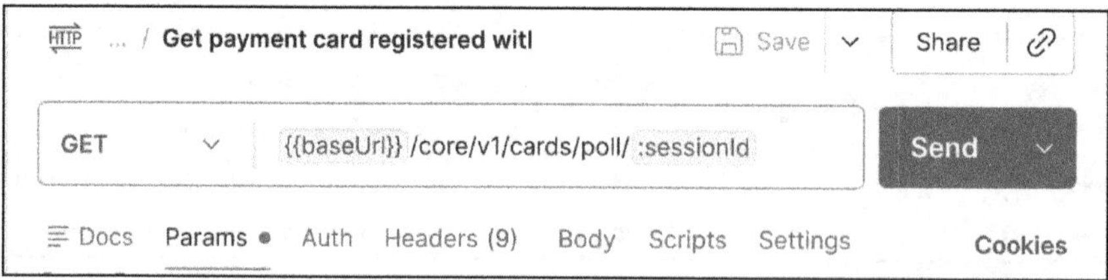


*Figure 8.16: Poll API Endpoint Signature in Postman*

The API endpoint will return the add-on-generated digital payments token number, masked credit card number, and a few other details, as shown in Figure 8.17.

This successful test proves that the add-on configuration and the PSP configuration are correct. Repeat the Postman tests as needed for all PSPs, currencies, and regions to validate the setup.

{} JSON ⌄ ▷ Preview Visualize ⌄


```json
{
    "DigitalPaymentTransaction": {
        "DigitalPaymentTransaction": "19bf80938bc8b25ca585447448cb66c7242b4fce2ba",
        "DigitalPaymentDateTime": "2026-01-26T02:01:52.573Z",
        "DigitalPaytTransResult": "01",
        "DigitalPaytTransRsltDesc": "Successful"
    },
    "PaytCardByDigitalPaymentSrvc": "T6SDCKNM4ZX2XBLOUYKX3VRS",
    "PaymentCardType": "DPVI",
    "PaymentCardExpirationMonth": "01",
    "PaymentCardExpirationYear": "2029",
    "PaymentCardMaskedNumber": "************4242",
    "PaymentCardHolderName": "DEMO Customer"
}
```


*Figure 8.17: Add-On Token and Masked Number*

After confirming the test results in Postman, proceed with the testing card registration process in the consumer applications—for example, in the Manage Business Partner Master Data app.

### 8.3.2 Authorization

Authorization testing validates that payment cards are properly authorized during sales order processing, with successful authorizations allowing order progression and declined authorizations appropriately blocking orders.

#### Successful Payment Card Authorization

In this scenario, the objective is to independently test the authorization API using Postman, separate from SAP ERP. Prerequisites include ensuring that Postman is configured with a valid OAuth token and that a test card token has been obtained through the card registration API. Follow these steps:

1. In Postman, navigate to the add-on API collection. Expand the **Authorizations** folder and select the **POST Authorize amount for payment card payment** (`/core/v1/authorizations`) endpoint (see Figure 8.18).

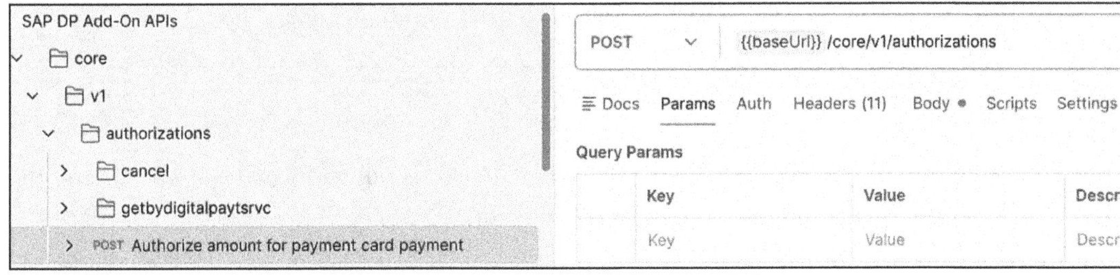


*Figure 8.18: Add-On Authorization API Endpoint*

2. Click the **Body** tab (see Figure 8.19), select the **raw** format, set the type to **JSON**, and enter the request payload shown in Listing 8.1.

```
{
  "Authorizations": [
    {
      "AmountInAuthorizationCurrency": "100",
      "AuthorizationCurrency": "USD",
      "CustomerAccountNumber": "3456345395",
      "DigitalPaymentCommerceType": "ECOMMERCE",
      "ReferenceDocument": "0001346385657226",
      "Source": {
        "Card": {
          "PaytCardByDigitalPaymentSrvc": "T6SDCKNM4ZX2XBLOUYKX3VRS"
        }
      },
      "_comment": "Sample 1: Payment card authorization request with
reference document and customer account number"
    }
  ]
}
```

*Listing 8.1: Example JSON Authorization Request for USD 100*

Docs Params Auth Headers (12) **Body** • Scripts Settings

raw JSON Schema

```
{
  "Authorizations": [
    {
      "AmountInAuthorizationCurrency": "100",
      "AuthorizationCurrency": "USD",
      "CustomerAccountNumber": "3456345395",
      "DigitalPaymentCommerceType": "ECOMMERCE",
      "ReferenceDocument": "0001346385657226",
      "Source": {
        "Card": {
          "PaytCardByDigitalPaymentSrvc": "T6SDCKNM4ZX2XBLOUYKX3VRS"
        }
```

*Figure 8.19: Example JSON for USD 100 Authorization*

3. Click **Send**. If successful, you will see HTTP status **200 OK** (see Figure 8.20), and the response payload will contain the authorization ID that can be used in subsequent settlement testing.

Body

200 OK 3.04 s 1.57 KB

{} JSON Preview Visualize


```
'Authorizations": [
    {
        "DigitalPaymentTransaction": {
            "DigitalPaymentTransaction": "19bf829a30e63a7885bf52349a18bb68793e7f4cf06",
            "DigitalPaymentDateTime": "2026-01-26T02:37:16.943Z",
            "DigitalPaytTransResult": "01",
            "DigitalPaytTransRsltDesc": "Successful",
            "ReferenceDocument": "0001346385657226"
        },
        "Authorization": {
            "AuthorizationByPaytSrvcPrvdr": "pi_3Stfg2J4PRGyzmc71buBcHAh",
            "AuthorizationByDigitalPaytSrvc": "YMK6HHLK4Y",
```


*Figure 8.20: Authorization API Response*

4. The response contains the `Authorizations` array, which contains the following:
   - `DigitalPaymentTransaction.DigitalPaytTransResult` = 01 (success)
   - `Authorization.AuthorizationByDigitalPaytSrvc` = DP authorization ID
5. Verify the complete JSON response and examine the various fields returned by the add-on. Repeat the tests as needed for different authorization amounts, currencies, and regions as per your implementation scope. After all the Postman tests are successful, test the authorization process from the consumer application either using the Manage Sales Orders app or Transaction VA01.
6. The results will be stored in table `FPLTC`, and you will see that the payment card plan shows a green traffic light for the authorization status. Check that table `FPLTC` contains an entry with the following fields:
   - `AUNUM`: Digital payments authorization ID
   - `AUTWR`: Order total amount
   - `CCWAE`: Order currency
   - `AUDAT/AUTIM`: Current date and time
   - `DP_TOKEN`: Populated with token value
   - `CCINS`: DPVI
   - `MERCH`: Configured merchant alias

#### Declined Payment Card Authorization

For declined payment card authorizations, there are a few possible variations that it is recommended you test. In the first scenario, the card registration is declined by the PSP, and in the second scenario the card registration is successful, but the authorization is declined.

### Card Registration Decline

You can simulate a card registration decline scenario by using one of the test cards that is provided under declined cards section on the Stripe test cards page at *https://docs.stripe.com/testing#declined-payments*. Get the registration URL by triggering the `getregistrationurl` API endpoint in Postman and launch the secure iFrame. Enter one of the test cards that is expected to be declined by the PSP (see Figure 8.21).

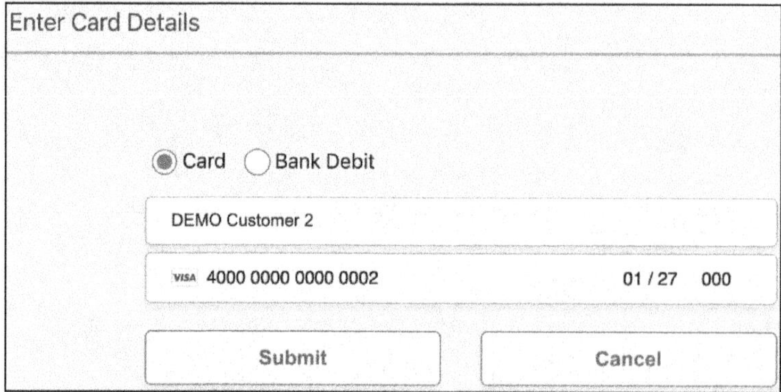


*Figure 8.21: Declined Registration Test Card Entry*

Click **Submit**. The add-on sends the card registration request to the PSP and, in this scenario, will get a decline message from Stripe. The add-on returns the failure message (see Figure 8.22) as expected to the consumer application.

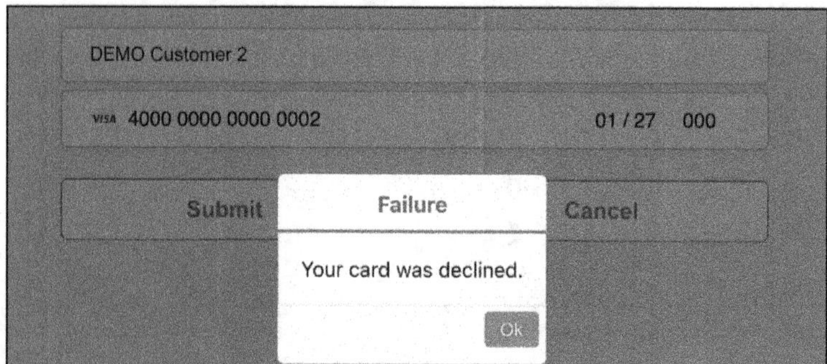


*Figure 8.22: Card Registration Decline Message*

Repeat the tests for all the various company codes and currencies where you intend to deploy the add-on for payment processing. After successful tests in Postman, start testing in the Manage Business Partner Master Data app and observe the behavior.

### Card Authorization Decline

In this scenario, the test card will allow a successful registration of the card, but when you attempt an authorization, it should be declined by the PSP. Trigger the `getregistrationurl` API endpoint in Postman and launch the URL in the browser. Enter the test card provided by Stripe for this specific test (see Figure 8.23).

Card
Bank Debit
DEMO Customer
VISA 4000 0000 0000 0341
01 / 29
000
Submit
Cancel

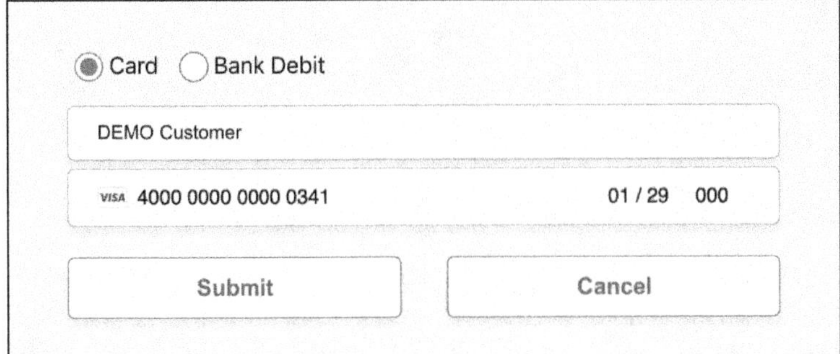

*Figure 8.23: Card Registration Secure iFrame*

Click **Submit**. You will see that the card registration was not declined (see Figure 8.24).

DEMO Customer
VISA 4000 0000 0000 0341
01 / 29
000
Success
Your request has been sent to the service provider.
Ok

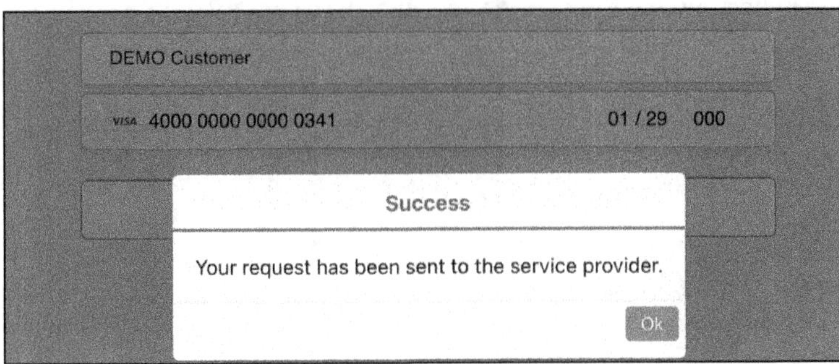

*Figure 8.24: Card Registration Success*

Copy the session ID from the response in Postman and trigger the poll API endpoint using the session ID. You will obtain the add-on token in the JSON response payload (see Figure 8.25).

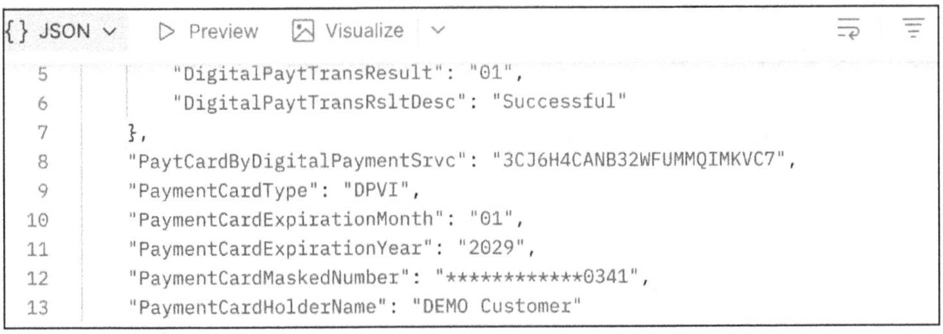

{} JSON
Preview
Visualize
"DigitalPaytTransResult": "01",
"DigitalPaytTransRsltDesc": "Successful"
},
"PaytCardByDigitalPaymentSrvc": "3CJ6H4CANB32WFUMMQIMKVC7",
"PaymentCardType": "DPVI",
"PaymentCardExpirationMonth": "01",
"PaymentCardExpirationYear": "2029",
"PaymentCardMaskedNumber": "************0341",
"PaymentCardHolderName": "DEMO Customer"

*Figure 8.25: Get SAP Digital Payments Token Using Poll API*

Use this token to obtain an authorization for USD 100 using the authorizations API endpoint (see Figure 8.26).

```
"Authorizations": [
  {
    "AmountInAuthorizationCurrency": "100",
    "AuthorizationCurrency": "USD",
    "CustomerAccountNumber": "3456345395",
    "DigitalPaymentCommerceType": "ECOMMERCE",
    "ReferenceDocument": "0001346385657226",
    "Source": {
      "Card": {
        "PaytCardByDigitalPaymentSrvc": "3CJ6H4CANB32WFUMMQIMKVC7"
      }
```

*Figure 8.26: Authorization Request Using Test Card*

You will observe that the authorization will be declined by the PSP, and a message will be returned to the add-on indicating that the card was declined (see Figure 8.27)

```
    },
    "Authorization": {
        "AuthorizationByPaytSrvcPrvdr": "pi_3Su2x2J4PRGyzmc70upRVb0j",
        "AuthorizationByDigitalPaytSrvc": "2ZRDQ554JE",
        "AuthorizedAmountInAuthznCrcy": "100.00",
        "AuthorizationCurrency": "USD",
        "AuthorizationStatus": "02",
        "DetailedAuthorizationStatus": "212",
        "AuthorizationStatusName": "Your card was declined.; code: card_declined;
        "MerchantAlias": "DPQAVIRNA",
        "DgtlPaytAuthznIsPrtlyChrgbl": false,
        "DigitalPaymentCommerceType": "ECOMMERCE"
```

*Figure 8.27: Authorization Decline Response*

You could attempt a similar test in your consumer application to observe the results and the system behavior based on the add-on responses.

### 8.3.3 Successful Settlement

This scenario tests settlement/capture API independently using Postman. As a prerequisite, you need an authorization ID from a previous authorization test. Follow these steps:

1. In Postman, select the **POST** `/core/v1/charges` endpoint (see Figure 8.28).

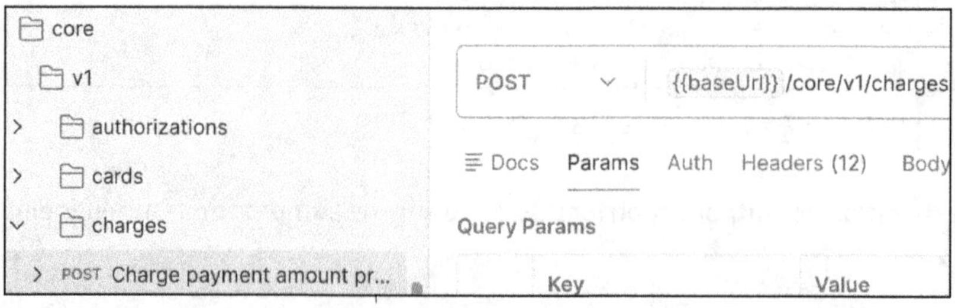

*Figure 8.28: Charges API Endpoint*

2. Enter the request payload shown in Listing 8.2.

```
{
  "Charges": [
    {
      "AmountInPaymentCurrency": "100",
      "Authorization": {
        "AuthorizationByDigitalPaytSrvc": "YMK6HHLK4Y",
        "AuthorizationCurrency": "USD",
        "AuthorizedAmountInAuthznCrcy": "100"
      },
      "PaymentCurrency": "USD",
      "PaymentIsToBeCaptured": false,
      "ReferenceDocument": "0001346385657226",
      "Source": {
        "Card": {
          "PaytCardByDigitalPaymentSrvc": "T6SDCKNM4ZX2XBLOUYKX3VRS"
        }
      },
    }
  ]
}
```

*Listing 8.2: Example JSON Capture Request Payload for USD 100*

3. Click **Send**. You should see an HTTP 200 OK result, and the response will contain a `Charge` **object with** `Charge.PaymentByPaymentServicePrvdr` **= PSP charge ID, showing funds successfully captured against the previous authorization (see Figure 8.29).**

```
            "ReferenceDocument": "0001346385657226"
        },
        "Charge": {
            "PaymentByPaymentServicePrvdr": "ch_3Stfg2J4PRGyzmc71YLXspF6",
            "AddlPaytByPaytSrvcPrvdr": "txn_3Stfg2J4PRGyzmc715yJtv0S",
            "PaymentByDigitalPaymentService": "19bfd94f0813381360b8e9d4fd0afaaf7d39bcd9ce5",
            "AmountInPaymentCurrency": "100.00",
            "PaymentCurrency": "USD",
            "PaymentDateTime": "2026-01-26T02:37:18Z",
            "PaymentStatus": "01",
            "DetailedPaymentStatus": "100",
            "PaymentStatusName": "The processing was completed successfully.",
```

*Figure 8.29: Successful Settlement Response*

Repeat the test for other company codes, currencies, and card types as per your implementation scope. After successful test results are obtained, you could start testing in SAP by creating a sales order and taking it all the way through settlement.

### 8.3.4 Refund Test Scenarios

Refund testing validates credit processing that returns funds to customer payment cards. In this scenario, we will use the `/core/v1/refunds/bypaymentbydigitalpaymentservice` refund API endpoint (see Figure 8.30).

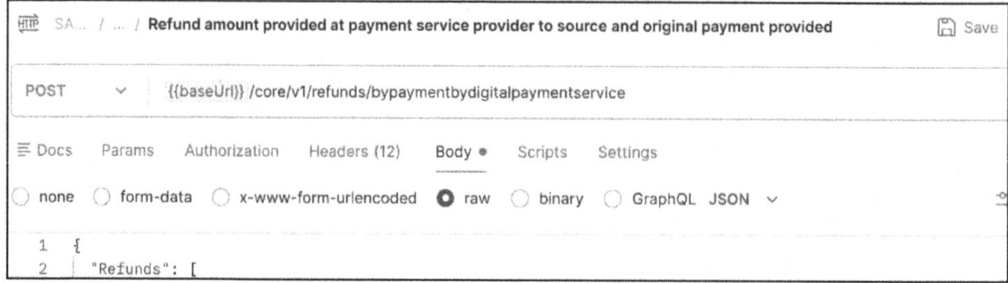


*Figure 8.30: Refunds API Endpoint*

Provide the digital payments transaction ID from the previous successful settlement test in the request payload, along with other details, as shown in Figure 8.31.

raw JSON Schema Beau

```
{
  "Refunds": [
    {
      "AmountInRefundCurrency": "100.00",
      "PaymentByDigitalPaymentService": "19bfd94f0813381360b8e9d4fd0afaaf7d39bcd9ce5",
      "ReferenceDocument": "0001346385657226",
      "RefundCurrency": "USD"
    }
  ]
}
```

*Figure 8.31: Refund Request Parameters*

You should see a refund success message in the JSON response, as shown in Figure 8.32.

Body Cookies Headers (15) Test Results 200 OK 2.05 s 1.7 KB

{} JSON Preview Visualize

```
        "Refund": {
            "RefundByPaymentServiceProvider": "re_3Stfg2J4PRGyzmc71nhcipng",
            "PaymentByPaymentServicePrvdr": "ch_3Stfg2J4PRGyzmc71YLXspF6",
            "AddlPaytByPaytSrvcPrvdr": "txn_3Stfg2J4PRGyzmc715yJtv0S",
            "AmountInRefundCurrency": "100.00",
            "RefundCurrency": "USD",
            "RefundDateTime": "2026-01-27T04:09:34Z",
            "RefundStatus": "01",
            "DetailedRefundStatus": "100",
            "RefundStatusName": "The processing was completed successfully.",
            "MerchantAlias": "DPQAVIRNA",
            "PaymentByDigitalPaymentService": "19bfd94f0813381360b8e9d4fd0afaaf7d39bcd9ce5"
```

*Figure 8.32: Refund Success Response JSON*

### 8.3.5 E-Commerce Integration

E-commerce testing validates that preregistered, preauthorized, or precaptured payments from external platforms integrate correctly with SAP ERP order processing.

#### Preregistered Card

This scenario is applicable for organizations operating e-commerce applications or self-service payment portals that allow customers to register payment cards within their profiles or accounts. Such applications may have been deployed prior to the implementation of the digital payments add-on in your environment. In these cases, direct integration with PSPs may have been established without utilizing the add-on API endpoints. When a customer submits an order or processes a payment using one of these registered cards, it may be necessary to integrate the card details with SAP. Given that SAP is connected to the add-on for payment processing, SAP will require the corresponding digital payments token and related information to store within the business partner record or the SAP order. For this purpose, the `getforpaymentcard` API endpoint should be used (see Figure 8.33).

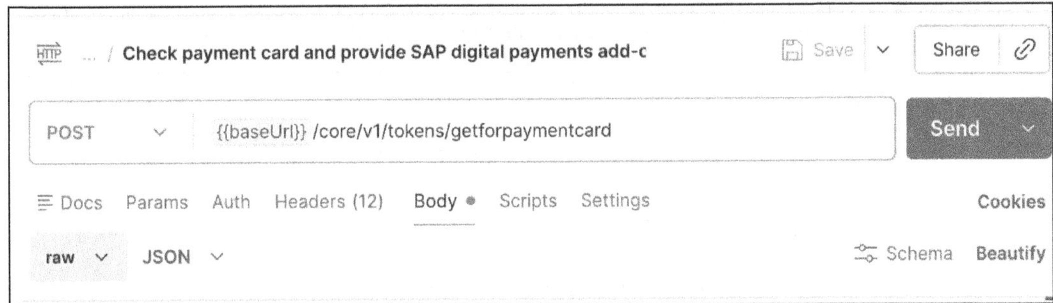


*Figure 8.33: getforpaymentcard API Endpoint*

Before accessing the `getforpaymentcard` API endpoint, you must first acquire the PSP code through the `serviceconfiguration` API endpoint (see Figure 8.34).

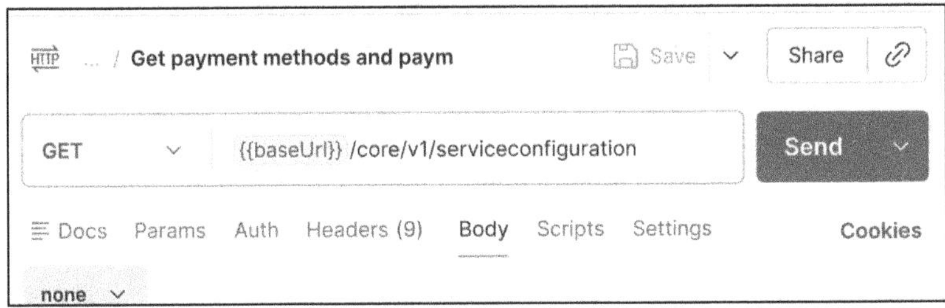


*Figure 8.34: serviceconfiguration API Endpoint*

When you click **Send**, you'll receive a JSON response listing the PSPs and their configured payment methods. Take note of the specific PSP code provided by the API (see Figure 8.35). The response also includes all enabled payment methods, such as external payments and payment cards, with their respective codes and descriptions. For each PSP set up in your system, you'll see both its PSP code and the associated list of enabled payment methods.

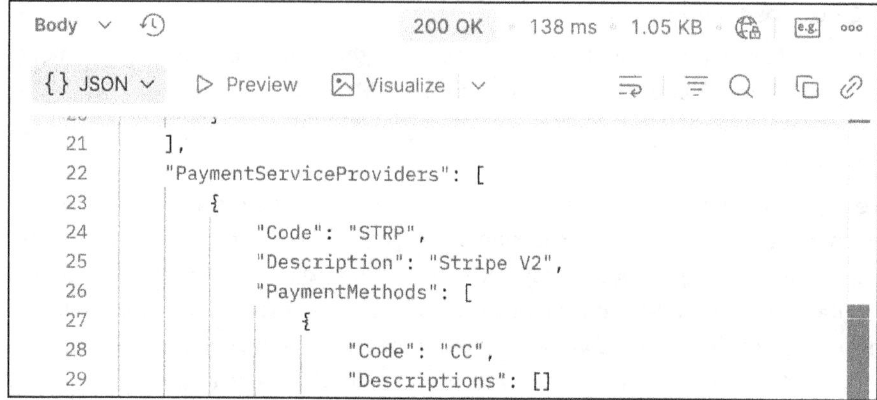

Figure 8.35: getserviceconfiguration Example Response Payload

In addition to the PSP code, the `getforpaymentcard` API needs the following details in the JSON request:

- `PaymentServiceProvider`: Provide the PSP code that was obtained.
- `PaytCardByPaytServiceProvider`: Provide the PSP token that was obtained through direct integration.
- `MerchantAccount`: Provide the PSP merchant account (e.g., the Stripe key).
- `PaytCardRegnLifeCycleType`: This code defines whether you intend to use the card once or more.

An example JSON request payload is shown in Listing 8.3.

```
{
 "PaymentCards": [
 {
 "PaymentServiceProvider": "STRP",
 "PaytCardByPaytServiceProvider": "pm_xxxxxxxxxxxxxxxxxxx",
 "MerchantAccount": "pk_test_xxxxxxxxxxxxx", // your PSP merchant
account key
 "PaytCardRegnLifeCycleType": "01".
 }
 ]
}
```

Listing 8.3: Example JSON Request Payload for Stripe PSP

You will obtain the corresponding digital payments token within the JSON response (see Figure 8.36) provided by the API. This response includes the digital payments token for the card used by the customer and registered with the PSP, as well as additional information such as card type, expiration month and year, and masked card number. These details are essential for storing the token in SAP.

Furthermore, the response contains supplementary data, including the PSP-generated token and PSP merchant account key, which can be retained according to your implementation requirements.


```
        "PaytCardByPaytServiceProvider": "pm_                    ",
        "PaymentServiceProvider": "STRP",
        "Source": {
            "Card": {
                "PaymentCardType": "DPVI",
                "PaymentCardExpirationMonth": "05",
                "PaymentCardExpirationYear": "2027",
                "PaymentCardMaskedNumber": "************4242",
                "PaytCardByDigitalPaymentSrvc": "AJO7JQS2VKZQB4W5N7IULW2P"
            },
            "Merchant": {
                "Account": "pk_test
```


*Figure 8.36: getforpaymentcard API Example Response*

**Preauthorized Order Integration**

In this scenario, customer orders are authorized within the e-commerce system and subsequently captured in SAP. When the e-commerce platform is integrated with the add-on APIs designated for payment processing, the testing process follows the process outlined previously in Section 8.3.2. Alternatively, if the e-commerce solution interfaces directly with the PSP, the `getforpaymentcardauthorization` API endpoint should be utilized, as illustrated in Figure 8.37.

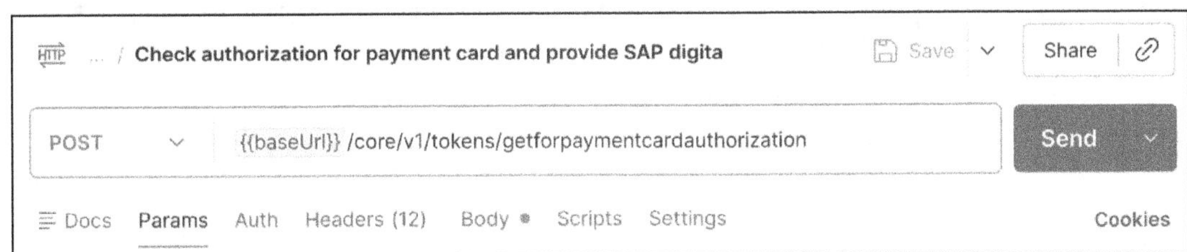


*Figure 8.37: getforpaymentcardauthorization API Endpoint*

To access this API endpoint, you'll need the following:

- An authorization ID provided by the PSP
- The digital payments add-on token
- The public key for your PSP merchant account
- The amount to authorize and its currency

An example JSON payload is shown in Listing 8.4.

```
{
 "Authorizations": [
 {
 "AuthorizationByPaytSrvcPrvdr": "pi_xxxxxxxxxxxxxxxxxxxxx",
 "PaytCardByDigitalPaymentSrvc": "BCS604LZVYMITCU3I4FUZ4EH",
 "MerchantAccount": "pk_test_xxxxxxxxxxxxxxxxxxxxxxxxxxxxxx",
 "AuthorizedAmountInAuthznCrcy": "12.99",
 "AuthorizationCurrency": "USD",
 "DigitalPaymentCommerceType": "ECOMMERCE",
 "DigitalPaymentSessionType": "OFFLINE"
 }
 ]
}
```


*Listing 8.4: Example JSON Request Payload to Request Add-On Authorization ID*

After you click **Send** in Postman, the add-on API will respond with a digital payments authorization ID. This ID is then used in SAP for further billing and settlement steps. Listing 8.5 shows an example of the JSON response returned by the API endpoint.


```
{
    "Authorizations": [
        {
            "DigitalPaymentTransaction": {
                "DigitalPaymentTransaction":
"19c06e57b2669a0a89814594337bd463655bd",
                "DigitalPaymentDateTime": "2026-01-28T23:17:08.518Z",
                "DigitalPaytTransResult": "01",
                "DigitalPaytTransRsltDesc": "Successful"
            },
            "PaytCardByDigitalPaymentSrvc": "BCS604LZVYMITCU3I4FUZ4EH",
            "MerchantAccount": "pk_test_xxxxxxxxxxxxxxxxxxxxx",
            "Authorization": {
                "AuthorizationByPaytSrvcPrvdr": "pi_
xxxxxxxxxxxxxxxxxxxxx",
                "AuthorizationByDigitalPaytSrvc": "XDKAD3A504",
                "AuthorizedAmountInAuthznCrcy": "12.99",
                "AuthorizationCurrency": "USD",
                "AuthorizationDateTime": "2025-06-26T09:42:04Z",
                "AuthorizationExpirationDateTme": "2025-07-
03T09:42:04Z",
                "AuthorizationStatus": "01",
                "DetailedAuthorizationStatus": "100",
                "AuthorizationStatusName": "Successful",
                "MerchantAlias": "xxxxxxxxxxxxx",
```

```
                "DgtlPaytAuthznIsPrtlyChrgbl": false
            }
        }
    ]
}
```

*Listing 8.5: Example JSON Response Payload with Add-On Authorization ID*

Conducting API tests in Postman can enhance your integration strategy, ensuring smooth connectivity between customer-facing platforms such as e-commerce systems and back-office solutions like SAP. The add-on APIs offer flexibility by allowing settlement of payment transactions not initially authorized via their APIs, eliminating the need for immediate full-scale integration. This enables organizations to adopt a phased approach when transitioning to the advanced digital payments infrastructure provided by the add-on.

**Presettled Order**

For organizations providing digital products and services, as no physical goods require shipment, funds can be captured immediately at checkout when a customer enters payment card details and places an order. In this context, the e-commerce platform used for purchasing may or may not interface directly with add-on APIs for payment processing. Typically, organizations centralize returns and refunds within their SAP system, where financial processes are administered. The add-on offers an API (see Figure 8.38) that facilitates retrieval of the digital payments transaction ID for presettled orders, enabling its storage in SAP. By recording the transaction ID from this API endpoint, refunds to customer credit cards can be processed directly through SAP.

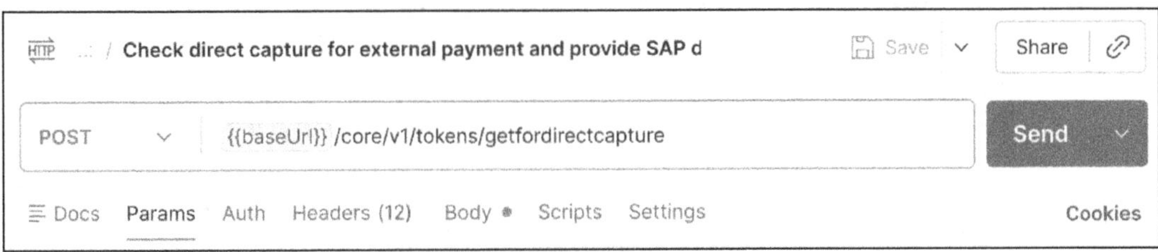


*Figure 8.38: getfordirectcapture API Endpoint*

## 8.4 Performance Testing

Performance testing validates that the digital payments integration handles expected transaction volumes without degradation in response times, system stability, or user experience. Unlike functional testing focusing on correctness, performance testing stresses the system with realistic and peak loads, ensuring production capacity requirements are met.

Digital payments performance testing addresses three primary concerns:

- API response time, ensuring that authorization and settlement complete within acceptable timeframes for user transactions
- Batch processing throughput, validating that Transaction FCC1 can settle daily transaction volumes within allocated batch windows
- System resource utilization, confirming that memory, CPU, and database capacity support the payment processing load

Load testing simulates realistic transaction volumes, validating that the system performs adequately under normal operational conditions. Consider the following scenarios:

- **Authorization load test**
  Define a clear objective, approach, and success criteria as shown in Table 8.2.

| Objective | Approach | Success Criteria |
|---|---|---|
| Validate that authorization processing handles concurrent sales orders without performance degradation | ■ Create concurrent Transaction VA01 sessions using load testing tools<br>■ Each session creates a sales order with a payment card authorization<br>■ Measure average authorization response time<br>■ Monitor system resources (Transaction ST03N or ST06) | ■ Authorization completes as per your business SLA<br>■ No system memory exhaustion<br>■ No HTTP timeout errors<br>■ All authorizations process correctly without data corruption |

*Table 8.2: Authorization Load Test*

- **Settlement batch performance**
  Define a clear objective, approach, and success criteria as shown in Table 8.3.

| Objective | Approach | Success Criteria |
|---|---|---|
| Validate that Transaction FCC1 processes daily settlement volume within acceptable timeframe | ■ Create test data: Billing documents requiring settlement matching expected daily production volumes<br>■ Execute Transaction FCC1 to observe the runtime and measure total processing time<br>■ Calculate throughput (in transactions per minute) | ■ Settlement of daily transactions should complete within your defined business SLA<br>■ No database locks or deadlocks<br>■ Memory consumption stays within allocated parameters |

*Table 8.3: Settlement Batch Load Test*

## 8.5 Common Issues and Resolution

This section presents frequently encountered issues during the implementation and operation of the digital payments add-on, detailing symptoms, root causes, and recommended resolution steps. For additional troubleshooting guidance, consult the relevant chapters addressing topic-specific challenges and their corresponding solutions.

Table 8.4 describes some common RFC destination and connectivity issues.

| Issue/Symptoms | Root Cause/Resolution/Prevention |
|---|---|
| OAuth token-retrieval failure:<br>■ Settlement or authorization fails with error retrieving OAuth token<br>■ The application log (Transaction SLG1) shows HTTP 401 Unauthorized on token endpoint or *authentication failed* message | RFC destination `DIGITALPAYMENTS_OAUTH` has incorrect credentials. Reenter `clientid` and `clientsecret` and test by executing report FAR_DP_UPDATE_PSP_PAYMENT_METH. |
| SSL Certificate Trust Error (RFC connection fails with SSL handshake error) | DigiCert Global Root G2 certificate not imported into SAP ERP trust manager. Download DigiCert Global Root G2 certificate from *https://www.digicert.com/kb/digicert-root-certificates.htm* and import into trust store. |

*Table 8.4: Common Connectivity Issues and Resolution Steps*

## 8.6 Log Analysis and Debugging

Effective troubleshooting requires understanding where the digital payments add-on logs information and how to analyze the logs to isolate root causes. SAP ERP provides multiple logging mechanisms that capture different aspects of payment processing, from application-level transaction logs to HTTP communication traces. Digital payments integration writes detailed processing logs to the SAP Application Logging service for SAP BTP, accessible via Transaction SLG1 (Application Log Display).

SAP's primary package for digital payments is `FINS_AR_DIGITALPAYMENTS`. Figure 8.39 displays its related subpackages in SAP ERP.

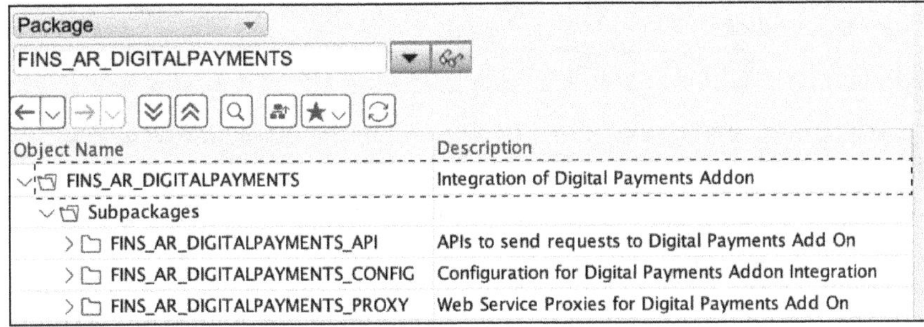


*Figure 8.39: Add-On Primary Package in SAP*

Figure 8.40 shows the ABAP classes included in the `FINS_AR_DIGITALPAYMENTS_API` subpackage, which are key for managing the API interactions with the add-on in SAP ERP.

| | |
|---|---|
| ∨ FINS_AR_DIGITALPAYMENTS_API | APIs to send requests to Digital Payments Add On |
| › Dictionary Objects | |
| ∨ Class Library | |
| ∨ Classes | |
| › CL_FAR_DP_ABAP_JSON | JSON mapper |
| › CL_FAR_DP_ADVICE | Utility Class for Advices |
| › CL_FAR_DP_GET_PSP_PAYMENT_METH | Get active PSPs and Payment Methods |
| › CL_FAR_DP_UTIL | DP Utility class |

*Figure 8.40: ABAP Classes in Add-On API Subpackage*

In SAP S/4HANA, SAP offers a significantly expanded selection of classes within the API subpackage (see Figure 8.41 for a partial list). These packages, their subpackages, and associated classes are available by default in SAP S/4HANA, eliminating the need to apply separate SAP Notes as was required in SAP ERP.

| | |
|---|---|
| FINS_AR_DIGITALPAYMENTS_API | APIs to send requests to Digital Payments Add On |
| › Package Interfaces | |
| › Dictionary Objects | |
| ∨ Class Library | |
| ∨ Classes | |
| › CL_FAR_DP_ABAP_JSON | Class to convert ABAP structure JSON format |
| › CL_FAR_DP_ADVICE | Utility Class for Credit Card Settlement |
| › CL_FAR_DP_AUTHORIZATIONS | Utility class for credit card authorizations |
| › CL_FAR_DP_AUTHORIZATIONS_V2 | Handles mass auth, preauth and renewal of expired auth |
| › CL_FAR_DP_AUTH_CANCEL | Utility class for cancelation of credit card authorization |
| › CL_FAR_DP_CARD_BASE | Base class for card operations |
| › CL_FAR_DP_CARD_BY_SESSIONID | Retrieve Payment Card data by Session ID |
| › CL_FAR_DP_CARD_BY_TOKEN | Retrieve Payment Card data by Token |
| › CL_FAR_DP_CARD_DELETE | Delete Payment Card by Token |

*Figure 8.41: SAP S/4HANA Add-On Classes*

When application logs don't provide sufficient detail, use ABAP debugging to trace the execution flow and inspect variable values during runtime. Set a debugging breakpoint in method `CREATE_BY_DESTINATION` of class `CL_HTTP_CLIENT` to intercept all HTTP calls to the digital payments add-on.

Follow these steps:

1. Run Transaction SE24, and enter "CL_HTTP_CLIENT" in the **Class/Interface** field (see Figure 8.42) to displace the corresponding class.
2. Navigate to the `CREATE_BY_DESTINATION` method (see Figure 8.43).
3. Set a breakpoint on the first executable line (see Figure 8.44)

Class Builder: Display Class CL_HTTP_CLIENT

Class/Interface CL_HTTP_CLIENT Implemented / Active

Properties Interfaces Friends Attributes Methods Events Types Aliases

Parameters Exceptions Sourcecode

Method Level Visibility Me..

*Figure 8.42: Class CL_HTTP_CLIENT*

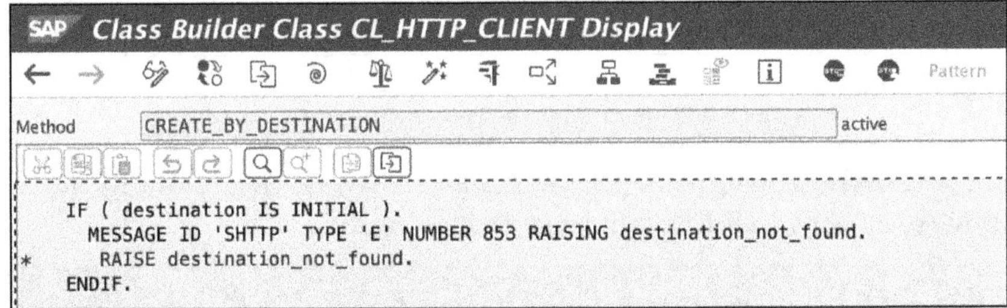


*Figure 8.43: CREATE_BY_DESTINATION Method*

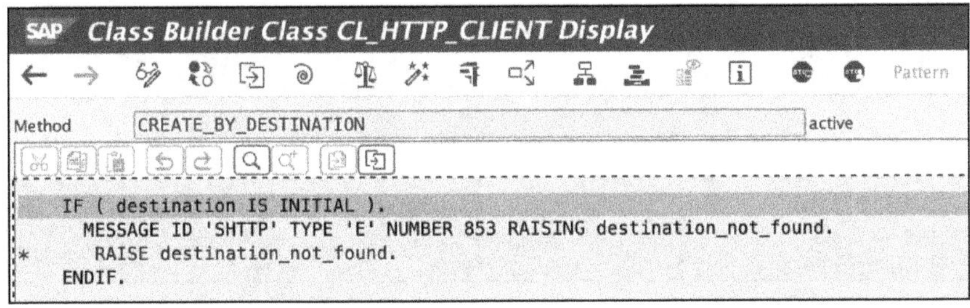


*Figure 8.44: Set Breakpoint Example*

4. Execute a transaction that triggers the add-on API call (payment card registration, sales order creation, settlement run, etc.). For this example, let's use report FAR_DP_UPDATE_PSP_PAYMENT_METH, which makes a digital payments API call to the `/core/v1/serviceconfiguration` API endpoint. When the breakpoint hits (see Figure 8.45), examine the call stack.

```
  IF ( destination IS INITIAL ).
    MESSAGE ID 'SHTTP' TYPE 'E' NUMBER 853 RAISING destination_not_found.
*     RAISE destination_not_found.
  ENDIF.

  DATA: scheme TYPE i VALUE schemetype_http.

*  HTTP Outbound for scenarios in Steampunk, where destinations from the destination ...
```

*Figure 8.45: Debug Breakpoint*

5. Navigate up the call stack to find the ABAP class that is responsible for making the API call, which is CL_FAR_DP_GET_PSP_PAYMENT_METH, and method GET_PSP_PAYMENT_METHOD (see Figure 8.46).

ABAP and Screen Stack

| Stac. | Stack. | S. Event Type | Event | Program | Navi. | Include | Line |
|---|---|---|---|---|---|---|---|
| → | 10 | METHOD | CREATE_BY_DESTINATION | CL_HTTP_CLIENT=============== | | CL_HTTP_CLIENT===============CM | 27 |
| | 9 | METHOD | INSTANTIATE_CLIENT | CL_FAR_DP_PROXY============= | | CL_FAR_DP_PROXY=============CM | 11 |
| | 8 | METHOD | GET_OAUTH_ACCESS_KEY | CL_FAR_DP_PROXY============= | | CL_FAR_DP_PROXY=============CM | 8 |
| | 7 | METHOD | SETUP_REST_CLIENT_FOR_REQ | CL_FAR_DP_PROXY============= | | CL_FAR_DP_PROXY=============CM | 56 |
| | 6 | METHOD | IF_FAR_DP_PROXY~SEND_GET_ | CL_FAR_DP_PROXY============= | | CL_FAR_DP_PROXY=============CM | 14 |
| | 5 | METHOD | GET_PSP_PAYMENT_METHOD | CL_FAR_DP_GET_PSP_PAYMENT_METHCP | | CL_FAR_DP_GET_PSP_PAYMENT_METHCM001 | 7 |
| | 4 | METHOD | GET_PSP_PAYMENT_METHOD | CL_FAR_DP_UPDATE_PSPPMT=======CP | | CL_FAR_DP_UPDATE_PSPPMT=======CM008 | 31 |
| | 3 | METHOD | IF_FAR_DP_UPDATE_PSPPMT~U | CL_FAR_DP_UPDATE_PSPPMT=======CP | | CL_FAR_DP_UPDATE_PSPPMT=======CM00H | 11 |

*Figure 8.46: ABAP Call Stack*

6. In GET_PSP_PAYMENT_METHOD, the lo_dp_proxy class (see Figure 8.47) is used for digital payments API calls. Double-click the **send_get_request** method.

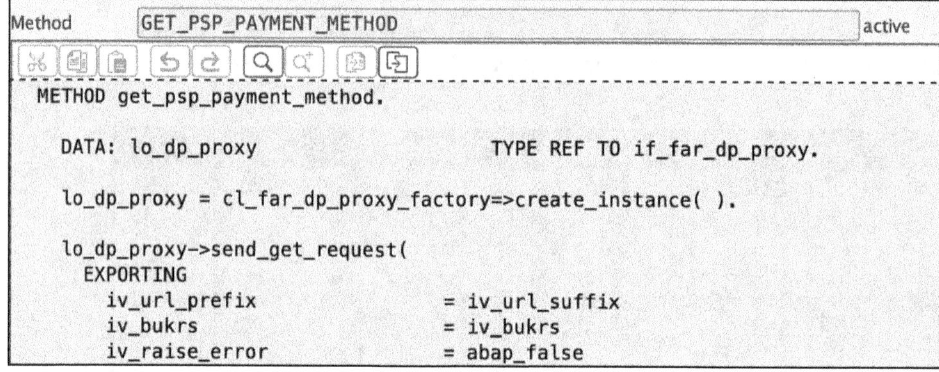

Method GET_PSP_PAYMENT_METHOD active

```
METHOD get_psp_payment_method.

  DATA: lo_dp_proxy                       TYPE REF TO if_far_dp_proxy.

  lo_dp_proxy = cl_far_dp_proxy_factory=>create_instance( ).

  lo_dp_proxy->send_get_request(
    EXPORTING
      iv_url_prefix                   = iv_url_suffix
      iv_bukrs                        = iv_bukrs
      iv_raise_error                  = abap_false
```

*Figure 8.47: lo_dp_proxy Class Used for API Calls*

7. Next, double-click the **setup_rest_client_for_request** method (see Figure 8.48).

Method IF_FAR_DP_PROXY~SEND_GET_REQUEST active

```
  TEST-SEAM create_client_get.
    setup_rest_client_for_request(
      EXPORTING
        iv_url_parameters             = lv_url_parameters
        iv_url_prefix                 = iv_url_prefix
        iv_bukrs                      = iv_bukrs
      IMPORTING
        eo_rest_client                = DATA(lo_rest_client)
      EXCEPTIONS
        http_client_failure           = 1
```

*Figure 8.48: Setup REST Client Method*

8. Inside the `setup_rest_client_for_request` method, you will see the `get_oauth_access_key` method (see Figure 8.49), which calls the add-on API to get the OAuth security token that uses the `DIGITALPAYMENTS_OAUTH` Transaction SM59 connection.

```
Method      SETUP_REST_CLIENT_FOR_REQUEST                                active

*   Get OAuth Access Key (JWT Token)
*    lv_oauth_access_key =
    get_oauth_access_key(
      EXPORTING
        iv_destination = lv_oauth_dest
        iv_url_suffix  = lv_url_suffix
      RECEIVING
        rv_token       = lv_oauth_access_key
      EXCEPTIONS
        http_client_failure     = 1
```

*Figure 8.49: Get OAuth Access Key Method*

9. The REST client object is created in the `instantiate_client` call (see Figure 8.50).

```
*   Get REST client for outbound call
    instantiate_client(
      EXPORTING
        iv_destination         = lv_destination
        iv_url_prefix          = lv_url_prefix
        iv_url_suffix          = iv_url_parameters
      IMPORTING
        eo_rest_client         = eo_rest_client
      EXCEPTIONS
        http_client_failure     = 1
        rfc_destination_missing = 2
        OTHERS                  = 3
```

*Figure 8.50: Instantiate Client*

10. At the end of the `setup_rest_client_for_request` method, the OAuth JWT bearer token is set in the **Authorization** header (see Figure 8.51).

```
    Set Header containing JWT-Token
    eo_rest_client->if_rest_client~set_request_header(
      EXPORTING
        iv_name  = 'Authorization' ##NO_TEXT
        iv_value = |Bearer { lv_oauth_access_key }|
    ).

  ENDMETHOD.
```

*Figure 8.51: Header Containing JWT Token*

11. The `setup_rest_client_for_request` method then returns the configured and authorized REST client object to the calling method, which in this case was `SEND_GET_REQUEST`, and this method makes the API call to the add-on (see Figure 8.52).

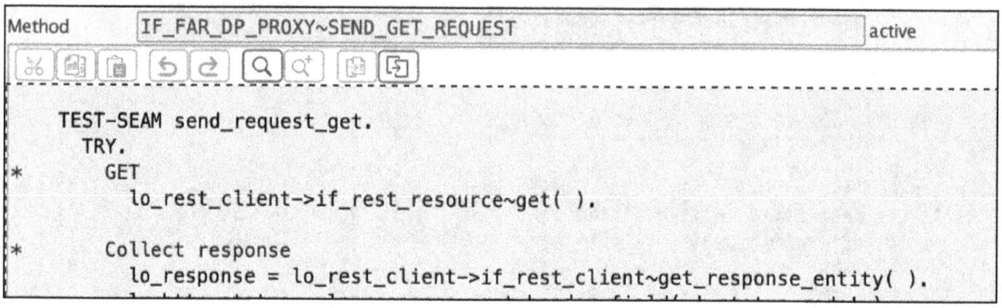

```
Method  IF_FAR_DP_PROXY~SEND_GET_REQUEST  active

    TEST-SEAM send_request_get.
      TRY.
*       GET
          lo_rest_client->if_rest_resource~get( ).

*       Collect response
          lo_response = lo_rest_client->if_rest_client~get_response_entity( ).
```

*Figure 8.52: Send Get Request Method*

## 8.7 Summary

Testing and troubleshooting the digital payments add-on integration requires a carefully structured and systematic approach that spans several testing phases, environments, and tools. The core objective is to ensure that both individual components and the broader end-to-end business processes function seamlessly and reliably, especially under production-like conditions. Establishing isolated test environments is essential to protect the production landscape from unintended impacts, while leveraging PSP-provided test cards allows teams to simulate a variety of authorization outcomes for thorough validation.

Key principles for successful testing include following a structured progression from unit testing through to user acceptance testing and ensuring that both successful processing and error scenarios are validated. It is equally important to document every test result and any issues encountered, creating a record for future reference. Systematic documentation aids in maintaining operational excellence and provides a valuable resource for ongoing support and troubleshooting.

Developing strong troubleshooting skills during the testing process is vital for future operational success. This involves understanding where critical processing logs are stored, how to debug ABAP code effectively, and which tables hold essential processing data. Effective digital payments testing provides confidence that payment processing will be stable and reliable in production, ultimately safeguarding revenue, ensuring customer satisfaction, and upholding compliance with financial and security regulations.

# Chapter 9
# Go-Live and Administration

*With the architecture detailed, configurations complete, and integrations built in the preceding chapters, the next critical step is transitioning your digital payments add-on implementation from project to production without disrupting revenue-impacting payment operations. This chapter provides the go-live playbook and ongoing administration guidance you need to launch successfully and keep the system running reliably.*

The go-live phase marks the completion of your SAP digital payments add-on project, when months of planning, setup, design, configuration, build, and testing finally result in launching the add-on for real use. Unlike standard SAP rollouts that only affect internal processes, a payment system go-live has an immediate impact on customers, revenue tracking, and financial compliance. If the payment system fails at launch, then sales can stop, customer transactions may be blocked, and the company's reputation could suffer. This makes it essential to carefully coordinate technical cutover steps, business continuity plans, and support preparedness.

This chapter covers the shift from implementation to operations, including cutover planning, legacy system migration, deployment, and post-go-live support for payment processing. It explains how to minimize business disruption during the go-live, run parallel legacy gateways, perform production cutover with rollback strategies, provide rapid issue resolution, monitor operational health, and improve long-term system performance.

## 9.1 Go-Live Planning and Preparation

Go-live planning begins during project initiation, not weeks before deployment. A successful digital payments cutover requires coordinated preparation across the technical teams deploying SAP configurations and RFC connections, business teams training users and updating procedures, integration teams validating e-commerce and portal connections, and support teams establishing help desk and escalation processes. The planning deliverable is a comprehensive cutover runbook documenting every activity, dependency, owner, and timing required for production deployment.

The cutover strategy defines the overall approach for transitioning from current state payment processing (legacy payment service provider [PSP] or manual operations) to digital payments add-on integration. Your strategy must address business-specific constraints including payment processing downtime tolerance (can you stop accepting payments for hours or days?), parallel operations requirements (run both legacy and new systems simultaneously), go-live scope (all company codes at once or phased rollout), and integration dependencies (e-commerce platforms, customer portals, and accounting systems).

### 9.1.1 Deployment Approaches

The first approach we'll discuss is a *big bang cutover*: On the go-live date, all company codes, payment types, and payment processing channels are transitioned at once to the digital payments add-on system (see Table 9.1). This means that rather than phasing the rollout, every part of the organization—including each distinct company code, every supported payment method, and all channels through which transactions occur (including e-commerce platforms, point-of-sale systems, and customer portals)—will begin using the new add-on integration simultaneously. This switch requires detailed planning and coordination between technical, business, integration, and support teams. This approach is designed to minimize the duration of the transition period, create consistency in the way payments are processed across the business, and simplify user training and change management efforts by having everyone move to the new system at the same time. However, this also means that any issues encountered during the cutover could have an immediate and widespread impact on payment operations and customer experience, making robust testing and contingency planning critical prior to launch.

| Approach | Advantages | Disadvantages | Recommended For |
|---|---|---|---|
| All company codes, payment types, and processing channels switch to the digital payments add-on simultaneously on the go-live date | Single cutover event, reducing project duration; uniform processing across organization; simplified training with one system | Higher risk, with no fallback to legacy processing; complex cutover requiring extensive coordination; significant business impact if issues occur | Smaller organizations with very few business units, limited payment volume enabling quick validation, or strong confidence from thorough testing |

*Table 9.1: Big Bang Cutover*

In a *phased rollout* approach, company codes or individual business units transition to the digital payments add-on in a controlled, step-by-step manner, with each group migrating over a period of weeks or even months. Rather than switching the entire organization at once, this method allows for a gradual adoption, enabling teams to validate processes, address potential issues, and adjust as needed during each phase. This sequential migration helps minimize operational risks by limiting the scope of changes at any given time and providing opportunities to learn from earlier deployments, thereby improving the success of subsequent rollouts and reducing the likelihood of widespread disruptions (see Table 9.2).

| Approach | Advantages | Disadvantages | Recommended For |
|---|---|---|---|
| Company codes or business units migrate to digital payments add-on sequentially over weeks or months | Reduced risk per deployment phase; ability to learn from early phases, improving later deployments; smaller coordination scope per phase | Extended project timeline; complexity involved in maintaining parallel systems; potential configuration drift between phases | Large organizations with complex e-commerce and back office operations, high transaction volumes requiring cautious validation, or first-time cloud payment processing implementations |

*Table 9.2: Phased Rollout*

### 9.1.2 Go-Live Readiness Assessment

A readiness assessment evaluates whether the implementation is prepared for production deployment using objective criteria across multiple dimensions. Establish go/no-go checkpoints at which a steering committee reviews the readiness status and either formally approves proceeding to the next phase or delays cutover until issues resolve. This section walks you through the key elements of readiness assessment, starting with the readiness categories that define objective validation criteria across application, configuration, integration, data, business, and technical dimensions.

From there, it covers the go/no-go decision framework for formally approving deployment, the development of a detailed cutover runbook with task-by-task instructions, risk-assessment and -mitigation strategies for common go-live failures, the communication plan for keeping internal teams, customers, and PSP partners informed, and rollback planning to define clear criteria and procedures for reverting to the previous state if critical issues arise.

#### Readiness Categories

This section defines assessment categories with specific validation criteria for digital payments implementations. Table 9.3 maps each readiness category to its specific validation criteria, the evidence required to confirm completion, and the team responsible for sign-off, giving your steering committee a clear checklist for evaluating go-live preparedness.

| Readiness Category | Validation Criteria | Evidence Required | Owner |
|---|---|---|---|
| Application readiness | ▪ All SAP Notes implemented<br>▪ All custom objects activated<br>▪ RFC destinations created and tested<br>▪ Report FAR_DP_UPDATE_PSP_PAYMENT_METH executes successfully | ▪ Transport logs showing successful imports, Transaction SM59 test results<br>▪ Transaction SE80 package review showing active objects<br>▪ FAR_DP_UPDATE_PSP_PAYMENT_METH log | ▪ Basis team<br>▪ ABAP developers |
| Configuration readiness | ▪ Payment card types defined<br>▪ General ledger accounts created and configured<br>▪ Virtual bank setup complete<br>▪ Account symbols assigned<br>▪ Posting rules configured<br>▪ Electronic bank statement settings | ▪ Master configuration document repository updated with latest configuration steps and screenshots<br>▪ Transaction FS00 account master data<br>▪ Transaction SPRO config screenshots<br>▪ Test transaction results | ▪ Order-to-cash configuration lead<br>▪ Finance configuration lead |
| Integration readiness | ▪ E-commerce test orders successful<br>▪ Payment portal integration tested<br>▪ Business partner master data sync validated<br>▪ Accounting document posting verified | ▪ Test execution logs<br>▪ Integration test results<br>▪ Sample transaction screenshots | ▪ Integration team |

*Table 9.3: Readiness Assessment*

| Readiness Category | Validation Criteria | Evidence Required | Owner |
|---|---|---|---|
| Data readiness | ■ Customer business partner records migrated with cards/tokens<br>■ Transaction data in table FPLTC/BSEGC for validation<br>■ Table FAR_DP_T100_MAP populated<br>■ Production PSP merchant accounts configured | ■ Table SE16N queries showing record counts<br>■ Data migration validation reports | ■ Data migration team |
| Business readiness | ■ User training completed with sign-off<br>■ Process documentation finalized<br>■ Help desk procedures established<br>■ Business stakeholder user acceptance testing (UAT) approval | ■ Training records<br>■ UAT sign-off documents<br>■ Process runbooks | ■ Change management |
| Technical readiness | ■ Production RFC destinations configured<br>■ SSL certificates installed (DigiCert)<br>■ Network firewall rules active<br>■ SAP BTP production tenant provisioned<br>■ PSP production merchant accounts active | ■ Transaction SM59 connection tests<br>■ Transaction STRUST certificate validation<br>■ Network connectivity tests<br>■ SAP BTP tenant confirmation | ■ Basis team, network team |

*Table 9.3: Readiness Assessment (Cont.)*

### Go/No-Go Decision Framework

The go/no-go decision requires participation from the steering committee, including the project sponsor, IT director, finance director, change management lead, and technical leads for SAP, SAP BTP, and the integration. Document decisions formally with signatures and the reasoning if go-live is postponed. Key criteria to be documented for the go/no-go decision include the following:

- All green statuses on all readiness categories
- Cutover simulation completed with all tasks validated, identified issues resolved, and runbook updated with lessons learned
- Stakeholder approvals obtained, support teams confirmed available, and rollback plan documented and approved
- Smoke tests passed, test transactions successful (authorization, settlement, refund), no critical defects, and monitoring dashboards operational

### Cutover Runbook Development

The cutover runbook provides task-by-task instructions for production deployment executed during the cutover window (typically a weekend for minimal business disruption). Table 9.4 organizes the cutover runbook into logical task groups, from initial setup and preparation through security, data migration, configuration, environment validation, end-to-end testing, and final go-live release, providing a structured sequence that your team can follow step by step during the cutover window.

| Task Group | Cutover Runbook Instructions |
| --- | --- |
| Setup and preparation | ■ Freeze all transports to production (with change control board approval)<br>■ Backup production system (database, file systems); verify all team members are available and on standby<br>■ Establish war room (physical or virtual meeting space); activate support ticketing system for go-live<br>■ Send communications to business users about cutover |
| Security | ■ Import DigiCert certificate to Transaction STRUST in production<br>■ Create production RFC destinations (`DIGITALPAYMENTS`, `DIGITALPAYMENTS_OAUTH`)<br>■ Configure OAuth with production service key credentials; test RFC connections<br>■ Create user authorizations for payment processing |

*Table 9.4: Cutover Runbook Task Groups*

| Task Group | Cutover Runbook Instructions |
|---|---|
| Data migration | ■ Execute report FAR_DP_UPDATE_PSP_PAYMENT_METH in production (synchronizes PSP list)<br>■ Execute report Z_FILL_FAR_DP_T100_MAP to populate error-mapping table<br>■ Migrate customer payment card tokens from legacy (if applicable)<br>■ Validate data loaded correctly with row counts and sample record verification |
| Configuration | ■ Assign settlement and authorization function modules<br>■ Configure account symbols (e.g., DPEXTCLEAR, DPEXTIN, DPEXTOUT)<br>■ Assign general ledger accounts to account symbols<br>■ Create posting rules and external transaction types<br>■ Validate configuration |
| Environment validation | ■ Basis and security: Verify transports imported, all objects active, no dumps in Transaction ST22<br>■ SAP BTP admin: Confirm production tenant accessible, service key valid<br>■ SAP digital payments add-on: Test API calls via Postman, verify `serviceconfiguration` returns production PSPs<br>■ Stripe (or your PSP): Confirm merchant accounts active, test mode disabled, webhook endpoints configured |
| End-to-end validation | ■ Execute first test authorization (Transaction VA01) with production test card<br>■ Verify authorization succeeds and table `FPLTC` populates<br>■ Create delivery and billing for test order<br>■ Execute Transaction FCC1 settlement<br>■ Verify settlement succeeds with table `BSEGC` updated<br>■ Execute test refund via FIN_DP_REFUND_PC<br>■ Verify all accounting documents post correctly<br>■ Review application logs (Transaction SLG1), confirming no errors |
| Go-live release | ■ Send go-live confirmation to business users<br>■ Enable production URLs in e-commerce platforms<br>■ Activate batch jobs for Transactions FCC1, VF04, FAR_DP_UPDATE_PSP_PAYMENT_METH<br>■ Begin monitoring dashboards<br>■ Transition to hypercare support mode |

*Table 9.4: Cutover Runbook Task Groups (Cont.)*

Each task in the runbook must specify a task number and sequence, a task description with detailed steps, an assigned team and owner, an estimated duration, any dependencies (predecessor tasks that must complete first), the success criteria and validation method, and the rollback steps to take if the task fails.

#### Risk Assessment and Mitigation

During this part of the assessment, you will identify risks (see Table 9.5) that could cause cutover failure or production issues and develop mitigation strategies to address each risk.

| Risk | Mitigation |
|---|---|
| RFC destination configuration error in production | ■ Impact: Authorization and settlement completely blocked, no payment processing possible<br>■ Mitigation: Create RFC destinations in production few days before cutover, test using FAR_DP_UPDATE_PSP_PAYMENT_METH, have configuration screenshots from QA for reference, and independently verify credentials |
| OAuth service key expired or incorrect | ■ Impact: All API calls fail with authentication errors<br>■ Mitigation: Obtain production service key a week before cutover, verify expiration date, store securely in password vault, and test in production during dress rehearsal |
| Missing table configuration | ■ Impact: Settlement function not called or errors display as generic messages<br>■ Mitigation: Checklist of all tables requiring manual entries in production, prepopulate data in Excel for quick Transaction SM30 entry, and validate immediately after configuration |
| First production transaction fails | ■ Impact: Business cannot process payments on go-live day, leading to potential revenue loss<br>■ Mitigation: Execute few test transactions during cutover validation, keep test customer and materials for production smoke testing, and have rollback plan ready if validation fails |
| Batch jobs not scheduled | ■ Impact: Settlements don't execute automatically; manual processing required<br>■ Mitigation: Precreate all batch job definitions before cutover, schedule with appropriate variants, test execution in simulation, and validate first execution |

*Table 9.5: Risk Assessment and Mitigation*

Document all identified risks in the risk register with severity ratings (High/Medium/Low), assign owners responsible for mitigation, and track their status through the cutover period, updating the status as risks are resolved or new risks emerge.

### Communication Plan

Stakeholder communication (see Table 9.6) ensures that business users, customers, and external partners understand cutover timing, impacts, and support procedures.

| Category | Recommended Communication Approach |
|---|---|
| Internal communications | ■ Email the sales, customer service, and finance teams, announcing the go-live date, explaining payment system changes, providing a training schedule, and identifying support contacts.<br>■ Send a follow-up communication with the cutover window details (expected system availability, alternative procedures if cutover extends beyond window, and emergency contact information).<br>■ Send a final reminder with go-live confirmation, war room dial-in details for key personnel, and support hotline activation.<br>■ Send go-live day status updates during the cutover window, including notification of critical issues and rollback decisions, or go-live success confirmation when validation completes. |
| External communications | ■ To customers (if payment portal downtime is required): Send notification before the cutover, explaining the brief maintenance window for payment system upgrades, providing alternative payment options during downtime, and confirming the time at which normal operations are expected to resume.<br>■ To PSPs: Coordinate production merchant account activation timing, verify that webhook URLs point to production endpoints, and confirm that production API credentials have been provided. |

*Table 9.6: Communication Approach*

### Rollback Planning

Despite thorough testing, the production cutover may encounter issues that require rolling back to a previous state. Define clear rollback criteria and procedures before the cutover begins. Execute the rollback if any of these conditions occur:

- Critical payment processing failure: Cannot authorize or settle any transactions
- Data corruption detected: Accounting documents posting to wrong accounts; amounts incorrect
- Integration failure: E-commerce orders not posting to SAP or payments failing
- PSP production merchant account issues: Account suspended; credentials invalid
- Time window exceeded: Cutover tasks not completed within allocated window; start of business at risk

The authority to make a rollback decision rests with the project sponsor and IT director. The cutover lead can recommend a rollback but cannot execute it without approval. Document rollback decisions with timestamps, approvers, and specific triggers that caused the rollback for lessons learned analysis.

## 9.2 Migration from Legacy Systems

The migration strategy addresses parallel operations, in which both systems run simultaneously; customer payment data migration from legacy to digital payments tokens; legacy transaction completion for in-flight orders; and knowledge transfer from legacy system experts to the new operations team.

*Parallel operations* enable gradual migration, reducing risk by maintaining legacy payment processing as a fallback while the digital payments add-on proves its operational stability in production. During implementation, configure your system with conditional logic, routing payment requests based on company code. You can maintain a parameter table that lists company codes activated for digital payment processing. This table-driven approach enables activating additional company codes without code changes as you can simply add a company code to the control table and restart processing.

Migrating customer payment card data from legacy systems to digital payments tokens requires careful handling given PCI DSS compliance and data sensitivity constraints. Legacy card data scenarios are as follows:

- **Legacy system stores full card primary account numbers**
  This has the highest security risk. Primary account numbers (PANs) in legacy databases must be tokenized through the digital payments add-on before legacy decommissioning. This migration requires PCI DSS–compliant processes. Card data must remain encrypted during extraction, transmission, and tokenization. Consider engaging specialized payment migration services for this scenario.
- **Legacy system uses PSP tokens**
  Legacy systems integrated with Stripe, PayPal, or other PSPs may already store tokens rather than PANs. These tokens may be reusable within the digital payments add-on depending on the PSP and token type. Validate that tokens are still valid with the PSP, use the digital payments add-on token exchange API to convert legacy PSP tokens to digital payments tokens, store add-on tokens in the business partner masters, and maintain a mapping table linking legacy tokens to digital payments tokens for reference.
- **Legacy transaction completion**
  Handle in-flight transactions in the legacy system to prevent revenue loss or customer confusion.

Sales orders authorized in the legacy system before cutover must complete processing in legacy system through billing and settlement. Do not attempt to transfer partially processed orders to digital payments midstream. Actions to take are as follows:

- Identify all open sales orders with payment card authorizations two weeks before cutover. Prioritize delivery and billing for these orders before the cutover date where possible.
- Document any orders that cannot complete before cutover and extend authorization validity if needed. Process these orders through the legacy system in the parallel operations period.

Billing documents posted but not yet settled through the legacy PSP must complete settlement before legacy system deactivation. Actions to take are as follows:

- Execute a final legacy settlement run the day before cutover. Verify that all billable items are settled and document any settlement failures requiring manual intervention.
- Process failed settlements through the legacy PSP during parallel operations.

Customer refunds in process through the legacy system must complete to avoid customer dissatisfaction. Actions to take are as follows:

- Process all pending refunds through the legacy system before cutover. If refunds cannot complete before cutover, document and process them manually.
- Ensure refund accounting is accurate in SAP even if it is processed through a legacy PSP.

After the parallel operations period confirms the digital payments add-on's stability (typically after 30–90 days), decommission the legacy payment gateway infrastructure. Deactivation steps are as follows:

- Disable legacy PSP API connections and remove legacy function modules or code from production.
- Archive the legacy configuration for reference, update disaster recovery plans to remove legacy dependencies, and cancel legacy PSP contracts and subscriptions.

Monitor for any residual legacy references or dependencies, validate the cost savings from eliminating the legacy system, update operational procedures to remove legacy content, and conduct a lessons learned session to document the migration experience.

## 9.3 Cutover Procedures

Cutover execution transforms planning into action through disciplined execution of the runbook during the defined cutover window. Successful cutover requires strict time management with defined task durations and start times, clear communication protocols to update stakeholders on progress, real-time issue tracking and resolution, and decision-making to resolve blockers immediately without lengthy approval chains.

Execute cutover simulation in the QA environment two to four weeks before production cutover, testing all procedures, timing, and coordination. The simulation identifies issues with task sequencing and dependencies, duration estimates that need adjustment, unclear ownership or missing resources, and gaps in the runbook documentation.

Schedule the simulation to mirror the production cutover timing. Assemble a complete cutover team for executing the assigned tasks. Execute every task from the runbook, documenting each task's actual duration versus its estimated duration, issues encountered and resolutions applied, and tasks requiring additional detail in the runbook. Update the runbook with simulation learnings, adjust task durations based on actual performance, add missing tasks discovered during execution, clarify ambiguous instructions, and reschedule the simulation if significant issues are identified.

Industry best practices recommend two dress rehearsals. The first rehearsal exposes major gaps in planning and procedures. The second rehearsal validates the remediation effectiveness and builds team confidence for production execution.

The *war room* serves as the command center during cutover: coordinating activities, resolving issues, and making decisions. War room staffing includes the cutover lead (overall coordination, decision authority), a Basis representative (transports, system administration), an ABAP developer (code issues, debugging), a finance configuration expert (accounting, general ledger accounts), a sales configuration expert (sales orders, billing), an integration specialist (e-commerce, APIs, interfaces), an SAP BTP administrator (tenant management, service keys), and a PSP liaison (merchant accounts, production credentials).

## 9.4 Post Go-Live Support

Post go-live support, commonly called *hypercare*, represents an intensive support period immediately following production deployment in which dedicated teams provide elevated support that ensures system stability, rapid issue resolution, and user confidence. Hypercare typically lasts 4–12 weeks, depending on implementation complexity, organizational change magnitude, and system stabilization rate. For digital payments implementations, hypercare is particularly critical because payment-processing failures directly impact revenue, customer experience, and business operations, requiring immediate resolution. As the hypercare period progresses and the team resolves production issues, it is equally important to document the lessons learned and create operational runbooks so that the business-as-usual support team can maintain the system independently after the hypercare team transitions off.

### 9.4.1 Hypercare Phase Structure

Hypercare operates differently from standard SAP production support, offering intensified support coverage with 24/7 availability during the initial weeks, reduced response times with critical issues addressed immediately, enhanced monitoring with real-time dashboards and automated alerts, daily standup meetings reviewing system health and open issues, on-site or dedicated remote support teams rather than a shared service desk, and direct escalation paths that bypass the normal support tier structures.

Table 9.7 illustrates typical timelines for hypercare and a checklist of key tasks.

| Hypercare Timeline | Key Tasks |
|---|---|
| Weeks 1–2:<br>Critical period | ■ 24/7 support coverage with war room staffing; all support team members on call<br>■ Daily standup meetings morning and evening; executive-level daily status reports; all-hands response for critical issues<br>■ Monitor every transaction, authorization, settlement, and refund |
| Weeks 3–4:<br>Stabilization | ■ 24/7 coverage continues, but with a reduced team size and daily standup meetings<br>■ Issue volume decreases; patterns emerge in common issues<br>■ Knowledge base develops from issue resolutions; transition planning begins |
| Weeks 5–8:<br>Transition period | ■ Business hours support; reduced standup meetings; issues resolved as per documented SLAs; users gain confidence and self-sufficiency<br>■ Support team documents operational runbooks |
| Weeks 9–12:<br>Handover to operations | ■ Standard support hours; weekly status meetings<br>■ Formal handover to business-as-usual support team; hypercare team transitions to other projects<br>■ Lessons learned documentation; final performance metrics review |

*Table 9.7: Hypercare Timeline and Key Tasks*

Establish a ticketing system (ServiceNow, Jira, or SAP Solution Manager) for logging, tracking, and resolving hypercare issues, including a structured severity classification and response time commitments.

Table 9.8 defines several severity levels.

| Severity | Definition | Response Time and Resolution Target | Escalation |
|---|---|---|---|
| Critical (P1) | Payment processing completely blocked; revenue loss occurring; system down or unusable | 15 minutes, 2–4 hours | Immediate executive notification |

*Table 9.8: Severity Levels and Resolution Targets*

| Severity | Definition | Response Time and Resolution Target | Escalation |
|---|---|---|---|
| High (P2) | Significant functionality impaired; workaround exists but is inefficient; multiple users affected | 1 hour, 8–24 hours | Management notification if not resolved in four hours |
| Medium (P3) | Minor functionality issue; single user or limited scope; workaround available | 4 hours, 2–5 business days | Team lead notification |
| Low (P4) | Enhancement request; cosmetic issue; documentation gap | Next business day, as resources are available | No escalation |

*Table 9.8: Severity Levels and Resolution Targets (Cont.)*

### 9.4.2 Knowledge Transfer and Documentation

The hypercare team develops operational knowledge that must be documented and transferred to the business-as-usual support team, ensuring continuity when hypercare ends. You'll want to create runbooks for common operational tasks based on the hypercare experience (see Table 9.9).

| Runbook | Common Operational Tasks |
|---|---|
| Daily payment processing checklist | ▪ Review overnight Transaction FCC1 settlement batch job results<br>▪ Check application log for errors requiring action<br>▪ Validate that settlement success rate meets threshold (>99%)<br>▪ Review any declined authorizations for patterns<br>▪ Check clearing account balances (open items aging report)<br>▪ Monitor payment advice processing if enabled |
| Authorization failure troubleshooting | ▪ Identify failure symptom (card declined, technical error, timeout)<br>▪ Check application log for message number and details<br>▪ Verify RFC destinations accessible<br>▪ Test OAuth token retrieval manually<br>▪ Review PSP dashboard for corresponding transaction<br>▪ Escalate to appropriate team based on root cause |

*Table 9.9: Runbooks*

| Runbook | Common Operational Tasks |
|---|---|
| Settlement processing issues | ■ Verify items exist for settlement<br>■ Check Transaction FCC1 selection criteria are not too restrictive<br>■ Validate table `TCCAA` configuration is present<br>■ Test the settlement function module manually<br>■ Review application log for API call failures<br>■ Check network connectivity to digital payments add-on |
| Monthly reconciliation procedures | ■ Extract settled transactions<br>■ Compare to PSP settlement reports (download from PSP dashboard)<br>■ Identify discrepancies (settlements in SAP not in PSP or vice versa)<br>■ Investigate missing settlements using `DP_TRANS_ID` correlation<br>■ Generate reconciliation report for finance team<br>■ Document any manual adjustments required |

*Table 9.9: Runbooks (Cont.)*

You'll also want to document resolutions to all P1 and P2 issues encountered during hypercare, creating a searchable knowledge base for future reference. Each article should include a problem description and symptoms, root cause analysis, a step-by-step resolution procedure, prevention measures, and related issues or variations.

Finally, define objective criteria (see Table 9.10) for ending hypercare and transitioning to standard support, preventing a premature transition while the system remains unstable.

| Exit Criteria | Checklist |
|---|---|
| System stability | ■ Zero P1 critical issues in past two weeks; fewer than five P2 high issues per week<br>■ Authorization success rate >99%; settlement success rate >99.5%<br>■ System performance within benchmarks; no recurring technical issues requiring workarounds |
| User confidence | ■ User-reported issues declining week over week; users able to resolve common scenarios independently<br>■ Positive feedback from business stakeholders; training completion >95% for payment-processing users; no escalations from users bypassing the hypercare team |

*Table 9.10: Exit Criteria Checklist*

| Exit Criteria | Checklist |
|---|---|
| Operational readiness | ■ All operational runbooks documented and tested; business-as-usual support team trained on digital payments troubleshooting<br>■ Knowledge base contains resolutions for common issues; monitoring dashboards are operational and reviewed daily; batch jobs are running successfully without intervention |
| Financial validation | ■ First month-end close completed successfully; reconciliation reports are accurate<br>■ All settled transactions match PSP statements; refunds are processing correctly with proper accounting; the finance team approves the transition |

*Table 9.10: Exit Criteria Checklist (Cont.)*

Hypercare exit requires formal approval from business and IT leadership to confirm readiness for standard support model transition.

## 9.5 Operational Monitoring

Operational monitoring provides visibility into digital payment processing health enabling proactive issue identification before business impact occurs. Unlike reactive support responding to user-reported problems, monitoring detects payment processing anomalies automatically through application log analysis, batch job monitoring ensuring critical settlement and conciliation jobs complete successfully, and automated performance measurement triggering alerts when thresholds are exceeded.

### 9.5.1 Application Log Monitoring

Application logs record detailed payment-processing events, including API calls to the digital payments add-on, authorization and settlement results, error conditions and exceptions, and data transformation activities. Regular log review identifies issues that require attention before they escalate.

Execute Transaction SLG1 (Application Log Display) with these selection criteria: **Object**: `FIAR` (Financial Accounting Receivables); **Subobject**: `FIAR_DP_ADVICE` (for settlement and advice); and **Date**: Yesterday's date (review previous day's processing). Execute to display the log entries. Review the entries, looking for the following:

- Error messages (red traffic light): Require immediate investigation and resolution
- Warning messages (yellow): May indicate configuration issues or approaching problems
- Success messages (green): Validate normal processing, confirm expected volumes

For each error, double-click to view the details, note the message class and number, review the message text and parameters, and check the timestamp identifying when the issue occurred.

In addition, consider developing a custom program to scan application logs automatically and email alerts when errors are detected. The program can be designed to read the application logs for the previous 24 hours using function modules `BAL_DB_SEARCH` (search logs), `BAL_LOG_HDR_READ` (read log headers), and `BAL_LOG_MSG_READ` (read log messages); filter for error and warning messages; count error frequency by message number and check if the error count exceeds a set threshold (e.g., >10 authorization failures); and send an email alert to the support team with the summary statistics, most frequent error messages, and affected transaction details.

### 9.5.2 Batch Job Monitoring

Payment processing relies on scheduled batch jobs that execute settlement, reconciliation, and synchronization functions. Job monitoring ensures that critical jobs complete successfully within allocated windows. Table 9.11 describes critical batch jobs for monitoring.

| Job | Monitoring/Validation/Alerting |
|---|---|
| Transaction FCC1 settlement (daily) | ▪ Schedule: Nightly after final Transaction VF04 billing run<br>▪ Monitoring: Execute Transaction SM37 (Job Overview) the next morning, filter via program RFCCSSTT, and verify the status is **Finished** (green checkmark)<br>▪ Validation: Check spool output for the settlement log, verify the line count matches the expected transaction volume, and review error messages in the log<br>▪ Alert criteria: Job status **Cancelled** or **Aborted** |
| Transaction VF04 billing (hourly or daily) | ▪ Schedule: Per business billing frequency<br>▪ Monitoring: Transaction SM37 filtering via program SDBILLDL<br>▪ Validation: Verify billing documents created and check for blocked billing documents requiring attention<br>▪ Alert criteria: Job failure; billing error rate greater than 5% |
| FAR_DP_UPDATE_PSP_PAYMENT_METH (weekly) | ▪ Schedule: Sunday mornings or other low-activity periods<br>▪ Monitoring: Transaction SM37 plus Transaction SE38 execution log review<br>▪ Validation: Log shows PSP list retrieved successfully; PSP count matches digital payments add-on configuration<br>▪ Alert criteria: Job failure; error with connection to add-on |

*Table 9.11: Batch Jobs*

| Job | Monitoring/Validation/Alerting |
|---|---|
| RFDP_ADVICE_V2 (daily if payment advice enabled) | ■ Schedule: After Transaction FCC1 settlement completion<br>■ Monitoring: Transaction SM37 plus application log review<br>■ Validation: Advice items processed; clearing entries posted<br>■ Alert Criteria: Job failure |

*Table 9.11: Batch Jobs (Cont.)*

### 9.5.3 Automated Performance Monitoring

Consider developing custom reports to execute hourly via background job collecting KPI measurements from tables FPLTC and BSEGC and batch job logs. Store the KPI history to enable trend analysis. Email alerts to the operations team when KPIs cross warning or critical thresholds.

Beyond payment-specific monitoring, standard SAP system health checks ensure that the infrastructure supporting payment processing remains stable (see Table 9.12).

| System Health Checks | Key Tasks |
|---|---|
| Transaction ST22 (ABAP Dump Analysis) | ■ Review dumps from past 24 hours; filter for programs related to payment processing (RFCCSSTT, Z_DP_*, RV*)<br>■ Investigate any dumps related to digital payments functions; document root causes and preventive measures |
| Transaction SM21 (System Log) | ■ Review system log for errors or warnings; filter for message IDs related to RFC, HTTPS, and database<br>■ Check for RFC destination communication failures; validate that there are no security audit messages related to payment processing |
| Transaction SM50 (Work Process Overview) | ■ Monitor for long-running dialog processes during payment processing; verify background work processes are available for batch jobs<br>■ Check for processes in PRIV mode, indicating exclusive lock issues; ensure that there are no payment-related deadlocks |
| Transaction DB02 (Database Performance) | ■ Verify that database tablespace usage is not approaching limits; check that buffer hit ratios are acceptable (>95%); review missing indexes on payment tables (FPLTC, BSEGC, BSIS, BSAS)<br>■ Validate database statistics are current for payment tables |

*Table 9.12: System Health Checks*

PSP dashboards (Stripe Dashboard, PayPal Manager, etc.) provide independent validation of the payment processing visible to the PSP, showing authorizations, captures, refunds, and settlements from their perspective.

We recommend a daily PSP dashboard review, in which you log into the PSP production dashboard (e.g., the Stripe dashboard at *https://dashboard.stripe.com*) and verify the following:

- Transaction volume: Matches SAP transaction counts (i.e., table `FPLTC` authorization count = PSP authorization count)
- Success rates: PSP approval rates match SAP success rates
- Disputes/chargebacks: Check for new disputes requiring investigation and resolution
- Failed charges: Review PSP failure details, supplementing the SAP error logs
- Webhook status: Verify webhooks are delivering events successfully (if configured)
- Account balance: Confirm settlement deposits are occurring per agreed-upon schedule

Discrepancies between SAP and PSP counts indicate data synchronization issues that require investigation.

Finally, monthly reconciliation ensures that SAP accounting records match PSP settlement statements and bank deposits, maintaining financial accuracy. Extract settled transactions for the month, then download the PSP settlement report from the PSP dashboard for the month. Compare the SAP settlement total to the PSP capture total and investigate any variance.

## 9.6 Maintenance and Optimization

Ongoing maintenance ensures that the digital payments integration continues operating reliably as business volumes grow, new requirements emerge, and underlying systems evolve. Optimization activities improve performance, reduce costs, and enhance user experience based on operational data and changing business needs. Establish routine maintenance schedules to prevent degradation and ensure long-term system health. We recommend focusing on the following:

- Tables `FPLTC` and `BSEGC` grow with every payment transaction. Consider archiving settled transactions older than two years (adjust as needed to meet compliance requirements). Consider using SAP's archiving framework (Transaction SARA) to create archiving objects for payment tables, and validate that archived data is accessible for audit via archive retrieval.
- Database indexes on payment tables (`BSEGC`, `BSIS`, `BSAS`) fragment over time. Execute index reorganization to improve query performance, schedule reorganization for extension indexes, and monitor index effectiveness after reorganization.
- SSL certificates expire, thus requiring renewal. Review the DigiCert Global Root G2 certificate expiration once per quarter. If expiration will occur within 90 days, download a renewed certificate and import it.
- Review users with payment-processing authorizations, remove access for users no longer requiring payment functions, validate that segregation of duties is maintained, and ensure that the principle of least privilege is followed.

- Check SAP Support Portal for updates to digital payments add-on pilot SAP Notes. Look for new versions or related SAP Notes, evaluate updated SAP Notes for bug fixes or enhancements, test SAP Note updates in the development system before production deployment, and plan a maintenance window for applying relevant updates.
- PCI DSS audits may request payment-processing evidence. Prepare documentation showing tokenization implementation to prevent PAN storage, application log samples to demonstrate an audit trail, and configuration screenshots of security controls. Review all logged payment transactions for the past year, and provide reconciliation reports that matching SAP to PSP settlements.
- SAP releases digital payments add-on updates on SAP BTP. Review release notes for new features, bug fixes, and breaking changes. Coordinate with your SAP BTP admin for production tenant upgrades. Schedule a test of the RFC connectivity after the upgrade (as API contracts may change), and update custom code if API changes require modifications.

## 9.7 Summary

Successful go-live and sustainable operations require disciplined execution of cutover procedures, postdeployment support during hypercare, and operational monitoring to ensure long-term system health. The transition from an implementation project to business-as-usual operations represents a critical phase in which technical capabilities translate into business value through reliable payment processing.

Key go-live success factors include comprehensive cutover planning with documented runbook and rehearsal validation, clear readiness criteria, a go/no-go decision framework, structured hypercare support with 24/7 coverage and defined exit criteria, a migration strategy addressing legacy systems and parallel operations, and operational monitoring providing visibility into payment-processing health. Organizations investing adequate preparation time and support resources during go-live achieve faster stabilization, higher user confidence, and fewer production issues than implementations rushing deployment without proper planning.

Operational excellence in digital payments administration combines proactive monitoring to detect issues before business impact, regular maintenance to prevent system degradation, continuous optimization to improve performance and efficiency, and capacity planning to ensure the infrastructure scales with business growth. The operational procedures, runbooks, and monitoring approaches established during hypercare become the foundation for long-term operational success.

As payment processing volumes grow and business requirements evolve, the digital payments add-on provides flexibility for expansion, including activating additional PSPs for new markets or payment methods, enabling new company codes or business units, integrating additional e-commerce platforms or customer portals, implementing payment advice processing for enhanced reconciliation, and adopting new digital payments add-on features as SAP releases updates.

Chapter 10
# Processing Digital Payments

*A successful payment modernization initiative depends not only on sound technical implementation but also on the ability of business users to operate the system effectively in their daily work. While previous chapters covered architecture, configuration, PSP integration, and SAP activation, this chapter shifts the focus to the people who will use the system every day: customer service representatives processing orders, finance teams managing settlements and refunds, and business stakeholders driving adoption. This practical, operations-oriented perspective is essential as even the best-configured digital payments environment will fail to deliver value if the teams relying on it cannot execute payment transactions confidently and resolve issues independently.*

This chapter addresses how business users interact with the SAP digital payments add-on through SAP transactions for daily payment operations. While previous chapters focused on technical implementation, configuration, and system administration, this chapter provides practical guidance for customer service representatives processing orders with payment cards, finance teams handling refunds and reconciliation, and business stakeholders ensuring successful user adoption through training and change management. The content serves as an operational reference for business teams and training foundation for end user enablement programs.

Unlike technical implementation knowledge requiring ABAP development or Basis expertise, the procedures in this chapter are intended for business users who need to execute payment transactions reliably without understanding the underlying technical architecture. The chapter emphasizes step-by-step transaction execution, error interpretation and resolution, and when to escalate issues to technical teams versus resolving independently. This practical focus ensures business teams can operate payment processing confidently, maintaining business continuity even when technical support resources are limited.

## 10.1 Customer Service Payment Operations

Customer service teams interact with payment processing primarily during sales order creation when payment cards are selected and authorized. Representatives must understand how to process orders with payment cards stored in customer master data, interpret authorization results and error messages, handle authorization failures requiring alternative payment methods, and respond to customer inquiries about payment status. Effective customer service payment operations balance providing a smooth customer experience with maintaining security and compliance discipline to protect sensitive payment information.

### 10.1.1 Processing Orders with Payment Cards

Sales order processing with payment cards follows the standard transaction flows, with additional payment-specific steps in the customer master data and sales order header enabling payment card selection and authorization validation.

#### Scenario: Assisting a Client by Phone

In this scenario, you are a customer service representative assisting a client over the phone who is prepared to place an order using a payment card. Launch the Manage Customer Master Data app from the SAP Fiori launchpad (see Figure 10.1).

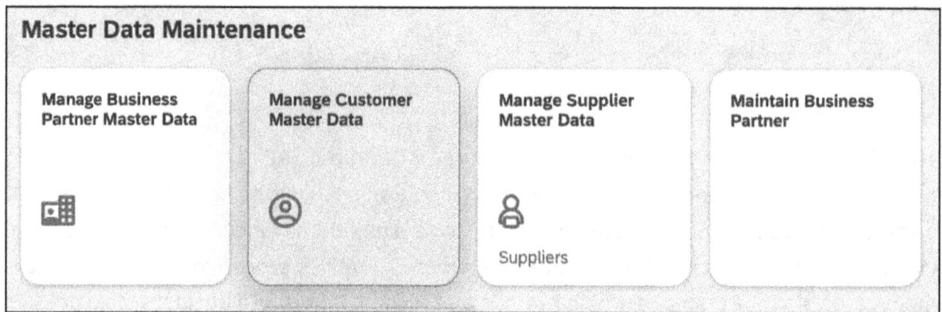


*Figure 10.1: Manage Customer Master Data*

Enter the customer's details into the search parameters (see Figure 10.2) and verify whether a customer master record already exists for the person currently on the phone with you.

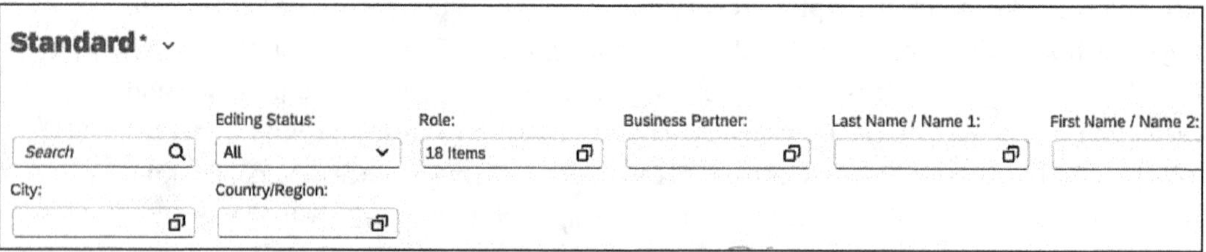


*Figure 10.2: Customer Master Search Parameters*

If the customer exists, verify the sales area for the order (**Sales Organization** field) and confirm that the **Payment Guarantee Procedure** field has the value **0002**, which means it is enabled for payment cards (see Figure 10.3).

Verify whether the customer's card information is already stored in the system. If it is not, switch to edit mode and select the **Create** option (see Figure 10.4) to open the secure iFrame for entering payment card details (**Cardholder Name, Card Number, Expiration Date**, and **CVC**). Once completed, store the digital payments add-on token in the customer's profile.

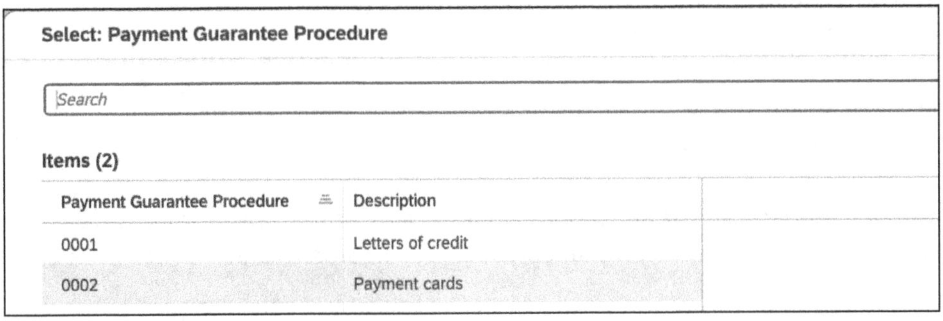

**Select: Payment Guarantee Procedure**

Search

**Items (2)**

| Payment Guarantee Procedure | Description |
|---|---|
| 0001 | Letters of credit |
| 0002 | Payment cards |

*Figure 10.3: Payment Guarantee Procedure*

**Payment Cards**

**Payment Cards** Standard ˅ Create

| Card ID | Payment Card Type | Card Number | Valid To | Payment Card Description |
|---|---|---|---|---|
| | | | No items available. | |

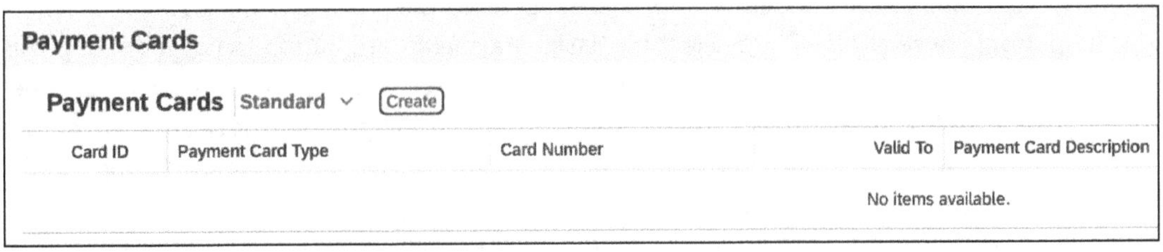

*Figure 10.4: Register New Payment Card*

After successfully saving the card information, proceed to the Manage Sales Orders app. Complete the **Order Type** field, sales area details (**Sales Organization, Distribution Channel, Division** fields), **Customer Reference** field, and the order line items (**Material, Order Quantity** and **Sales Unit** fields) that the customer has requested. Then, access the **Electronic Payments** section within the order header (see Figure 10.5). The **Total** field will reflect the order price once all mandatory parameters have been successfully entered in accordance with your business requirements.

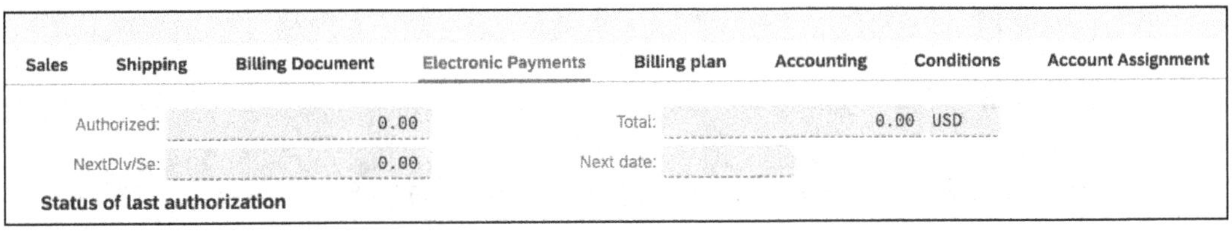

Sales | Shipping | Billing Document | Electronic Payments | Billing plan | Accounting | Conditions | Account Assignment

Authorized: 0.00 Total: 0.00 USD

NextDlv/Se: 0.00 Next date:

**Status of last authorization**

*Figure 10.5: Order Header: Electronic Payments Tab*

Press F4 for search help in the **Card Number** field (located next to the **Type** field), select the registered card, and assign it in the first row of the **Electronic Payments** table (see Figure 10.6).

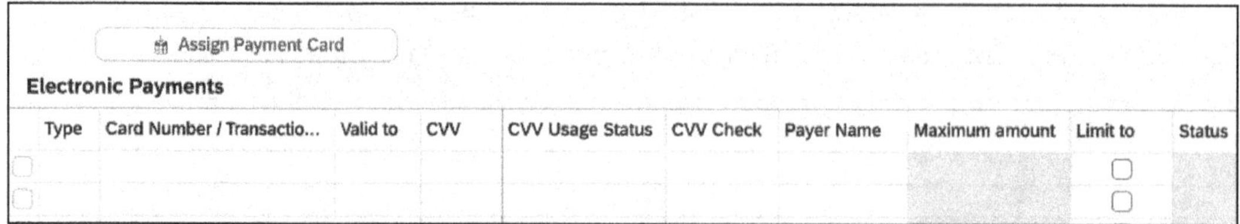


*Figure 10.6: Payment Card Selection*

Submit the order to initiate SAP system authorization, with the results recorded in the electronic payments section. Provide the customer with the order number for future reference.

### Scenario: Creating a Sales Order with a One-Time Payment Card

In this situation, the customer chooses not to store their card information in their profile; they want to use the card only for this transaction. The process is similar, but in the **Electronic Payments** tab, you select the **Assign Payment Card** button (see Figure 10.7). This action opens a secure iFrame for one-time registration with the payment service provider (PSP), allowing you to get a token and use that add-on token during the sales order process. After you enter the card details and submit to the PSP successfully, press **Confirm** in the dialog window to save the order.

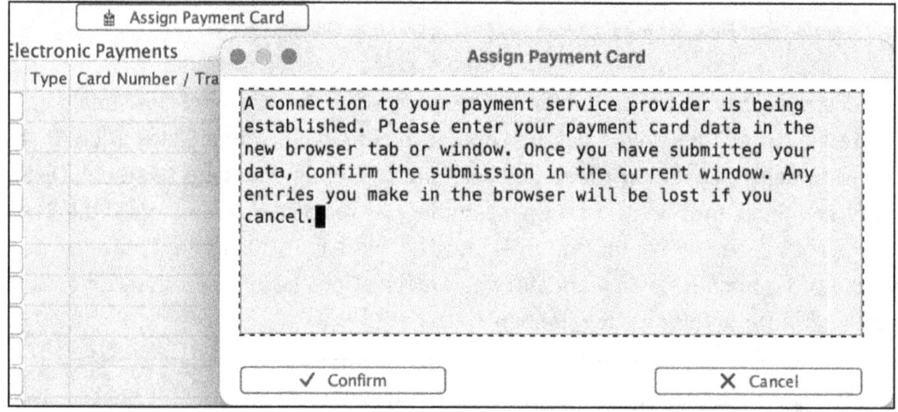


*Figure 10.7: Assign Payment Card Options*

### Scenario: Creating Sales Order with Stored Payment Card

If the customer payment card already exists, review the selected card and ensure the following is true:

- Expiration date hasn't passed (system issues a warning for expired cards but may allow saving)
- Card type matches customer's intended payment method (Visa vs. Mastercard vs. Amex)
- No blocking reason is displayed (blocked cards show a lock icon or blocking text)

After selecting the payment card and completing all order data entry, save the sales order. During save processing, the system executes automatic authorization, and the system displays one of the following outcomes:

- **Successful authorization**
  - Message **"Sales order [number] has been saved"** displays in the status bar
  - System assigns sales order number
  - Payment card plan shows green traffic light icon in **Authorization Status** column
  - Authorized amount displays matching order total
  - **Authorization** section shows success indicator
- **Authorization Failure**
  - Error message displays with specific decline reason
  - Payment card plan shows red traffic light in **Authorization Status**
  - Sales order saves, but with blocks preventing delivery
  - Overall status shows **Not approved** (Status B)

After successful save, verify the authorization completed correctly by navigating to **Header** and the **Electronic Payments** tab in the saved sales order. Click the **Authorization Items** section to expand its details. Review the authorization line item, and note the following:

- **Authorization Amount**: Should match order total
- **Authorization Number**: Digital payments authorization ID populated
- **Authorization Date/Time**: Current date and time
- **Authorization Result**: Success indicator or code
- **PSP Reference**: External authorization reference if provided

If the authorization amount doesn't match the order total, contact your supervisor or IT support. A discrepancy indicates a configuration issue or authorization calculation error that requires technical investigation.

### Scenario: IVR Card Tokenization

Some companies opt for solutions like CardEasy alongside the digital payments add-on to securely process card transactions. This approach reduces the risk of card fraud, minimizes manual entry mistakes during payment processing for customers, accelerates payment procedures, and boosts customer trust in the security of their card information. For further details, refer to the *Digital Payments Add-On Administration Guide* at *http://s-prs.co/v629405*.

At a basic level, CardEasy operates by offering a virtual terminal in which card details are submitted to the merchant's PSP, which then returns a PSP token to the add-on. This token is subsequently used by the add-on to finalize the transaction directly with the merchant's PSP.

Integrating CardEasy into your organization's telephony voice call system is a necessary requirement. Integration details for CardEasy (now part of Eckoh) are available at *https://www.eckoh.com/integrations*.

### 10.1.2 Handling Payment Failures

Payment authorization failures occur for business reasons (customer's card declined by issuing bank) or technical reasons (system communication errors). Customer service representatives must distinguish between these scenarios, providing appropriate customer communication and taking correct resolution actions.

If you are running SAP ERP in your landscape, consider using Transaction VCC1 (Payment Cards: Worklist) to review and manage payment card processing. Transaction VCC1 (Program RV21A001) is the standard SAP tool specifically designed for systematically reviewing and reprocessing sales documents that are on credit hold due to failed or insufficient payment card authorizations. This transaction should be used instead of generic credit release transactions (like Transaction VKM1) because the orders lack a valid authorization required for further processing. Table 10.1 illustrates key features that customer service representatives managing payment card processing can be trained on to handle payment failures.

| Feature | Description | Agent Action |
|---|---|---|
| Identification of issues | Generates a work list of sales orders and deliveries blocked due to payment card authorization failures, displaying the reason for the block. | Run the report regularly to identify problem orders. |
| Targeted selection | Allows agents to filter the list using specific criteria to quickly find relevant orders. | Use specific filters to address customer inquiries efficiently. |
| Reauthorization trigger | The primary function is to retrigger the authorization process with the payment processor. | Select the blocked order(s) and trigger reauthorization from the toolbar. |
| Integration with Transaction VA02 | Provides direct navigation to the Change Sales Order transaction (Transaction VA02) for manual review or updates if needed. | Double-click an order or select **Process** to view/edit details if reauthorization fails or requires manual intervention. |
| Processing flexibility | Agents can select multiple orders from the list and attempt reauthorization for all of them simultaneously. | Use checkboxes to manage multiple blocked orders at once. |

*Table 10.1: Transaction VCC1 Key Features*

Customer service agents can use a wide range of selection criteria to review the sales documents and deliveries that the agent is responsible for via the **Sales Organization** and **Created By** filters (see Figure 10.8).

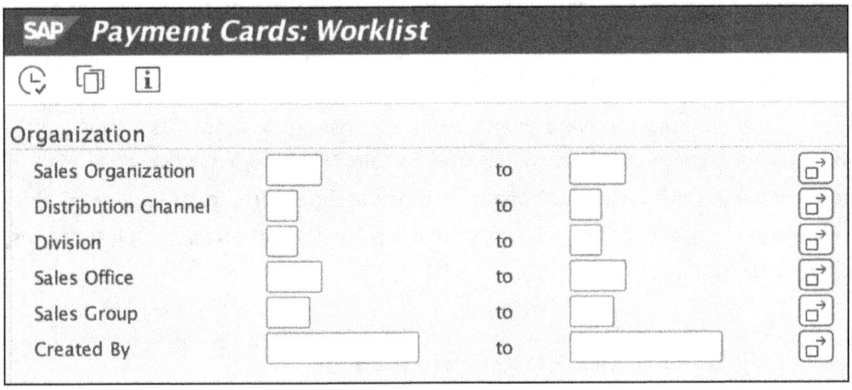

*Figure 10.8: Selection Criteria: Payment Cards Worklist*

In addition, you can specifically filter by specific **Sales Documents, Overall processing status**, and other options, as shown in Figure 10.9.

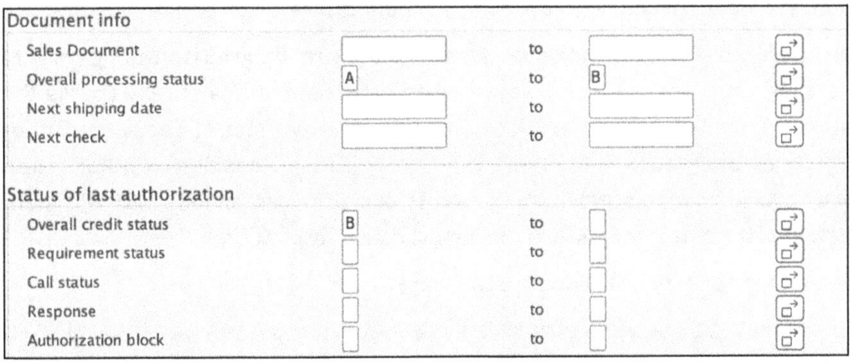

*Figure 10.9: Additional Filter Criteria*

Once the appropriate selection criteria have been entered, generate the report to display a list of sales orders and/or deliveries with pending or failed authorizations. Select the relevant order, review and update payment card details as necessary, and then click the **Forward for authorization** button located on the application toolbar (see Figure 10.10).

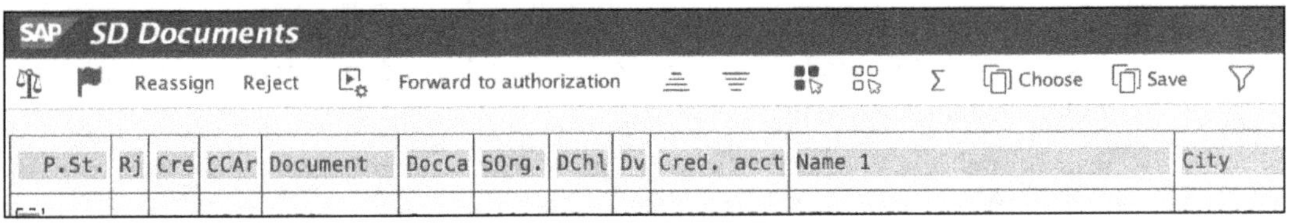

*Figure 10.10: Report Output*

The selected items will be grayed out, and the system will attempt a new authorization. Click the **Save** button to finalize the reauthorization request. The system will return a log or list with the updated authorization status. A successful authorization (typically indicated by a green light) means the block is removed, and the order can proceed for delivery and billing. When authorization failed previously, the sales order saved but the system set up blocks to prevent delivery creation and goods issue.

For technical errors, do not modify payment card data; instead, wait for IT support confirmation that the technical issue is resolved, then navigate to the payment card plan in Transaction VA02, and remove the authorization block manually (if the option is available) or simply save order without changes. The system will retry the authorization. If it succeeds, the block is removed automatically.

### 10.1.3 Managing Authorization Expiration in SAP ERP

Customer service agents handling payment processing need training on dealing with expired payment card authorizations. The options available vary depending on whether your system is SAP ERP or SAP S/4HANA.

SAP provides manual and automated processes within SAP ERP to handle expiring payment card authorizations, which typically have a validity period of seven days.

Agents can be proactive in identifying orders before settlement issues arise. During delivery creation, system warnings will be displayed when the system automatically checks the authorization age. If an authorization will expire soon, a warning message appears: **Authorization for sales order [Order Number] expires on [Date]. Consider reauthorization before delivery.** For periodic monitoring, agents can use Transaction VCC1 (Program RV21A001) to find the oldest orders with the largest risks of expiration.

Manual reauthorization is typically done in two ways:

- Via Transaction VCC1 (Mass Processing): Use this transaction to see a worklist of blocked orders. Select the order(s) and click **Forward for authorization**, then save the transaction.
- Via Transaction VA02 (Single Order): If the **Authorization** section shows a yellow (approaching expiration) or red (expired) traffic light warning, make a minor change (e.g., add a note) or update a new payment card, then click **Save** to trigger a new authorization call automatically.

Automation is key to preventing manual workload and delivery delays, as follows:

- Automated background job: Schedule program RV21A001 as a regular background job (e.g., nightly). This program automatically identifies and reauthorizes expiring orders without manual intervention.
- Expedite delivery: Encourage the warehouse/shipping team to process deliveries quickly to settle transactions within the initial authorization validity period.

### 10.1.4 Managing Authorization Expiration in SAP S/4HANA

SAP S/4HANA leverages the same core logic as SAP ERP, but it offers a better user experience and tighter integration with modern tools like the SAP digital payments add-on.

SAP S/4HANA offers modern, SAP Fiori–based identification alongside classic methods, as follows:

- Resolve Payment Card Issues app: This is the primary tool for customer service agents. It provides a visual, role-based dashboard to filter, monitor, and manage blocked sales orders and deliveries efficiently. The SAP Fiori app uses the same underlying logic as Transaction VCC1.
- Display Payment Card Data app: This SAP Fiori app also provides multiple features to review and manage payment card processing effectively.

Reauthorization is streamlined through the SAP Fiori interface, as follows:

- Resolve Payment Card Issues app: Agents select the blocked order(s) in the app and use the **Reauthorize** or **Process** function, which triggers the underlying program logic to request a new authorization.
- Manage Sales Orders app: Like in SAP ERP, manual changes and saving the sales order in the Manage Sales Orders app will trigger reauthorization.

### 10.1.5 Customer Payment Enquiries

Customers contact customer service to request payment transaction information, including authorization confirmation, settlement status, refund status, and payment method on file. Representatives must provide accurate information while maintaining PCI DSS compliance and protecting sensitive payment data.

#### Checking Settlement Status

Transaction FCCR (Payment Card Evaluations) is the primary transaction (see Figure 10.11) for checking the status and errors of settlement runs. It displays logs that include successful captures and specific error codes if a settlement failed. Enter the number of a settlement run that you want to review as input (e.g., enter "18" in the **Settlement run** field).

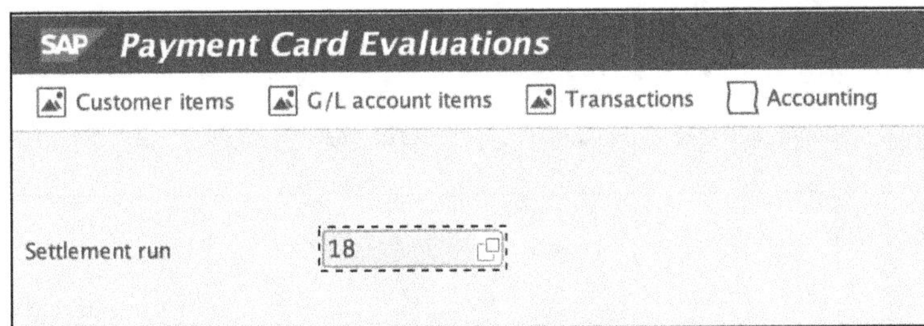


*Figure 10.11: Transaction FCCR: Payment Card Evaluations*

In the next step, you have a few options to choose from in the dialog window (see Figure 10.12).

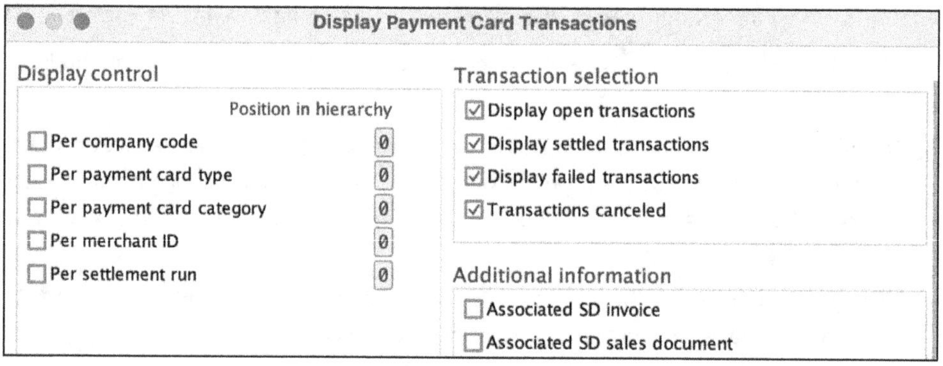

**Display Payment Card Transactions**

Display control

Position in hierarchy

- ☐ Per company code 0
- ☐ Per payment card type 0
- ☐ Per payment card category 0
- ☐ Per merchant ID 0
- ☐ Per settlement run 0

Transaction selection

- ☑ Display open transactions
- ☑ Display settled transactions
- ☑ Display failed transactions
- ☑ Transactions canceled

Additional information

- ☐ Associated SD invoice
- ☐ Associated SD sales document

*Figure 10.12: Filter Criteria for Transactions*

You could add additional filters to narrow down your search (see Figure 10.13) to a specific **Company Code, Payment card type, Merchant ID**, and so on, depending on your needs.

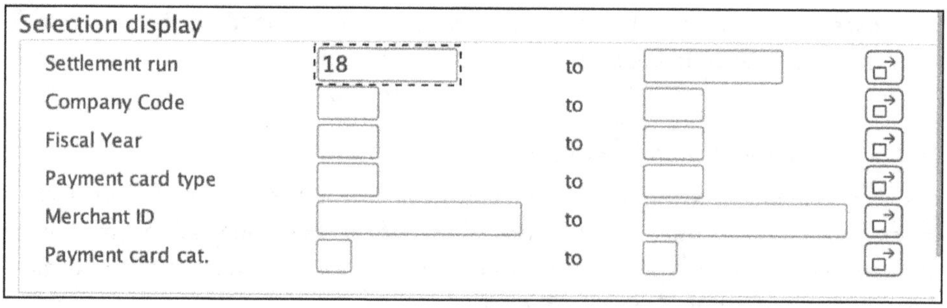

Selection display

| | | | |
|---|---|---|---|
| Settlement run | 18 | to | |
| Company Code | | to | |
| Fiscal Year | | to | |
| Payment card type | | to | |
| Merchant ID | | to | |
| Payment card cat. | | to | |

*Figure 10.13: Additional Filter Options*

You will see a list of transactions for which to review the settlement status (see Figure 10.14).

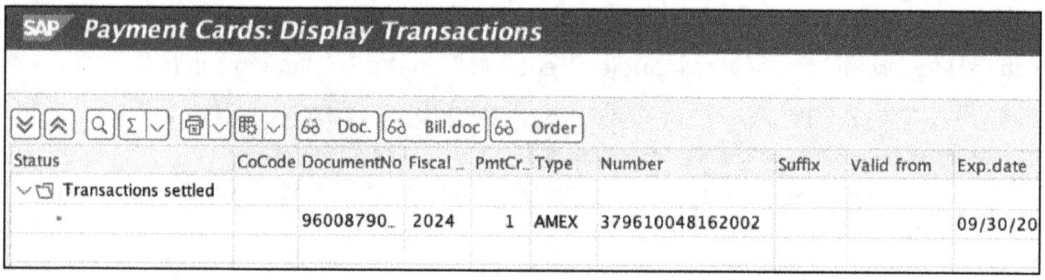

**Payment Cards: Display Transactions**

Doc. Bill.doc Order

| Status | CoCode | DocumentNo | Fiscal ... | PmtCr... | Type | Number | Suffix | Valid from | Exp.date |
|---|---|---|---|---|---|---|---|---|---|
| Transactions settled | | | | | | | | | |
| • | | 96008790... | 2024 | 1 | AMEX | 379610048162002 | | | 09/30/20 |

*Figure 10.14: Settlement Transactions List*

If Transaction FBL5N (Customer Line Item Display) is enabled for customer service agents, you also can check if the credit card clearing account line item has been cleared.

### Refund Status Inquiries

For customers requesting a refund status, you must check both the credit memo document creation and the refund settlement processing.

Execute Transaction VF03 (Display Billing Document) and enter the credit memo number the customer provides, or use menu option **Environment • Display Document Flow** from the original invoice to find the related credit memo. The existence of the credit memo (see Figure 10.15) confirms that a refund was initiated in the SAP system.

SAP Document Flow

Status overview | Display document | Service documents

**Business partner**
**Material** 000000005

| Document | Quantity | Unit | Ref. value | Currency | On | Time | Status |
|---|---|---|---|---|---|---|---|
| / 10 | 1 | EA | 49,00 | USD | 01/06/2026 | 06:25:03 | Open |
| 54 / 10 | 1 | EA | 49,00 | USD | 01/06/2026 | 06:25:06 | Completed |
| Invoice / 10 | 1 | EA | 49,00 | USD | 01/06/2026 | 06:40:06 | FI doc. generated |
| → Credit Memo Request / 10 | 1 | EA | 49,00 | USD | 01/09/2026 | 13:47:45 | Completed |
| Credit Memo / 10 | 1 | EA | 49,00 | USD | 01/14/2026 | 05:34:42 | |

*Figure 10.15: Credit Memo: Document Flow*

Refunds are typically settled via Transaction FCC1 (used for both charges and refunds) or through the dedicated refund transaction, Transaction FIN_DP_REFUND_PC. Settlement timing may depend on batch job schedules, such as the daily Transaction FCC1 run, or manual processing times. You can check the document flow and confirm if the credit memo has settled successfully (see Figure 10.16).

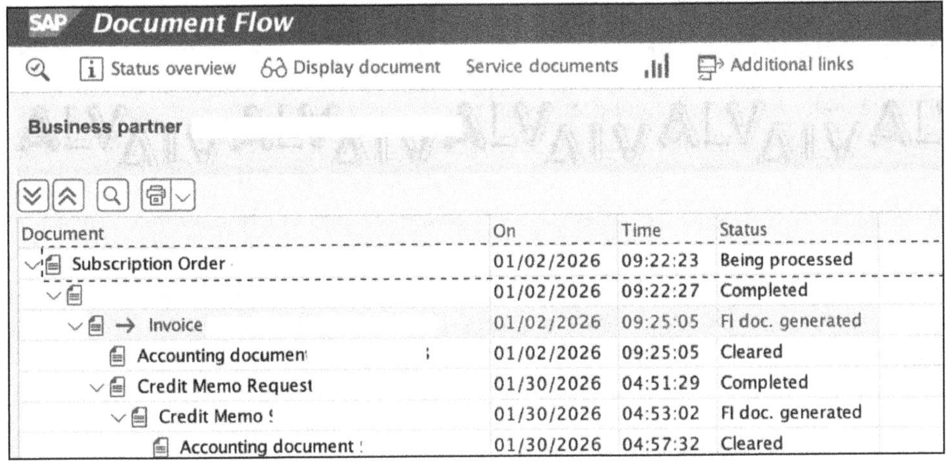
SAP Document Flow

Status overview | Display document | Service documents | Additional links

**Business partner**

| Document | On | Time | Status |
|---|---|---|---|
| Subscription Order | 01/02/2026 | 09:22:23 | Being processed |
| | 01/02/2026 | 09:22:27 | Completed |
| → Invoice | 01/02/2026 | 09:25:05 | FI doc. generated |
| Accounting document | 01/02/2026 | 09:25:05 | Cleared |
| Credit Memo Request | 01/30/2026 | 04:51:29 | Completed |
| Credit Memo | 01/30/2026 | 04:53:02 | FI doc. generated |
| Accounting document | 01/30/2026 | 04:57:32 | Cleared |

*Figure 10.16: Successfully Settled Credit Memo*

When communicating with customers, it is important to clarify that refund timing is determined by the customer's bank. This can help manage expectations and prevent any misdirected frustration.

**Privacy and Compliance Considerations**

PCI DSS compliance and privacy regulations restrict what payment information customer service representatives can access and share. Before discussing payment details, authenticate the customer's identity using established procedures:

- Verify the account number and customer name
- Confirm the billing address or shipping address
- Ask a security question, if configured
- Never share payment details with anyone who cannot properly authenticate

If the customer cannot authenticate, direct them to written inquiry channels (email, secure portal) where identity verification processes can be properly followed. Phone-based payment inquiries carry a higher risk of social engineering attacks.

The following are some circumstances in which it may be appropriate to escalate to the finance team:

- Settlement status is unclear or appears incorrect
- Accounting discrepancies exist between SAP and the customer's bank statement
- Refund is not appearing on customer's card after 15 business days
- Duplicate charge claims that require investigation
- Chargeback or dispute notifications

## 10.2 Finance Operations

Finance teams handle payment-processing activities after sales order completion, including settlement execution capturing authorized funds, refund processing returning payments to customers, reconciliation validating that SAP records match PSP statements and bank deposits, dispute and chargeback management responding to customer transaction challenges, and month-end procedures ensuring payment accounts close accurately. Finance payment operations require more detailed accounting knowledge and system access than customer service roles, with responsibilities for financial accuracy and audit compliance.

### 10.2.1 Refund and Credit Memo Processing

Refund processing returns funds to customer payment cards for returned merchandise, service failures, pricing errors, customer goodwill, or order cancellations after billing. SAP provides a few mechanisms for payment card refunds:

1. The Create Refunds for Digital Payments app
2. The credit memo request workflow, which provides approval controls
3. Processing refunds via Transaction FCC1
4. Direct refund processing via Transaction FIN_DP_REFUND_PC, enabling immediate refunds for settled transactions

### Create Refunds for Digital Payments App

Launch the Create Refunds for Digital Payments app. The app would typically be enabled in the **Accounts Receivable** pages in the SAP Fiori launchpad. You can also search for it in the SAP Fiori launchpad if you don't have it saved. Based on your permissions, you can use this app to do the following:

- Search for refundable payments using different criteria (see Figure 10.17).

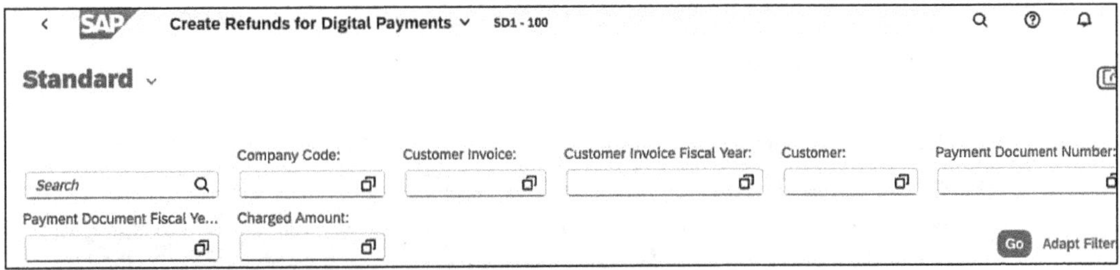


*Figure 10.17: Refund Search Criteria*

- Create full or partial refunds, but only if the previous refund for that payment is complete. Complete refunds via the Payment Card Settlement job in the Schedule Accounts Receivable Jobs app.
- View refund process flows by selecting the arrow icon on a payment row to access its details page, including related customer invoices and statuses.
- For payments covering multiple invoices, you can display one invoice in the process flow.

Refunds have four possible statuses:

- **Pending Settlement**: Recently created, not settled
- **Settled**: Refund completed
- **Reversed**: Refund cancelled
- **Failed**: Errors occurred during processing

You can view refund creation logs in the payment's detail page. Once a refund is created, its amount posts to the originating accounts receivable account.

### Credit Memo Request Workflow

Credit memo requests represent sales documents requiring approval before refund processing occurs, providing financial controls over customer credit issuance.

Execute Transaction VA01, select the **Order Type** (e.g., **ZCR** for a credit memo request), enter organizational data (**Sales Organization**, **Distribution Channel**, **Division**), and click the **Create with Reference** button (see Figure 10.18).

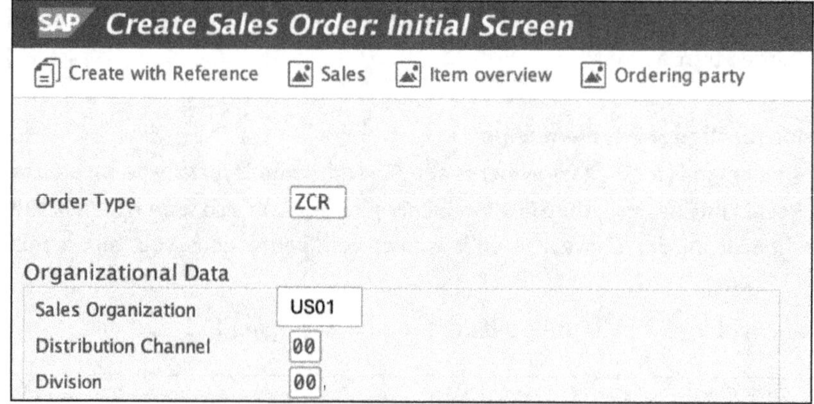


*Figure 10.18: Create Credit Memo*

The **Create with Reference** dialog appears. Enter the **Billing Document** number (e.g., 930010786) in the appropriate field (see Figure 10.19) and click the **Copy** button.

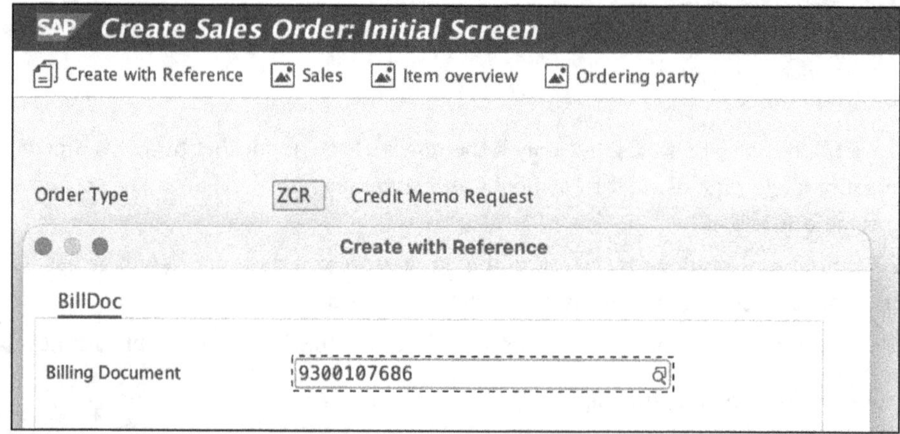


*Figure 10.19: Billing Document Reference*

The system displays the **Create Credit Memo Request** screen with data copied from the reference document, including the following:

- Customer information (sold-to, ship-to, payer)
- Line items with materials, quantities, and prices
- Pricing conditions
- Payment card information (see Figure 10.20)

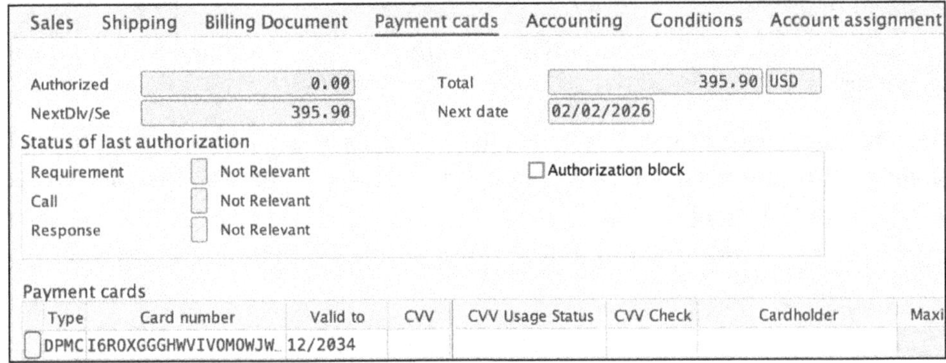


*Figure 10.20: Payment Card Information Copied Over*

Modify quantities for partial refunds by changing the **Target Quantity** field to the refund amount, leaving items at zero quantity for those not being refunded. Navigate to the **Header • Sales** tab and configure the following:

- **Order Reason**
  Select a reason code for the refund (customer complaint, defective goods, pricing error) from the dropdown (see Figure 10.21).

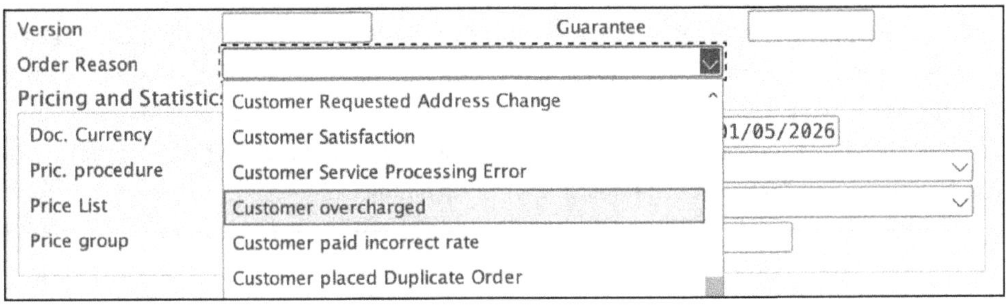

*Figure 10.21: Order Reason Dropdown Values*

- **Billing Block**
  The system may automatically set a block (see Figure 10.22) requiring approval (typically block reason 01 or 02).

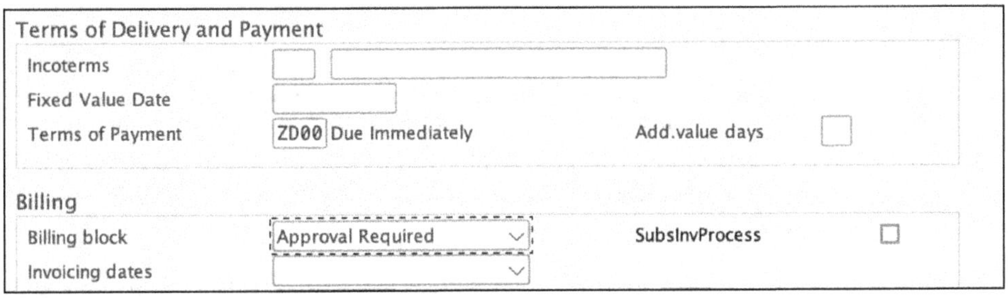

*Figure 10.22: Billing Block*

- **Pricing Date**
  Use the current date for refund pricing or the original order date to preserve the original pricing.

Save the credit memo request. The system generates a document number and displays the **Credit Memo Request [number] has been saved** message once the refund authorization is successful (see Figure 10.23).

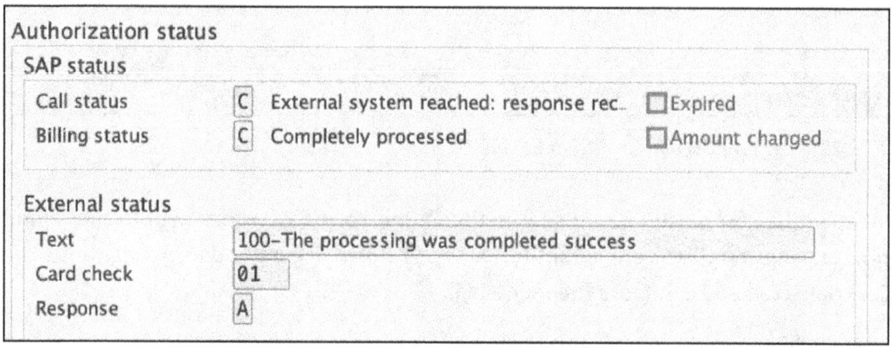

*Figure 10.23: Refund Authorization Successful*

For credit memo requests with billing blocks, an authorized approver must remove the block before billing can occur. Approval workflows vary by organization (manual release, workflow engine, external approval system integration). For manual approval, execute Transaction VA02, enter the credit memo request number, navigate to the **Header • Sales** tab, and clear the **Billing block** (see Figure 10.24) field (delete the block reason code). Save the document. Block removal enables credit memo billing.

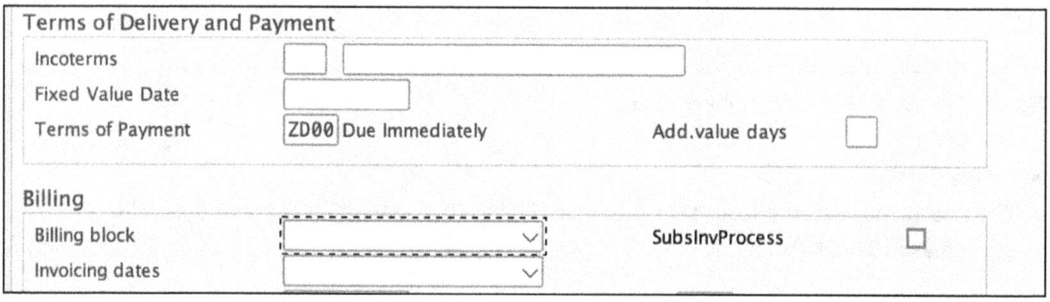

*Figure 10.24: Remove Billing Block*

For workflow approval, the approver receives the workflow notification (via email, the SAP Business Workplace inbox, or an external system), reviews the credit memo request details and business justification, approves or rejects the memo via the workflow interface, and the system automatically removes the billing block if approved.

After the credit memo request is approved (billing block removed), create a credit memo billing document via Transaction VF01. Enter the credit memo request number as a reference.

The system displays billing data copied from the request. Save the billing document. The system posts the following accounting entries:

- Debit: Revenue account (reversing original revenue recognition)
- Credit: Customer account (reducing customer receivables)
- Debit: Customer account (offsetting credit)
- Credit: Payment card accounts receivable clearing account (reversing the original accounts receivable posting)

### Processing Refund via Transaction FCC1

Prior to the settlement run, the table BSEGC entries will show the status as **Settled** = blank (see Figure 10.25).

| Group description | Cell Content |
|---|---|
| Name of cardholder | |
| Entry mode | |
| Authorized amount | 395.90 |
| Currency | USD |
| Settled | |
| Authorization no. | KYDWNVSMFA |
| Auth.refer.code | |
| Authorization date | 02/02/2026 |
| Authorization time | 00:14:29 |
| Merchant ID | |
| Point of receipt | |
| Terminal | |
| Settlement run | |

*Figure 10.25: Settled Field Shows Blank Value*

A successful settlement run will update the refund settlement status. Check the settlement log in Transaction SM37 by clicking the **Job Log** button to display the logs (see Figure 10.26).

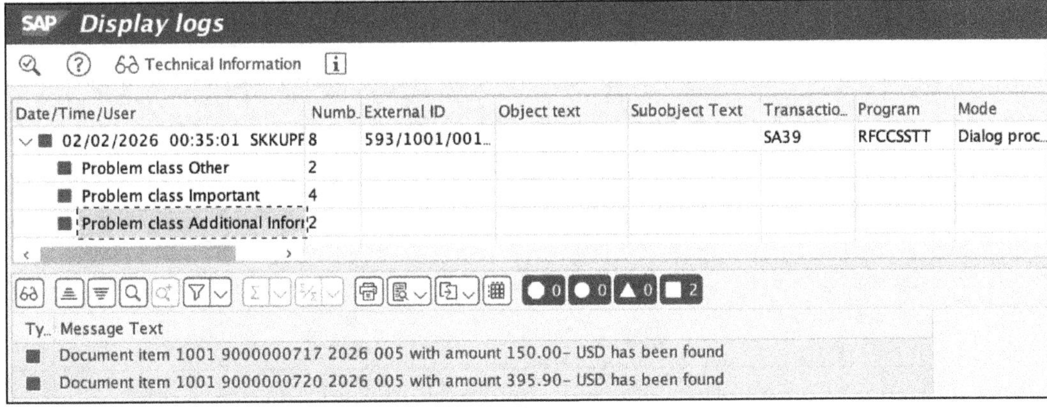


*Figure 10.26: Settlement Log Example*

The system updates the settlement status (see Figure 10.27) after a successful run, indicating the refund has been issued. In Transaction SE16, enter "BSEGC" as the table name. In the settlement run search filter, provide your settlement run number (in this example, 0000000593) and click **Display**.

| | |
|---|---|
| Authorized amount | 395.90 |
| Currency | USD |
| Settled | X |
| Authorization no. | KYDWNVSMFA |
| Auth.refer.code | |
| Authorization date | 02/02/2026 |
| Authorization time | 00:14:29 |
| Merchant ID | |
| Point of receipt | |
| Terminal | |
| Settlement run | 0000000593 |

*Figure 10.27: Table BSEGC with Settled Status as X*

Refunds return to the customer's original payment card. Timing depends on the card issuer's processing times (typically 5–10 business days from the settlement date to funds appearing on the customer's card statement).

### Direct Refund Processing via FIN_DP_REFUND_PC

Transaction FIN_DP_REFUND_PC provides an alternative refund mechanism for finance users who need to process immediate refunds without credit memo request workflow delays. Execute Transaction FIN_DP_REFUND_PC. A selection screen appears with the following criteria:

- **Company Code** (required)
- **Fiscal Year** (defaults to current year)
- **Document Number** (optional; narrows to specific billing document)
- **Posting Date** range (see Figure 10.28)

**Digital Payments: Refund Payment Card Payments**

| | | | |
|---|---|---|---|
| Company Code | | to | |
| Fiscal Year | 2026 | to | |
| Posting Date | | to | |
| Document Type | | to | |
| Document Number | | to | |
| Reference | | to | |

*Figure 10.28: Refund Card Payments: Search Criteria*

Select **Execute**. The system selects cleared customer line items from table BSAD for which digital payments fields are populated (PAYS_PROV, PAYS_TRAN, DP_PAYMENT_TYPE, and DP_TRANS_ID are not initial). Select line items to refund by checking the appropriate selection boxes. Enter a refund amount in the **Refund Amount** column for each selected item (partial or full amount up to the maximum shown). Click the **Post** button.

The system creates a credit memo document in financial accounting, posts refund accounting entries per configuration in view FARV_DP_REFUNDPC, calls the digital payments add-on refund API, and displays a success message that includes the posted document number.

This direct refund approach bypasses the sales credit memo request workflow and is suitable for finance-initiated refunds (accounting errors, goodwill adjustments, pricing corrections). Merchandise return scenarios are better handled through the sales credit memo process.

### 10.2.2 Dispute and Chargeback Management

Chargebacks are a form of forced transaction reversal in which a customer's issuing bank pulls funds back from a merchant's account. Although this is a necessary consumer protection, chargebacks represent a significant financial loss, extending beyond lost revenue to include fees, operational overhead, and potential penalties if chargeback ratios exceed card brand thresholds. An effective management strategy must prioritize both prevention and efficient dispute resolution.

The chargeback process involves multiple parties (merchant, customer, issuing bank, acquiring bank, and card networks like Visa/Mastercard) and typically spans multiple weeks, from the initial dispute to final resolution. The core steps include the following:

- **Customer disputes transaction**
  The cardholder contacts their bank (issuer), claiming an issue such as fraud, nonreceipt of goods, or quality issues. The issuer provides a provisional credit and assigns a reason code.
- **Issuer files chargeback**
  The issuer sends the chargeback through the card network to the merchant's bank (acquirer/PSP), including the reason code and claim details.
- **Merchant notification and provisional debit**
  The acquirer notifies the merchant and immediately debits the disputed amount, along with a chargeback fee, from the merchant's account.
- **Merchant response (representment)**
  The merchant has a limited window (typically 20–45 days) to either accept the chargeback or dispute it through the representment process by submitting compelling evidence.
- **Issuer review and decision**
  The issuer reviews the merchant's evidence. If the evidence is compelling, the chargeback is reversed; otherwise, the customer keeps the funds, and the merchant permanently loses the revenue and fees.

- **Arbitration (optional)**
  If either party disagrees with the issuer's decision, the case can proceed to arbitration with the card network, which makes a final, binding decision. This is costly and typically a last resort.

The most effective strategy is to stop chargebacks from occurring in the first place. Prevention falls into three main categories (see Table 10.2), divided by chargeback triggers.

| Trigger Category | Description | Prevention Best Practices |
|---|---|---|
| Merchant errors | Mistakes in billing, poor customer service, unclear policies, or duplicate charges | ■ Provide clear billing descriptors, accessible customer service contact information, and transparent return/refund policies. |
| Unauthorized fraud | Unauthorized transactions using stolen card information | ■ Implement fraud detection tools like an address verification service (AVS) and CVV checks.<br>■ Use 3D Secure authentication for online sales to shift liability. |
| Customer fraud | Customers making a legitimate purchase but filing a chargeback due to buyer's remorse, forgetfulness, or an attempt to get a free item | ■ Maintain detailed records of all transactions, including IP addresses, delivery confirmations, and customer communication logs.<br>■ Send clear order confirmations and shipping updates. |

*Table 10.2: Chargeback Prevention Strategies*

When a chargeback is received, you must decide whether to fight it (representment). Success rates can vary, but a well-documented case is crucial. Key steps for successful representment are as follows:

- **Respond Promptly**
  Adhere strictly to the deadlines provided by the acquirer. Missing the deadline results in an automatic loss.
- **Analyze the Reason Code**
  Each card network (Visa, Mastercard, Amex, Discover) uses specific reason codes that dictate the type of evidence required.
- **Gather Evidence**
  The evidence must directly address the reason code. For example, for Merchandise Not Received (Visa 13.1), you need shipping confirmations, signed delivery receipts, and tracking numbers. For Not as Described (Visa 13.3), provide detailed product descriptions, images, and proof the customer agreed to the terms.

- **Prepare a formal rebuttal Letter**
  Submit a professional, concise cover letter summarizing why the claim is invalid and listing all attached evidence.
- **Submit the Package**
  Send the complete documentation to your acquiring bank by the deadline.

Chargebacks affect financial statements and require specific accounting entries to ensure accuracy. When a chargeback is filed, the original revenue transaction is reversed. You record a debit to accounts receivable (specific to chargebacks) and a credit to the bank account (cash outflow). The associated fees from the payment processor must be recorded as *bank fees* expense or equivalent general ledger account.

If a dispute is settled in your favor, the accounts receivable entry is cleared, and the funds are returned to the bank account. Otherwise, the accounts receivable balance must be written off to a *bad debt* expense account. It should not be recorded as a cost of goods sold or a simple refund.

The following tasks ensure that disputes are handled within required deadlines and don't lapse into automatic losses:

- Daily PSP portal review: Monitor the PSP portal daily for new dispute notifications and chargebacks.
- Acknowledge and assign: Log all new chargebacks immediately in an internal tracker and assign responsibility for investigation and representment.
- Gather evidence: For all disputable items, gather evidence (shipping proof, customer communications, AVS/CVV data) and prepare the rebuttal package.
- Submit representment: Ensure all rebuttal packages are submitted to the acquiring bank before the hard deadline (typically 14–30 days).

These tasks occur during the final days of the month, often concurrently with the general ledger clearing and fee-accrual processes:

- Run PSP reports: Generate a month-to-date report from your PSP showing all new chargebacks initiated and all resolved chargebacks (both wins and losses).
- Reconcile accounts receivable (chargeback account): Compare the PSP report data against your designated accounts receivable general ledger account used for tracking disputes.
- Target: Ensure every transaction recorded has a corresponding status (pending, won, lost).

Consider implementing SAP Collections and Dispute Management to manage disputes. This is a specific application within financial supply chain management designed to process receivables-related dispute cases efficiently. It offers the following advantages:

- It uses a case-management approach, assigning a unique case ID to each dispute. This allows all relevant documents, communication logs, and internal notes to be centrally documented and tracked under one ID, which is a major enhancement over manual tracking in general ledgers alone.

- Dispute cases can be created directly from identifying underpayments during bank statement processing or during manual clearing of open items.
- It leverages built-in SAP workflows to automatically route dispute cases to the correct department (e.g., sales, logistics, accounts receivable team) for investigation and approval, reducing processing time.

### 10.2.3 Month-End Procedures

The month-end financial close requires validating that payment processing accounts balance correctly, all transactions are posted to proper periods, and reconciliation is complete before closing the books. Your month-end payment-processing checklist might look like this:

- **Execute Transaction FCC1 settlement run**
  - Transaction code: Use SAP Transaction FCC1 (Program RFCCSSTT) to initiate the settlement of credit card transactions.
  - Selection criteria: Set the **Posting Date** range from the first day of the current month through the last business day to ensure no pending billings are missed.
  - Process pending settlements: Ensure all authorizations used to post invoices have been captured in accounting documents so that they are available for this final run.
- **Verify results and logs**
  - Success verification: Review the settlement log to confirm the number of successfully settled items.
  - Check for failures: Identify any items flagged with errors. These often require manual intervention, such as reauthorizing an expired card or correcting account assignments. Several common error codes may appear in a settlement log, including payment card expired, card declined, no authorization, duplicate authorization, authorization expired, and refund exceeds original amount.
- **Postsettlement actions**
  - Reconcile subledgers: Confirm that the settlement run has credited the credit card receivable account and debited the appropriate clearinghouse reconciliation account.

There are multiple ways to validate settlement completeness, including reports and SAP Fiori apps that we have discussed in the prior sections. Use them as a guide.

Execute Transaction FAGLL03 (General Ledger Account Line Items) or FS10N (General Ledger Account Balances) for each payment-processing clearing account: the PSP clearing account and accounts receivable clearing accounts. Review open items and account balances:

- PSP clearing should have a minimal balance (only items between settlement and advice processing). Accounts receivable clearing should have a minimal balance (only current day billings awaiting settlement). Maintaining minimal balances confirms that transactions have successfully moved from the initial clearing phase to the final bank settlement and reconciliation phase.

- Aged items (more than seven days old) indicate clearing failures requiring investigation. Items older than a set threshold indicate an operational or system failure in the automatic clearing process that requires investigation and manual correction.

Use available data from your internal systems (total settled amount and transaction count) to estimate the expense based on the known PSP contract rates. Use Transaction FB01 (Post Accrual Entry) to record the estimated expense via a manual journal entry in accounting. The PSP fees are an operating expense, and the credit goes to a liability account for expenses not yet paid.

In the subsequent month, once the official PSP statement is received, reverse the original accrual entry to clear the temporary liability. Post the actual, precise fees from the statement. This ensures your final bank reconciliation is accurate. This process helps management have a more accurate view of expenses and cash flow before the final bank statement.

Once settlements are executed, the next critical month-end task is reconciling those settlements with your actual bank activity:

- Import bank statements: Use Transaction FF_5 to upload electronic bank statements (e.g., in MT940 or BAI2 format).
- Match transactions: The system attempts to match settlement totals from Transaction FCC1 against the deposits listed in the statement.
- Investigate variances: Identify items that were settled in SAP but never deposited (e.g., chargebacks) or deposits that have no matching SAP record.
- Final posting: Resolve discrepancies and post any adjusting entries to ensure the general ledger matches the bank statement.

## 10.3 Training and Change Management

For digital payments to be adopted successfully, it is important to provide thorough training so that business users learn the new payment processes. Change management should be used to address any resistance within the organization and to build users' confidence. Ongoing support must also be available to help users when they face challenges. Investing in training is closely linked to higher user adoption rates, fewer support tickets, and improved operational efficiency after launch. This section covers three key aspects of preparing your organization for digital payments add-on adoption: role-based training that tailors content to each user's specific responsibilities, best practices for driving user adoption across the organization, and common challenges that users encounter. We also examine with self-service support procedures that help users resolve issues.

### 10.3.1 Role-Based Training

Generic SAP training is ineffective as roles vary in responsibilities, system access, and learning requirements. Role-based training delivers relevant add-on content tailored to users' specific job duties. Table 10.3 defines training audiences for the digital payments add-on with role-specific learning objectives.

| Role | Primary Responsibilities | Training Focus |
|---|---|---|
| Customer service representative | Create sales orders; select payment cards; handle authorization failures | Payment card selection; authorization interpretation; error handling; customer communication |
| Finance clerk | Execute settlement; process refunds; basic reconciliation | Settlement processing; refund execution; table queries; document display |
| Finance manager | Month-end close; reconciliation; chargeback handling | Advanced reconciliation; PSP statement analysis; chargeback response; month-end procedures |
| IT operations | Monitor batch jobs; troubleshoot errors; maintain configuration | Application log analysis; batch job monitoring; RFC troubleshooting; performance monitoring |
| Help desk | First-level support; ticket triage; basic troubleshooting | Read-only transaction access; error message interpretation; escalation criteria |

*Table 10.3: Training Audience Segmentation*

Develop training materials specific to each role, using the following approaches:

- **Hands-on simulations**
  Provide sandbox access or a dedicated training system for users to practice transactions with realistic test data. Simulations should mirror actual business scenarios users will encounter. Examples include the following:
  - Customer service reps practice creating a few sales orders with various payment cards and handling a few decline scenarios
  - Finance clerks practice executing Transaction FCC1 settlement with test billing documents and processing a few refunds
  - IT operations practice investigating application log errors and troubleshooting simulated failures
- **Job aids and quick reference guides**
  Create job aids as PDF documents for common tasks. Examples include Quick Guide: Creating Sales Order with Payment Card (Transaction VA01 steps with screenshots), Quick

Guide: Handling Declined Authorization (error message interpretation and resolution steps), and Quick Guide: Processing Refunds (Transaction FIN_DP_REFUND_PC flow).

- **Screen recordings and video tutorials**
  Record transaction walkthroughs using screen-capture software, showing actual SAP screens and providing narration that explains each step. Videos enable self-paced learning and on-demand reference when users forget procedures. Store videos in a learning-management system or a shared repository with clear naming for easy discovery.
- **Process documentation**
  Document complete business processes from end to end, not just individual transactions. For example, create a Order-to-Cash with Payment Cards process document covering creating a sales order with a payment card (Transaction VA01), handling a authorization decline if one occurs, creating a delivery after goods are picked (Transaction VL01N), creating a billing document (Transaction VF01), verifying that settlement occurs, and responding to customer payment inquiries.

### 10.3.2 Best Practices for User Adoption

User adoption measures to what degree business users accept and consistently follow digital payments add-on processing procedures, as well as how valuable they find these methods compared to previous ones. When adoption is high, there are usually fewer support tickets, transactions are processed accurately, and users report higher satisfaction. In contrast, low adoption often leads to users bypassing the system with alternative methods, making frequent mistakes that need fixing, and showing reluctance toward new procedures. Consider these key factors to boost user adoption:

- **Early user involvement**
  Engage representative business users during the implementation project, not just before go-live. Include customer service representatives in design workshops to provide input on sales order screen flows. Involve finance users in settlement testing to validate batch job procedures. Gather feedback on usability and incorporate improvements before deployment. Users who participate in the design feel ownership and advocate for adoption among their peers.
- **Clear communication of benefits**
  Users are more likely to adopt new processes when they see direct advantages. The add-on streamlines work in the following ways:
  - For customer service: Authorizing payments during order entry stops invalid orders, reducing cancellations and disputes.
  - For finance: Automated settlement cuts manual processing, shortens month-end closing, and boosts cash flow visibility.
  - For all users: Tokenization removes sensitive card data from SAP, enhancing security and easing compliance.

- **Super user network**
  Building a super user network involves selecting and training two to three super users from each business area before rolling out general user training. These super users undergo in-depth instruction on how the system operates, troubleshooting steps, and how to deliver training. After the system goes live, they support their peers by answering questions, demonstrating processes, and forwarding complex problems to IT if necessary. Ideal super users are well-regarded team members that coworkers feel comfortable approaching, and they actively participate in implementation testing to build system knowledge.
- **Sandbox practice environment**
  Offer continuous access to a sandbox training system so that users can practice safely without affecting production data. Quarterly updates with recent production data keep the environment realistic, allowing users to build confidence through risk-free experimentation.
- **Digital adoption platforms**
  Implement in-application guidance tools (SAP Enable Now, WalkMe, or similar), providing contextual help directly within SAP screens. Digital adoption platforms overlay guidance on the live system, with pop-up instructions guiding users through task steps, tooltips explaining field meanings, and automated data entry for repetitive portions.

### 10.3.3 Common Challenges and Support Procedures

Despite thorough training, users will encounter challenges that require support assistance. Establishing clear support procedures and empowering users with self-service troubleshooting guidance minimizes the support burden while ensuring that users receive help when they genuinely need it.

This section covers common user challenges, steps for self-diagnosis before raising a support ticket, and most likely resolution steps.

#### Payment Cards Tab Not Visible in Sales Order

A user creates a sales order but cannot find the **Payment Cards** tab to enter card information. The user should perform a self-diagnosis:

- Check the order type. Is this a document type configured for payment cards?
- Verify successfully saved prior orders to confirm the behavior.

**For the resolution**, for order types for which payment cards must be supported, your SAP configurator defines the correct configuration in Transaction VOV8 (see Figure 10.29).

The **Payment Cards** tab (see Figure 10.30) appears based on this order type configuration.

In SAP S/4HANA 1909, the tab was renamed to **Electronic Payments** (see Figure 10.31), but the functionality is the same.

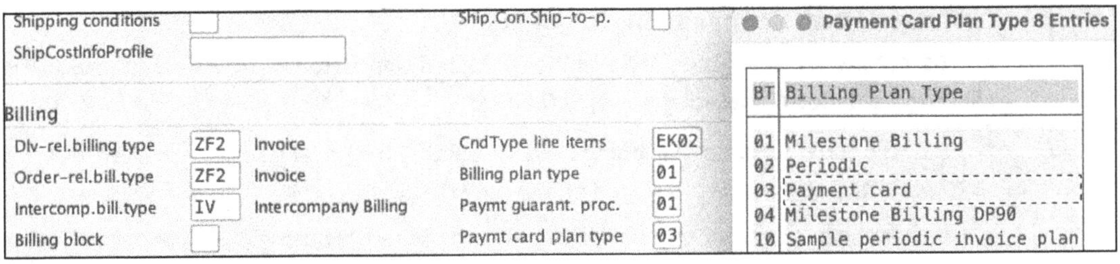

*Figure 10.29: Transaction VOV8 Configuration Screen*

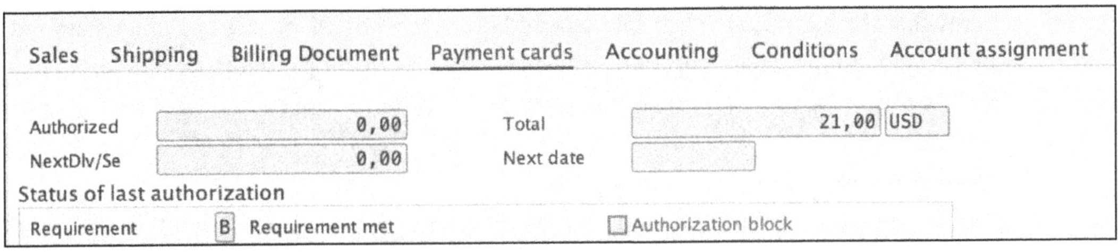

*Figure 10.30: Payment Cards Tab in Transaction VA01 in SAP ERP*

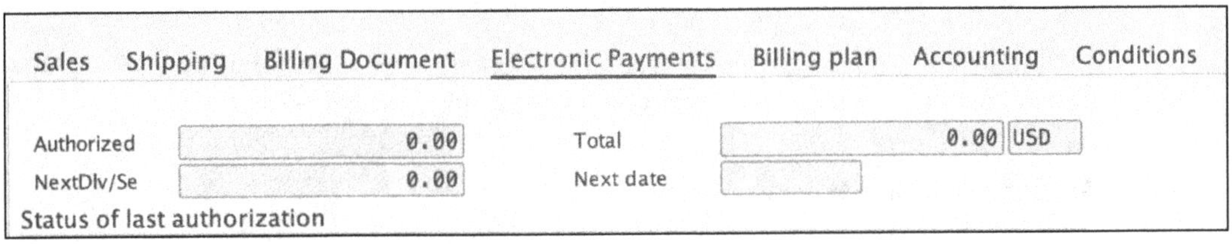

*Figure 10.31: Electronic Payments Tab in SAP S/4HANA*

If the order type is correct and the tab still does not appear, escalate to the help desk. The likely cause is that the sales document type is missing a payment card plan type assignment in the Transaction VOV8 configuration. If the **Payment Cards/Electronic Payments** tab is not needed, then it can be hidden with a setting tied to the sales order type. Keep in mind that if this setting is removed (see Figure 10.32), then the document type cannot use payment cards to settle the payment.

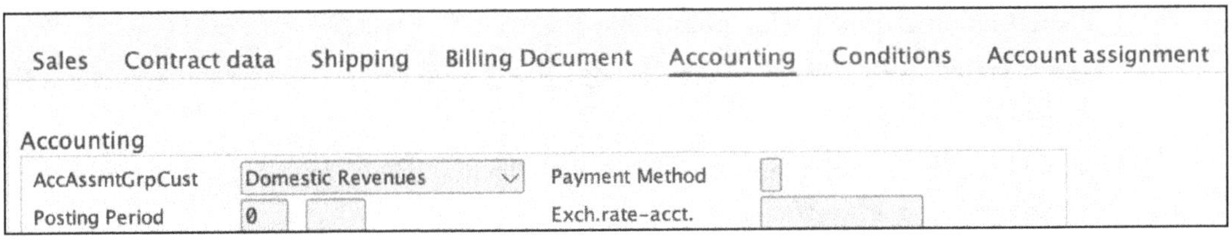

*Figure 10.32: Missing Payment Cards Tab*

#### Customer's Card Not Appearing in F4 Help

The user presses F4 in the **Card Number** field, but the customer's card is not listed. The user should perform a self-diagnosis, as follows:

- Verify the correct customer is being used. Check that the payer partner is in the order, not the sold-to party (cards are stored on the payer).
- Confirm the business partner has payment cards. View the business partner record in display mode and check the **Payment Transactions** tab.

If the payer is correct and the business partner has cards but they are still not appearing, verify that the business partner role selected in Transaction BP enables payment card visibility (field grouping configuration). This may require selecting a different role to see payment cards. If the cards are still unavailable, escalate to IT support.

#### Authorization Taking Very Long Time

The order's **Save** button is pressed, but authorization processing exceeds 30 seconds without response. The correct user action is as follows: Do not press **Save** multiple times; this creates duplicate authorization attempts. Note the order number, customer, and time of attempt for troubleshooting and escalate to IT support.

Likely causes include a network connectivity issue with the add-on, PSP service degradation or timeout, or a code-performance issue requiring IT investigation.

#### Settlement Didn't Process Overnight

A finance user checks prior-day settlement results but sees no settled items. The user should perform a self-diagnosis, as follows:

- Check the batch job status. Open Transaction SM37, filter by program RFCCSSTT, and verify that the job ran and finished successfully.
- Review the settlement log. If the job ran, check the spool output for error messages.
- Verify the billing documents exist. Open Transaction SE16N and enter table name as "BSEGC" and leave the **SETTL** field blank and check **Record Count**. If records exist, then the settlement did not process successfully.

For resolution, there are a few possibilities:

- If the batch job didn't run, check the job scheduling in Transaction SM36. Verify the job is active.
- If the job ran but zero items processed, the selection criteria may be too restrictive or no items are pending settlement.
- If the job failed with errors, review the application log (Transaction SLG1), identify the error messages, and escalate to IT support with the error details.

## 10.4 Summary

Processing digital payments in daily business operations requires business users to master payment-specific transactions, interpret authorization and settlement results, handle exceptions gracefully, and maintain financial accuracy through reconciliation and compliance procedures. This chapter provided practical guidance for customer service teams creating sales orders with payment cards and handling authorization failures and finance teams executing settlement processing and refunds. We also addressed dispute management for responding to chargebacks, and training approaches that can help ensure user adoption success.

Key operational capabilities include payment card selection from the business partner master, authorization validation for interpreting success and decline messages appropriately, settlement execution for capturing authorized funds, refund processing through credit memo workflows or directly through Transaction FIN_DP_REFUND_PC, and reconciliation procedures that validate SAP records match PSP statements and bank deposits. Business users equipped with these capabilities can independently handle standard payment processing scenarios, reducing dependency on technical support and maintaining operational continuity.

The transition from technical implementation to business operations is a critical success factor for SAP digital payments add-on deployments. Organizations investing in comprehensive role-based training, establishing clear support escalation procedures, and measuring adoption through objective metrics achieve higher user confidence, lower error rates, and faster return on implementation investment. The operational procedures documented in this chapter become the foundation for training materials, user reference guides, and help desk support scripts, ensuring a knowledge transfer from the implementation team to business operations teams.

As payment-processing volumes grow and user proficiency increases, organizations should continuously optimize operational procedures based on user feedback and performance data. Common optimization opportunities include streamlining payment card selection for high-volume users, automating reconciliation reporting to reduce manual effort, enhancing error message clarity to improve user self-diagnosis, and implementing digital adoption platforms to provide just-in-time guidance directly in SAP screens. These optimizations maintain user satisfaction and operational efficiency as the business scales.

Successful digital payments operations deliver business value through reliable revenue capture, accurate financial reporting, compliant payment processing, and positive customer experiences. The combination of well-trained users, clear operational procedures, effective support mechanisms, and continuous improvement culture transforms the digital payments add-on from a technical integration to a strategic business capability that enables digital commerce growth and payment processing excellence.

# The Author

**Saravana Kumar Kuppusamy** is a senior technology leader with more than 24 years of professional SAP expertise in consulting, enterprise architecture, process and systems integration, artificial intelligence, and technology leadership. He has directed large-scale SAP implementations and global rollout programs for Fortune 500 clients in the manufacturing, energy, utilities, publishing, and consumer products industries. His project experience encompasses end-to-end SAP implementations, technology integrations related to mergers and acquisitions, enterprise integration platform modernization, and multi-country deployment initiatives. His expertise in SAP ERP, cloud ERP, the SAP digital payments add-on, and SAP BTP gives him deep, hands-on experience with the technologies and business processes at the heart of this book.

He holds a master's degree in computer applications from the College of Engineering, Anna University, India, and a bachelor of science degree in computer science from Bharathiar University, India. He is based in New Jersey and can be reached via LinkedIn at *www.linkedin.com/in/saravanakumark*.

# Index

## D



## E



## F



## G

## P



## Q



## R

## T

## U



## V



## W



## X



## Z

- Manage credit, disputes, and collections with SAP S/4HANA
- Configure and integrate standard and advanced receivables management with step-by-step instructions
- Run key receivables processes with classic transactions and new apps

Chokshi, Mohapatra, Galal

## Receivables Management with SAP S/4HANA

Does your organization manage receivables in SAP S/4HANA? This book is your comprehensive guide for both standard and advanced processes! Once you've set up your master data, configure and use both basic AR and specialized tasks. Master credit, collections, and dispute management to successfully manage incoming payments. With step-by-step instructions and screenshots, this is your all-in-one receivables resource!

623 pages, pub. 04/2022
**E-Book:** $84.99 | **Print:** $89.95 | **Bundle:** $99.99

**www.sap-press.com/5408**

**Interested in reading more?**

Please visit our website for all new book and e-book releases from SAP PRESS.

**www.sap-press.com**